A Comprehensive Review of Biology
with a Special Section on

The College Board Achievement Test in Biology ©

Author:

Wayne H. Garnsey
John Jay High School

Illustrations and Graphics:

Eugene B. Fairbanks
Wayne H. Garnsey

Cover Design and Artwork:

Eugene B. Fairbanks

N & N Publishing Company, Inc.
18 Montgomery Street   Middletown, New York 10940
(914) 342 - 1677

Dedicated to my students, with the sincere hope that this

N&N's Science Series© — *Biology*

will further enhance their education and better prepare them
with an appreciation and understanding of their role
in the care and protection of the Earth's biosphere.

Special Credits

Thanks to the many teachers that have contributed their knowledge, skills, and years of experience to the making of this review text.

To these educators, a sincere thanks
for their assistance in the preparation of this manuscript:

Douglas Bean
Cindy Fairbanks
Virginia Page
Patrick Ryan
Gloria Tonkinson
Donald West
Mary Edler
Alan M. Gardner

N&N's Science Series© — *Biology* has been produced on a Macintosh Quadra 840AV and LaserWriter Pro 630 printer.

The applications *MacWrite II* and *Canvas* were used to produce the text and graphics. Original line drawings were reproduced with *VersaScan* on a Microtek MSF-300ZS scanner and modified with *Photoshop* by Adobe. Formatting, special designs, graphic incorporation, and page layout were accomplished with *Ready Set Go* by Manhattan Graphics.

Special technical assistance was provided by Frank Valenza and Len Genesee of Computer Productions, Newburgh, New York.

To all, thank you for your excellent software, hardware, and technical support.

N & N Publishing Company, Inc.
SAN # 216 - 4221 ISBN # 0935487 16 6

Printed in the United States of America
8 9 0 BMP 2000 1999 1998

Table Of Contents

Unit I
Unity and Diversity Among Living Things

Unit II
Maintenance in Living Things

Unit III
Human Physiology

Unit IV
Reproduction and Development

Unit V
Transmission of Traits

Unit VI
Evolution

Unit VII
Ecology

Unit 1 Unity and Diversity Among Living Things

Objectives

The student should be able to:

- Define life in terms of the functions performed by living organisms.
- Describe some of the methods by which organisms are classified.
- Recognize the role of the cell as the basic unit of structure and function of living things.
- Identify major biochemical compounds and some of the metabolic reactions in which these compounds are involved.
- Recognize that a unity of pattern underlies the diversity of living things.

Life – Terms & Concepts

Life Functions
- Nutrition
- Transport
- Respiration
- Excretion
- Synthesis
- Regulation
- Growth
- Reproduction

Metabolism

Homeostasis

Cell Instrumentation
- Measurement
- Techniques

Kingdoms (5)
- Monera
- Protista
- Fungi
- Plant
- Animal

Nomenclature
- Kingdom
- Phylum
- Class
- Order
- Family
- Genus
- *Genus species*

Cell Theory/History

Cell Organelles
- Cytoplasm
- Nucleus
- Nucleolus
- Endoplasmic Ret.
- Ribosome
- Mitochondrion
- Golgi Complex
- Lysosome
- Vacuole
- Centriole
- Chloroplast
- Cell Wall

I. Concept of Life

A. Definition of Life

Modern scientists have been unable to agree upon a single definition of life. However, most biologists characterize life by the functions that living organisms perform.

B. Life Functions, *key to "living"*

The life functions include the many activities carried on by all living things. Whether the life functions are the activities of cells or an organism composed of cells, these functions are basically the same.

1. **Nutrition** includes the activities of an organism by which it *obtains* materials (for structure) and energy (for life activities) from its environment and *processes* them for its use.

2. **Transport** involves the *absorption* (taking in) and *distribution* (circulation) of materials by an organism.
3. **Respiration** includes those processes which provide the energy (through the conversion of stored energy in food) necessary for the maintenance of life functions.
4. **Excretion** involves the removal from the organism of waste products manufactured by the organism.
5. **Synthesis** involves those chemical activities by which large molecules or structures are built from smaller ones.
6. **Regulation** involves the control and coordination of the various activities of an organism through responses to change both around and within the organism.
7. **Growth** involves an increase in cell size and/or cell numbers. This process utilizes the products of synthesis.
8. **Reproduction** involves the production of new individuals from preexisting ones. The reproduction of cells is also responsible for the growth of an organism and the repair of tissue. While the individual's survival is *not* dependent on reproduction, *species* survival *is* dependent upon the reproduction of individuals.

C. Metabolism, *key to "survival"*

Metabolism is the total of all the life activities required to sustain life. Each life process is chemical (metabolic) in nature and may be affected by both physical and chemical conditions within its internal and external environment. **Anabolism** is a constructive (building) activity, whereas, **catabolism** is a destructive (breaking down) process.

D. Homeostasis, *key to the "quality of life"*

Life functions are carried out by an organism in an integrated manner that results in the maintenance of a stable internal environment (internal dynamic equilibrium). This maintenance is known as **homeostasis**. Homeostasis refers to the organism's stability (normal balance) and is dependent upon the organism's ability to coordinate and control its many chemical reactions.

Questions

1 For survival, a hummingbird uses a considerable amount of energy. This energy most directly results from the life activity of
1 transport 2 excretion 3 regulation 4 respiration

2 Which life function provides substances that may be used by an organism for its growth and for the repair of its tissues?
1 excretion 3 nutrition
2 reproduction 4 regulation

3 The term that represents all the chemical activities that occur in an organism is
1 synthesis 2 regulation 3 metabolism 4 homeostasis

4 Which life function includes processes responsible for an animal being able to react in a coordinated manner to changes in its environment?
1 reproduction 3 nutrition
2 synthesis 4 regulation

5 A paramecium absorbs materials from its environment and circulates these materials throughout its cytoplasm. Which life function is described by these activities?
1 synthesis 3 respiration
2 reproduction 4 transport

6 Organisms release stored chemical energy from nutrients by the process of
1 assimilation 3 respiration
2 transport 4 ingestion

7 Which term is used to represent all of the physiological activities being maintained in a balance?
1 regulation 2 metabolism 3 synthesis 4 homeostasis

8 In plants, the production of poisons, drugs, waxes, and fibers is a direct result of
1 respiration 2 digestion 3 hydrolysis 4 synthesis

9 The control of all physiological activities of an organism is necessary to maintain that organism's stability in its environment. This life activity is known as
1 nutrition 2 transport 3 respiration 4 regulation

10 Which life activity is not required for the survival of an individual organism?
1 nutrition 2 reproduction 3 respiration 4 synthesis

II. Diversity of Life

A. Necessity for Classification

In order to study more easily the unity and diversity of living organisms in an organized manner, biologists classify organisms. Physical structure is often the primary basis for biological classification. However, similarities in biochemistry, genetics, embryology, and cytology are also considered.

B. Scheme of Classification

There is some disagreement as to the best classification system. In one modern system, organisms are grouped into five kingdoms. Each kingdom is divided into phyla. Phyla are categories which indicate major differences in structure among organisms of a kingdom.

The complete classification system, from broad groupings to more narrowing and specific (more diverse) groupings, includes: Kingdom, Phylum, Class, Order, Family, *Genus*, and *species*.

The science of classification is called **taxonomy**.

The classification system used in this book utilizes five kingdoms: **Monera, Protista, Fungi, Plant**, and **Animal**.

This five kingdom system is based on the following criteria:

- The presence (eukaryotic organism) or absence (prokaryotic organism) of a nuclear membrane within the cell.
- Unicellularity versus multicellularity.
- Type of nutrition — autotrophic vs. heterotrophic.

This five-kingdom system of classification is based on the idea that Monerans are the most primitive. Most classification systems suggest relationships among organisms which may indicate **common ancestry**.

Five Major Kingdoms and Phyla

KINGDOMS	Characteristics	*Examples*
I. MONERA	Primitive Cell Structure, Prokaryotic - Lack Nuclear Membrane and All Organelles, Circular DNA	
Blue Green Algae Cyanobacteria	Photosynthetic, Contain - Chlorophyll A, Ribosomes, Produce Oxygen	*Nostoc oscillatoria*
Bacteria Schizophyta	Mostly Heterotrophic, Decomposers, Pathogens, Cell Walls, Some Photosynthetic	*Proteus vulgaris, Escherichia coli*
II. PROTISTA	Eukaryotic - Contain Nuclear Membranes and Organelles	
Protozoa	Mostly Unicellular, Animal-like in Movement and Mode of Nutrition	*Ameba, Paramecium, Foraminifera*
Algae	Unicellular - Colonies Photosynthetic, Chloroplasts, Cell Walls	*Spirogyra, Kelp, Diatoms, Sargassum*
Slime Mold	Similar to Huge Amoeba, but Reproduces by Sporangium and Spores	*Physarum polycephalum*
III. FUNGI	Eukaryotic - Cells Organized into Branched, Multi-nucleated Filaments, External Digestion	*Yeast, Breadmold, Mushroom*
IV. PLANTS	Eukaryotic - Multicellular, Photosynthetic Organisms	
Bryophytes	Lack Vascular Tissue, Xylem, and Phloem No True Roots, Stems, or Leaves	*Mosses, Liverworts*
Tracheophytes	Contain Vascular Tissue, Xylem, and Phloem Have True Roots, Stems, and Leaves	*Ferns, Pines, Oaks, Tulips, Carrots*
V. ANIMALS	Eukaryotic - Multicellular Organisms, Heterotrophic Nutrition, Ingestion, and Muscular Contraction	
Coelenterates	Sessile, Body - 2 Cell Layers, Tentacles, Nematocysts Sac - Like Digestive System	*Hydra, Jellyfish*
Annelids	Segmented Worms, Closed Circulation, Tubular Digestive System with Mouth and Anus, Nephridia	*Earthworm, Leech, Sandworm*
Arthropods	Jointed Appendages, Chitinous Exoskeleton Body Plan - Head, Thorax, and Abdomen	*Grasshopper, Spider, Daphnia*
Chordates	Hollow Notochord, Dorsal Nerve Cord, Tail - Posterior to Anus	
(Vertebrates)	Dorsal Nerve Cord, Enclosed by Vertebrae Composed of Cartilage or Bone	*Mammals, Frog, Birds, Fish, Shark*

[Note: This chart is not a complete representation of the five kingdoms. The cells of Monerans are prokaryotic. The term multi-nucleated is used in place of multicellular since well-defined cell partitions are lacking.]

C. Nomenclature

The modern system of naming organisms is based upon **binomial nomenclature** devised in the 18th century by Carolus Linnaeus. In this system the first part of an organism's name, in Latin, is its *Genus*, and the second part of the name, in Latin, indicates its *species*.

A **species** is a group of organisms that are similar in structure and can mate and produce *fertile* offspring.

Questions

1 *Ursus horribilis*, the scientific name for the grizzly bear, refers to the bears
 1 kingdom and phylum
 2 kingdom and species
 3 genus and phylum
 4 genus and species

2 *Homo erectus* and *Homo sapiens* are classified in the same
 1 kingdom, phylum, genus, and species
 2 kingdom, phylum, and genus
 3 phylum, genus, and species
 4 genus and species

3 Which group of organisms is arranged in order of increasing complexity according to a modern classification system?
 1 protozoa, coelenterates, annelids, arthropods
 2 coelenterates, protozoa, arthropods, birds
 3 protozoa, arthropods, mammals, coelenterates
 4 amphibians, protozoa, coelenterates, arthropods

4 In a modern system of classification, two organisms would be most closely related if they were classified in the same
 1 kingdom 2 phylum 3 genus 4 species

5 The scientific name of the common housefly is *Musca domestica.* This name indicates the housefly's
 1 genus and species
 2 phylum and genus
 3 kingdom and phylum
 4 kingdom and species

6 Based on the five-kingdom system of classification, which two groups of organisms are included in the plant kingdom?
 1 protozoa and slime molds
 2 algae and fungi
 3 bryophytes and tracheophytes
 4 sponges and coelenterates

7 Organisms that are very similar in structure and in the manner in which they perform life functions, but are not capable of interbreeding, could be classified in the same
 1 genus, but different species
 2 species, but different phyla
 3 phylum, but different kingdoms
 4 genus, but different kingdoms

8 Based on the modern system of classification, unicellular organisms that contain both chloroplasts and flagella would be classified as
 1 annelids 2 protists 3 chordates 4 tracheophytes

9 In which kingdom is an organism classified if it lacks a membrane separating most of its genetic material from its cytoplasm?
 1 protist 2 plant 3 monera 4 animal

10 Blue-green algae lack a membrane separating their nuclear material from their cytoplasm. On this basis these organisms are classified as
1 fungi 2 monerans 3 protists 4 plants

III. Unity of Life

A. Structure of Living Organisms

1. Cell Theory

The unity of life, or how living things are alike, can be supported by the two main statements of the cell theory:

1. Cells are the **basic units of structure and function** of living things.
2. Cells **come from preexisting cells.**

Historical Background

The improvement of the microscope and microscopic techniques throughout the last four centuries has allowed scientists to observe cells and to better understand living things. The following is a listing of some historical scientists and their major discoveries which contributed to the establishment of the cell theory:

- **Anton van Leeuwenhoek** (1632 - 1723) discovered one-celled organisms in water by using a simple one lens microscope.
- **Robert Hooke** (1635 - 1703) used a two lens microscope to examine the bark of a tree. The empty cork cells led him to discover the cell wall and name the empty structures - "cells."
- **Robert Brown** (1773 - 1858) discovered a body, which he called a nucleus, at the center of most cells.
- **Matthias Schleiden** (1804 - 1881) studied many types of plants and determined that all plants were composed of cells.
- **Theodor Schwann** (1810 - 1882) studied many types of animals and determined that all animals were composed of cells.
- **Louis Pasteur** (1822 - 1895) proved that tiny microscopic organisms (microbes) caused disease.
- **Rudolph Virchow** (1821 - 1902), after studying previous discoveries, concluded that all cells arise from preexisting cells.

Louis Pasteur

Methods of Cell Study

Continuing advances in techniques and instrumentation have enabled biologists to increase their understanding of cell structures and functions.

Instrumentation

The **compound light microscope** is the major tool for the study of cells. It is most likely used in your school laboratory and can produce magnification, generally of up to 500x. Although some compound light microscopes can produce a magnification of 2000x, there is a limitation to the magnification, due to the **resolving power** of the lens. Therefore, it is primarily useful for viewing large cell organelles, such as nuclei and chloroplasts. Parts of the compound light microscope include:

- **Ocular lens** (eyepiece) is used to "look through" and generally magnifies at 10x.
- **Objective lenses**, found closest to the object to be studied, provide the rest of the magnification of the light compound microscope and generally magnify at 10x to 50x. For example, a 10x ocular and a 40x objective produce a total **magnification** of 400x.
- **Stage** is the flat surface (platform) on which the specimen (slide) to be studied is placed.
- **Stage clips** are used to hold the specimen (slide) in place.
- **Coarse adjustment** (the larger knob) is used to focus under low power. It causes large movement for focusing.
- **Fine adjustment** (the smaller knob) is used to focus under the higher powers. It causes small movement for focusing.
- **Light source** provides the light which must pass through the object to be studied in order to provide an image in the objective lens.
- **Diaphragm**, found under a stage, controls the amount of light passing through the specimen and into the objective lens.

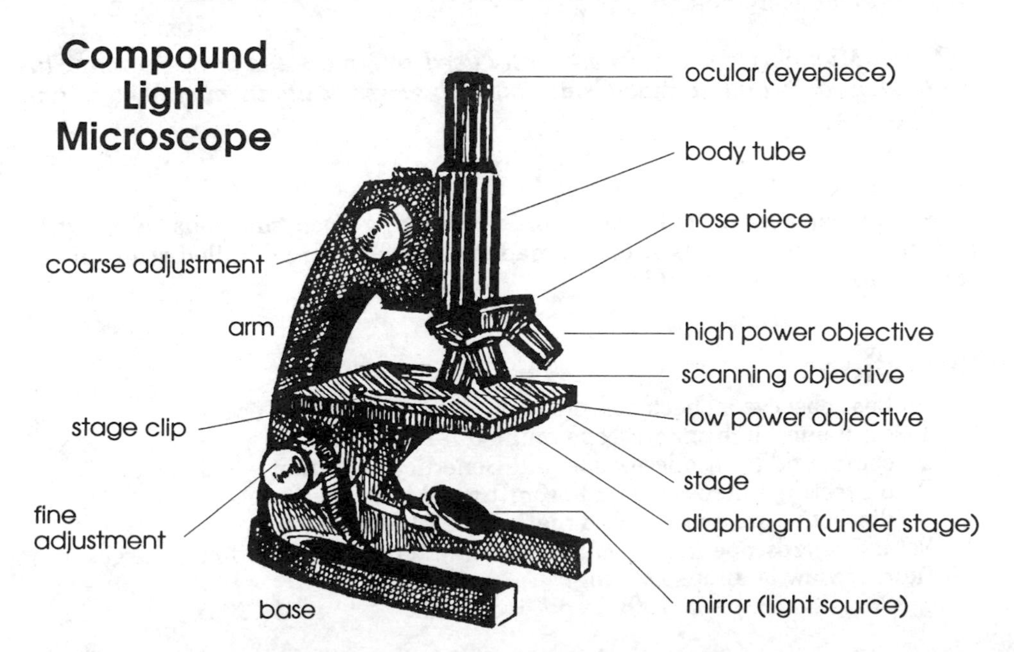

Electron Microscope

Other microscopes used by scientists include the **electron microscope,** which can produce a magnification of greater than 200,000x and is useful for studying the small cell organelles and detail structures of the cell, such as the mitochondria and chromosomes. A third, low-powered microscope is the **dissecting microscope.** It is used during dissection work at a magnification of 5x to 20x.

The **ultracentrifuge** is a tool that separates cell parts according to their relative densities by spinning the cellular materials very fast.

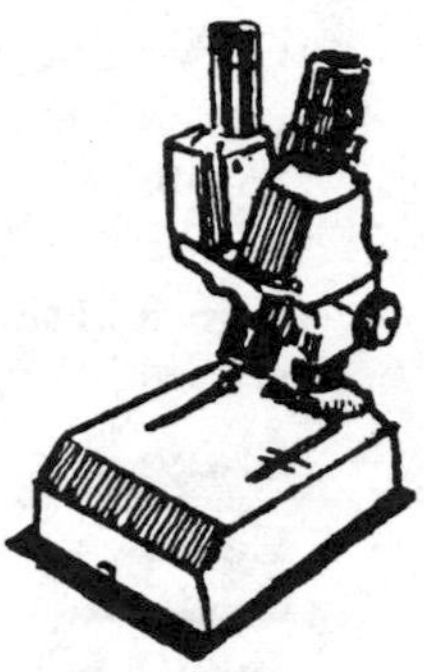

Dissecting Microscope

Microdissection instruments help the biologist to perform very small dissection of cells. These instruments are also used to transfer large cell organelles between cells.

Measurement

The very small size of most cells requires the use of a small unit for measurement. This unit is known as the **micrometer** (**µm**). One thousand micrometers is equal to one millimeter (mm). Cells often have diameters of only 10 to 50 micrometers.

[Note: Micrometer is a term currently used to replace the term micron. Students should be able to make conversions between units in micrometers and millimeters.]

Techniques

The development of staining techniques, using such solutions as **Lugol's iodine and methylene blue,** has made possible a more detailed study of cell structures.

Questions

1 To best observe a cheek cell nucleus, a student should use a
  1 compound light microscope and iodine solution
  2 compound light microscope and Benedict's solution
  3 dissecting microscope and bromthymol blue solution
  4 dissecting microscope and methylene blue solution

2 Which microscope magnification should be used to observe the largest field of view of an insect wing?
  1 20x 2 100x 3 400x 4 900x

3 A student observed an ameba under the low-power objective (10x) of a compound microscope and noted that the organism occupied one-fourth of the field of view after it was centered. When the student changed to the high-powered objective (40x), how much of the field of view would the ameba most likely occupy?
1 0%
2 50%
3 75%
4 100%

4 A student viewing a specimen under low power of a compound light microscope switched to high power and noticed that the field of view darkened considerably. Which microscope part should the student adjust to brighten the field of view?
1 diaphragm
2 coarse adjustment
3 fine adjustment
4 eyepiece

Base your answers to questions 5 and 6 on the diagram of the compound light microscope at the right and on your knowledge of biology.

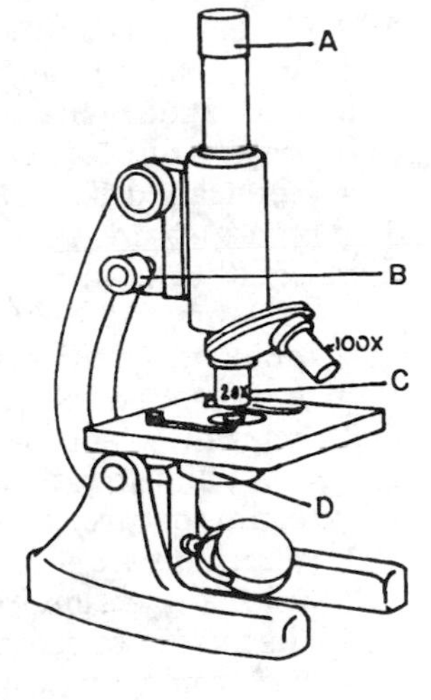

5 Which microscope part regulates the amount of light that reaches the objective lens?
1 A
2 B
3 C
4 D

6 If the ocular (eyepiece) of this microscope is marked 10x and the high power objective is marked 100x, what is the greatest total magnification of this microscope?
1 100x
2 200x
3 1000x
4 2000x

7 When a student uses a compound microscope and switches from low power to high power, the diameter of the field of view will
1 decrease 2 increase 3 remain the same

8 When the microscope is changed from low power to high power, the distance between the objective lens and the cover slip of the slide will
1 decrease 2 increase 3 remain the same

9 While focusing a microscope on high power, a student crushed the cover slip. The student probably
1 closed the diaphragm
2 rotated the eyepiece
3 turned up the light intensity
4 used the coarse adjustment

10 A student using a compound microscope measured the diameter of several red blood cells and found that the average cell length was 0.008millimeter. What is the average length of a single red blood cell in micrometers?
1 0.8 2 80 3 8 4 800

11 An instrument used to collect ribosomes for chemical analysis is
1 a compound microscope
2 an ultracentrifuge
3 a scalpel
4 an electron microscope

12 Which substance, when added to a wet mount containing starch grains, would react with the starch grains and make them more visible?
1 litmus solution
2 iodine solution
3 distilled water
4 bromthymol blue

13 Which technique is commonly used to separate the pigments found in a coleus leaf?
1 electron microscopy
2 ultracentrifugation
3 microdissection
4 amniocentesis

14 To collect mitochondria from cells and study their structure in fine detail, which instruments would a scientist most likely use?
1 microdissection apparatus and phase-contrast microscope
2 ultracentrifuge and electron microscope
3 microdissection apparatus and dissecting microscope
4 ultracentrifuge and compound light microscope

15 An important function of the ultracentrifuge is to
1 increase the magnification of different parts of cells
2 slice specimens embedded in wax into thin sections
3 transplant nuclei from one cell to another
4 separate different cell organelles by their density

16 Which instrument would provide the most detailed information about the internal structure of a chloroplast?
1 a compound light microscope
2 an electron microscope
3 a phase contrast microscope
4 an interference microscope

17 Of the following, which instrument is most commonly used to observe the external features of a grasshopper's abdomen?
1 ultracentrifuge
2 microdissection instrument
3 dissecting microscope
4 electron microscope

18 The best instrument to use to separate the liquid portion from the solid portions in whole blood is
1 a centrifuge
2 a compound microscope
3 an electron microscope
4 a microdissection instrument

19 To transplant a nucleus from one cell to another cell, a scientists would
1 use an electron microscope
2 use staining techniques
3 use an ultracentrifuge
4 use microdissection instruments

20 Which piece of equipment represented below would most likely be used to separate the red blood cells from blood plasma?

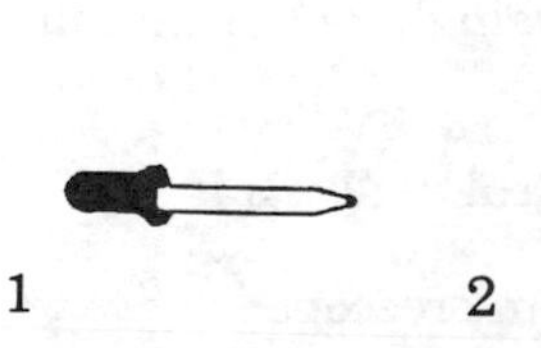
1

2

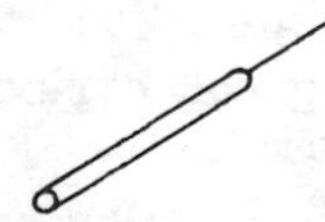
3

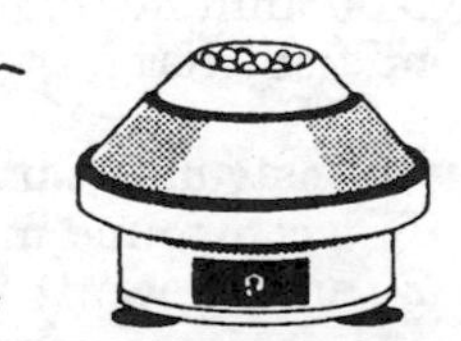
4

Plant Cell

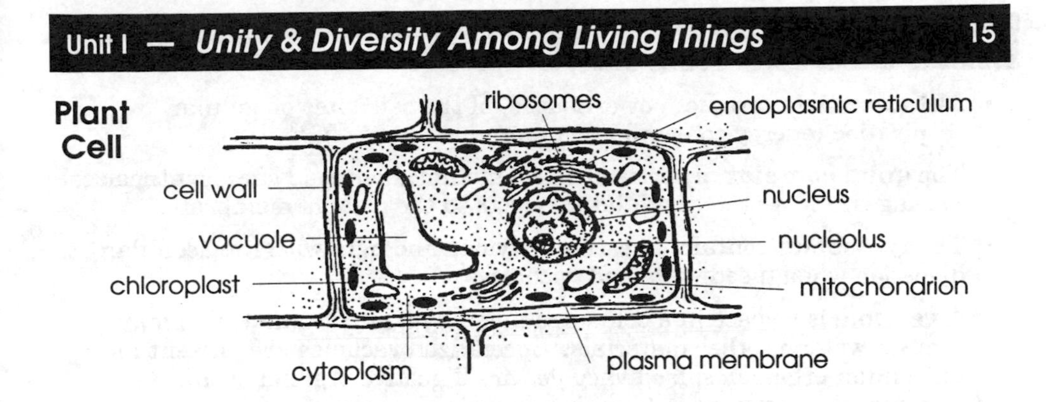

Cell Organelles

In cells, various specialized functions occur in subcellular structures known as organelles. Some major organelles and their functions are:

Organelle and Function

- The **plasma membrane** is the outer membrane of the cell and regulates the transport of certain materials into and out of the cell. It is selectively permeable (semipermeable).
  [The Fluid-Mosaic Model will be taught in Unit II.]
- **Cytoplasm** is a fluid-like environment between the nucleus and the plasma membrane in which other organelles are suspended and within which diffusion and many biochemical processes occur.
- The **nucleus** is spherical, often found in the center of the cell, and surrounded by the nuclear membrane. It contains genetic information (within **chromosomes**) and directs cell activities.
- The **nucleolus**, found within the nucleus, is involved with the synthesis of ribosomes (mRNA) which manufacture proteins.
- The **endoplasmic reticulum** is a series of interconnecting channels (membranes) associated with storage, synthesis, and the transport of substances (mainly proteins) within the cell.
- **Ribosomes** are the sites of protein (enzyme) synthesis, either occurring as free bodies or attached to membranes of the endoplasmic reticulum.

Animal Cell

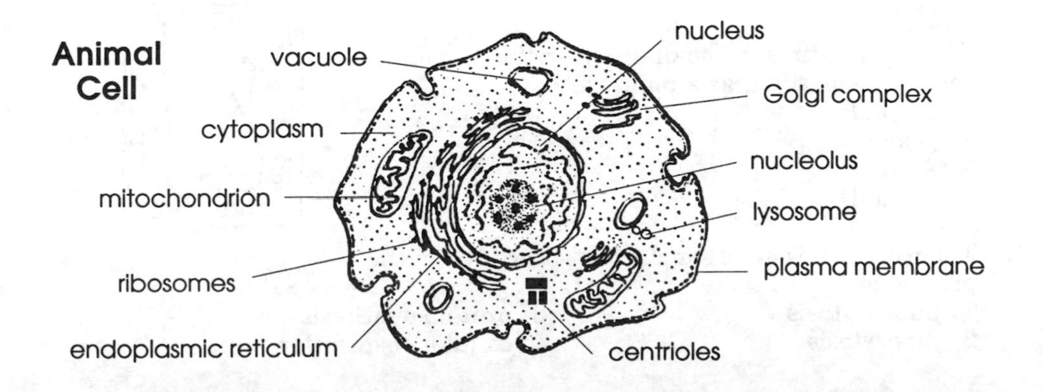

- **Mitochondria** are the "powerhouses" of the cell, sites of cellular respiration (energy conversion) and production of ATP.
- The **golgi complex,** a series of small membrane-bound sacs, synthesizes, packages, and secretes cellular products to the plasma membrane.
- The **lysosomes** contain digestive enzymes and help with intracellular digestion when fused with food vacuoles.
- A **vacuole** is a space in a cell surrounded by a membrane which may contain water or other materials. Specialized vacuoles are present in unicellular organisms: food vacuoles are digestive organelles and **contractile vacuoles** help maintain water balance.
- **Centrioles** are cylindrical structures found in the cytoplasm which appear to function during cell division. Although commonly found in animal cells, they are rarely found in plant cells.
- **Chloroplasts,** primarily found in plant and algae cells, are pigment containing structures which serve as a site for photosynthesis. Their main pigment is **chlorophyll.**
- The **cell wall,** made of **cellulose,** is a nonliving structure which surrounds, protects, and supports a plant cell.

2. Exceptions to the Cell Theory

There are several exceptions to the cell theory:

- Since all cells must come from pre-existing cells, what produced the first cell? The first cell could not have arisen from a non-existing (previously existing) cell.
- Viruses are not composed of cells and do not carry on most cellular activities, except for reproduction and synthesis (only within the host cell). However, viruses do contain genetic material and can "take over" life activities in the host cell.
- Although considered organelles, mitochondria and chloroplasts contain their own genetic material and can independently reproduce in the presence of a cell.

Questions

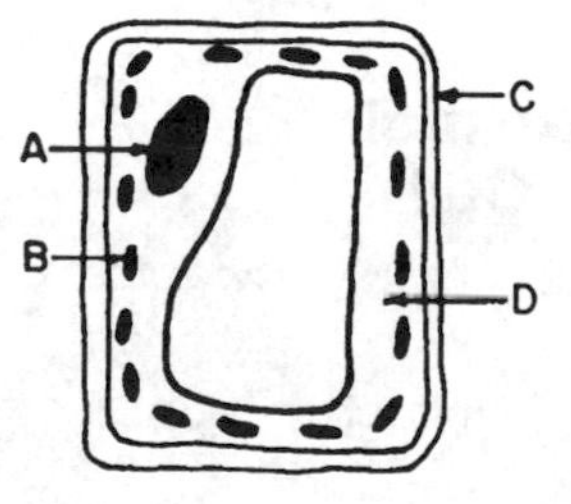

1 Which structures in the diagram enable the observer to identify it as a plant cell?
1 A and B
2 B and C
3 A and C
4 B and D

2 The ribosome is an organelle that functions in the process of
1 phagocytosis
2 pinocytosis
3 protein synthesis
4 cellular respiration

3 Which organelles can be observed only with the aid of an electron microscope?
 1 ribosomes 3 nuclei
 2 chloroplasts 4 cell wall

4 Intracellular transport of materials is most closely associated with which cell organelle?
 1 cell membrane 3 cell wall
 2 ribosome 4 endoplasmic reticulum

5 Some scientists disagree on whether or not viruses are alive. A major reason for this disagreement is that viruses
 1 cannot manufacture food
 2 are not composed of units of structure known as cells
 3 do not contain nucleic acids
 4 do not contain the element carbon

6 Enzyme synthesis in a living cell occurs at which cytoplasmic organelles?
 1 centrioles 3 vacuoles
 2 ribosomes 4 chromosomes

7 Which cell organelle is involved most directly in the digestion of large particles brought into a cell's vacuole?
 1 ribosome 3 lysosome
 2 mitochondrion 4 nucleolus

8 Which structure includes all of the others?
 1 nucleolus 3 chromosomes
 2 nucleus 4 genes

9 An organelle found in most plant cells, but absent from animal cells, is the
 1 contractile vacuole 3 chloroplast
 2 centriole 4 Golgi complex

10 In which organelles are enzymes for intracellular digestion stored?
 1 centrioles 3 nucleoli
 2 lysosomes 4 chloroplasts

11 According to the cell theory, which statement is correct?
 1 Viruses are true cells.
 2 Cells are basically unlike in structure.
 3 Mitochondria are found only in plant cells.
 4 Cells come from preexisting cells.

12 Most cellular respiration in plants takes place in organelles known as
 1 chloroplasts 3 stomates
 2 ribosomes 4 mitochondria

13 Pathways for the transport of materials within a living cell are provided by the
 1 centrosome 3 ribosome
 2 endoplasmic reticulum 4 cell membrane

14 Which organelle contains hereditary factors and controls most cell activities?
 1 nucleus 3 plasma membrane
 2 vacuole 4 endoplasmic reticulum

15 Centrioles are cell structures involved primarily in
 1 cell division 3 enzyme production
 2 storage of fats 4 cellular respiration

B. Chemistry of Living Organisms

Biochemistry – Terms & Concepts

Chemical Elements
- C - H - O - N
- S - P - Mg - I - Fe
- Ca - Na - Cl - K

Organic Compounds
- Carbohydrates
- Lipids
- Proteins

Inorganic Compounds
- Water - Salts
- Acids - Bases

pH scale

Dehydration Synthesis

Hydrolysis

Enzymes
- Active Site

Enzyme-Substrate-
- Complex

"Lock & Key" Model

Enzyme Actions
- pH
- Temperature
- Substrate Amount
- Enzyme Amount

1. Chemical Elements in Living Matter

The cell is a complex "Chemical factory" composed of some of the same elements found in the nonliving environment. Of all the elements present in living matter, carbon (**C**), hydrogen (**H**), oxygen (**O**), and nitrogen (**N**) are present in the greatest percentages. Some examples of elements found in smaller quantities are sulfur (S), phosphorus (P), magnesium (Mg), iodine (I), iron (Fe), calcium (Ca), sodium (Na), chlorine (Cl), and potassium (K).

2. Chemical Compounds in Living Matter

Organisms consist of inorganic and organic compounds. A **compound** is a substance made up of two or more different elements bonded together. When the *transfer* of electrons between the atoms of a compound is involved, the bond is called **ionic**. When there is a *sharing* of electrons between the atoms, the bond is called **covalent**.

Atomic Structures and Compounds

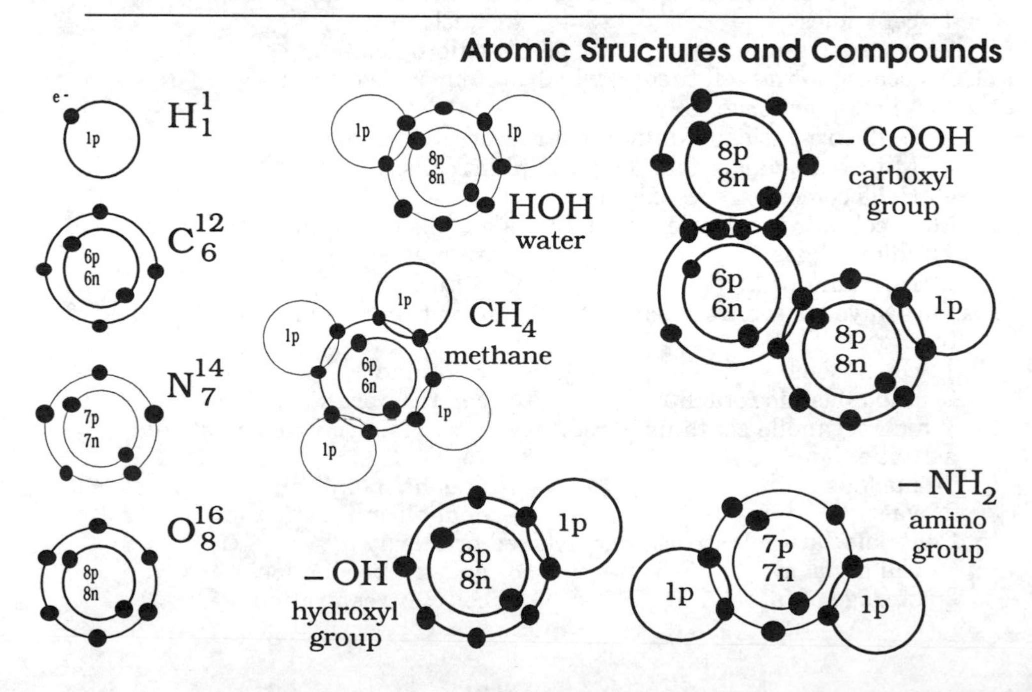

Inorganic Compounds

Inorganic compounds are chemical compounds that *lack the combination* of the elements carbon and hydrogen. The principal inorganic compounds in living organisms include water, salts, inorganic acids, and bases.

- **Water** makes up from 60 percent to 98 percent of a living organism and is necessary for transport and chemical activities.
- **Salts** are chemical compounds of ionically bonded metallic and nonmetallic particles. For living organisms, salts help maintain cellular osmotic (fluid - water) balance and supply ions which are necessary for cellular chemical reactions.
- **Acids and bases** are used by cells to maintain a balance (acidity/alkalinity — pH) of hydrogen-ion concentration. A **pH scale** is used to show the relative amounts of H^+ and OH^- ions present in a solution. A pH of 7 represents a neutral substance (equal amounts of H^+ and OH^- ions). On the pH scale, numbers 6.9 (weak) to 0 (strong) represent acids. Bases are between 7.1 (weak) to 14 (strong).

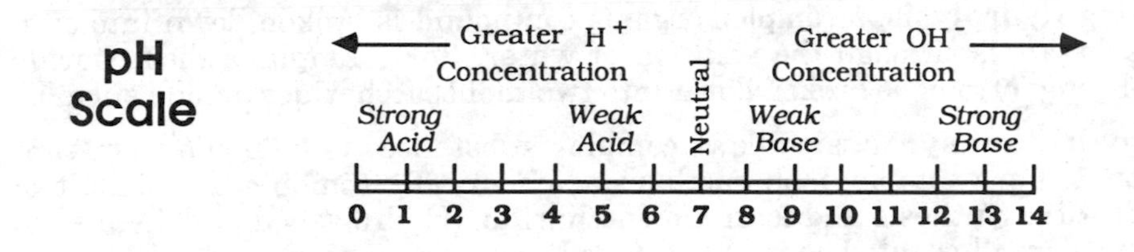

Organic Compounds

Organic compounds are compounds which *contain both* the elements carbon and hydrogen.

The major categories of organic compounds found in living things are **carbohydrates** (sugars and starches), **lipids** (fats, oils, and waxes), **proteins** (functional and structural), and **nucleic acids** (RNA and DNA).

Carbohydrates

Composition of Carbohydrates

Carbohydrates consist of the elements carbon (C), hydrogen (H), and oxygen (O). Hydrogen and oxygen atoms are most often present in a 2:1 ratio. The basic unit of the carbohydrate is the simple sugar or **monosaccharide.** A pentose sugar has 5 carbon atoms. A hexose sugar has 6 carbon atoms. Glucose, $C_6H_{12}O_6$, is an example.

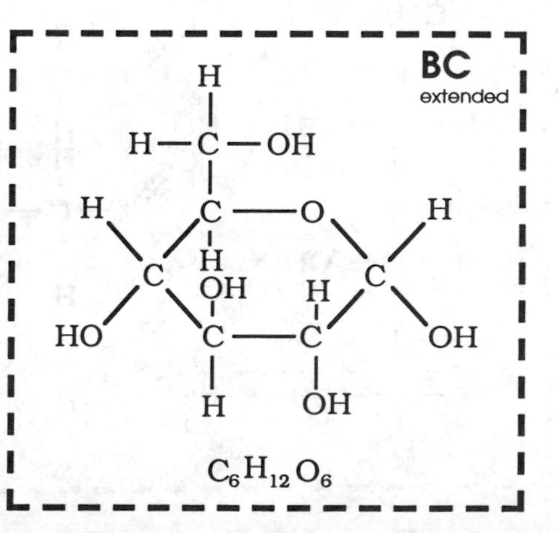

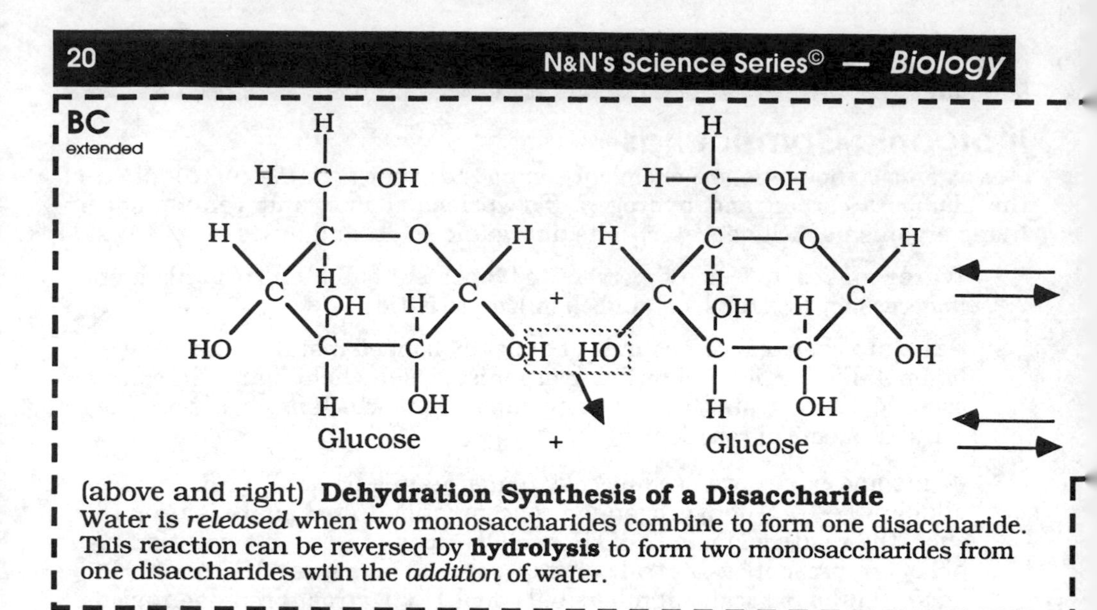

(above and right) **Dehydration Synthesis of a Disaccharide**
Water is *released* when two monosaccharides combine to form one disaccharide. This reaction can be reversed by **hydrolysis** to form two monosaccharides from one disaccharides with the *addition* of water.

Hydrolysis and Dehydration Synthesis

During **hydrolysis**, a complex organic compound is broken down into simpler substances through the addition of water. For example, a disaccharide (double sugar) may be broken down into two monosaccharides (single sugar).

Dehydration synthesis allows complex substances to be made from simpler ones. For example, monosaccharides chemically combine as a result of dehydration synthesis and form a disaccharide. (Hydrolysis and dehydration synthesis are illustrated on the next page.)

Further dehydration synthesis of monosaccharides will result in the formation of a polysaccharide such as starch. Conversely, starches can be broken

(below and right) **Dehydration Synthesis of a Triglyceride**
Three molecules of water are *released* when one glycerol combines with three fatty acids to form a lipid. This reaction can be reversed by **hydrolysis** with the *addition* of three water molecules.

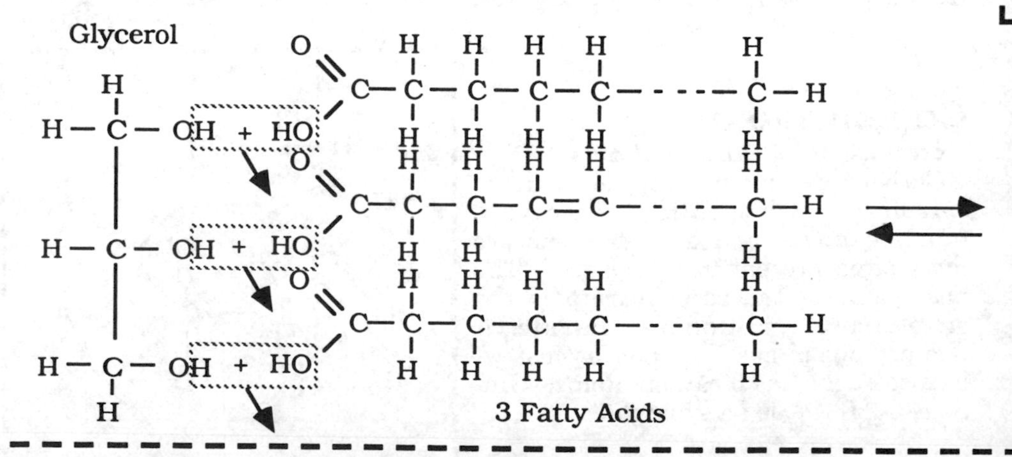

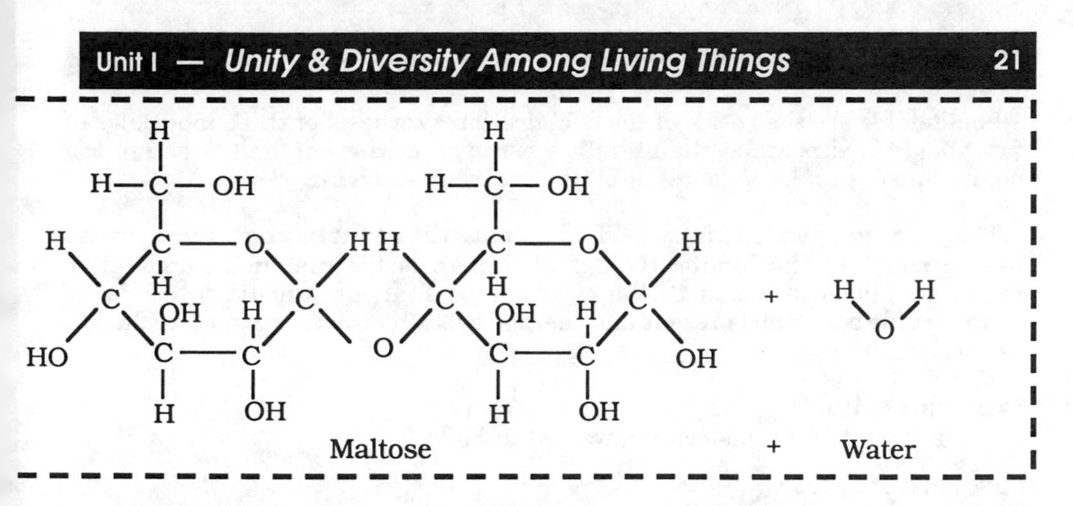

down into monosaccharides through repeated hydrolysis. (Note: the reaction arrow is reversed to show both dehydration synthesis (to the right) and hydrolysis (to the left).

Examples of Carbohydrates

Carbohydrates include all sugars and starches. The names of sugars end in "-ose." Some examples of common carbohydrates follow:

- simple sugars (monosaccharides) — glucose, fructose, and galactose
- double sugars (disaccharides) — maltose, lactose, and sucrose
- polysaccharides — starches, cellulose, and glycogen

Functions of Carbohydrates

In organisms, carbohydrates are used primarily as sources of energy, starch in plants and glycogen (animal starch) in animals. Also, carbohydrates are components of cell structures such as the cell wall, cellulose.

Lipids

Composition of Lipids

Lipids, like carbohydrates, contain carbon, hydrogen, and oxygen. However, the ratio of hydrogen to oxygen is *much greater* than 2:1 and is not the same from one lipid to another.

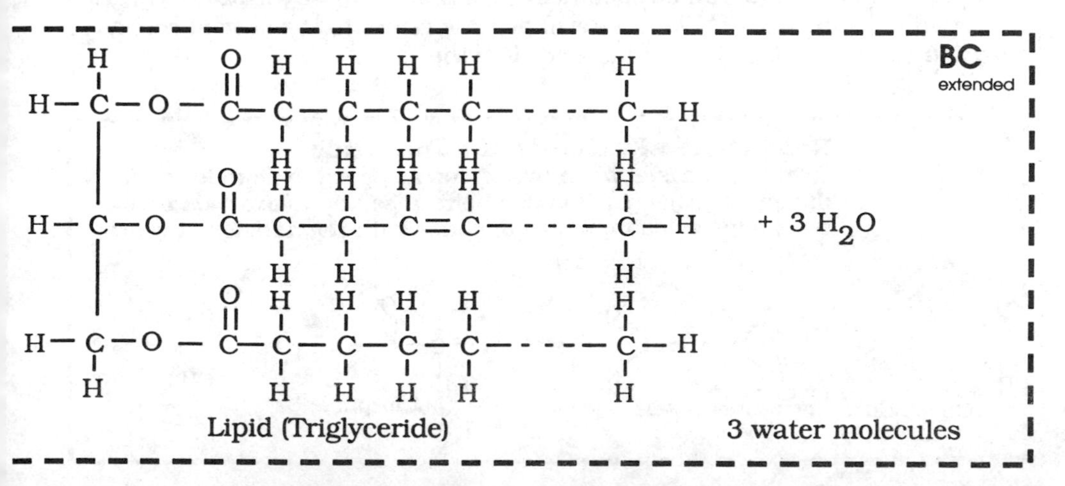

Some lipids are the result of the dehydration synthesis of three molecules of fatty acids (hydrocarbon chains with a carboxyl acid group on the end of the chain) and one molecule of glycerol (three carbon alcohol).

There are two types of fatty acids. In a **saturated fatty acid**, there are *no* double bonds in the hydrocarbon chain. There is the maximum amount of hydrogen atoms attached to the carbon atoms. In an **unsaturated fatty acid**, double bonds are present and there are less hydrogen atoms attached to the carbon atoms.

Examples of Lipids

Lipids include fats, cholesterol, waxes, and oils.

Functions of Lipids

Lipids are used primarily as sources of stored energy and also as components of cell structures such as the nuclear and plasma membranes.

Proteins

Composition of Proteins

Proteins contain carbon, hydrogen, oxygen, nitrogen, and, in many instances, sulfur. A protein is composed of building blocks known as **amino acids**. An amino acid contains an amino group (—NH_2), a carboxyl group (—COOH), and another group (R = variable sidechain) which differs in each amino acid and gives the amino acid its special properties.

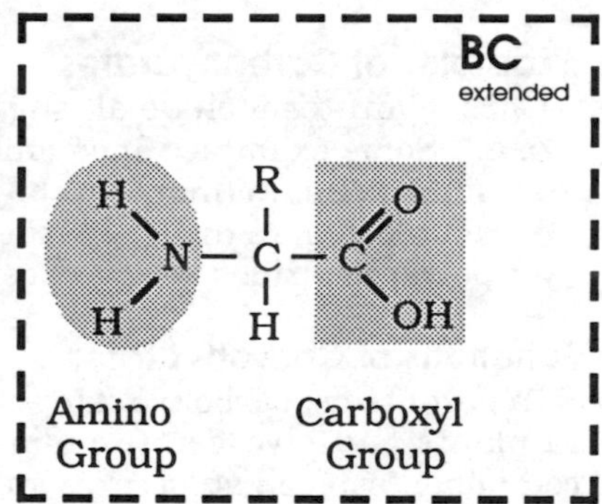

Twenty amino acids are usually found in living systems. Some proteins contain special amino acids that supplement the basic set of 20 amino acids. All the special amino acids are formed by modification of a common amino acid.

Two amino acid units can be chemically combined by dehydration synthesis and form a **dipeptide**. [Note: the term peptide refers to the special *carbon to nitrogen bond* that joins two amino acids together.]

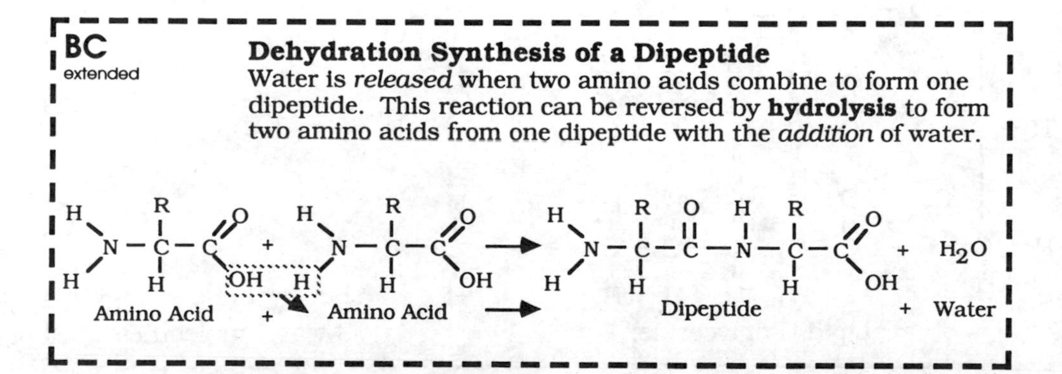

BC
extended
Dehydration Synthesis of a Dipeptide
Water is *released* when two amino acids combine to form one dipeptide. This reaction can be reversed by **hydrolysis** to form two amino acids from one dipeptide with the *addition* of water.
Amino Acid + Amino Acid → Dipeptide + Water
OH H
H_2O

Polypeptides result from the dehydration synthesis of many amino acids. Most proteins are large and complex and are usually composed of one or more polypeptide chains. There is an extremely large number of different proteins. The basis for variability includes differences in the number or kind and the sequence of amino acids present.

Examples of Proteins

Proteins include insulin (a hormone that regulates sugar levels), hemoglobin (oxygen carrier in red blood cells), and enzymes (necessary for most all cellular chemical reactions).

Functions of Proteins

Proteins are components of cell structures and organelles (structural proteins) and enzymes (functional proteins).

Questions

1 Which element is present in living cells and in all organic compounds?
 1 potassium 2 sulfur 3 nitrogen 4 carbon

2 A biochemist was given a sample of an unknown organic compound and asked to determine the class of organic compounds to which it belonged. The chart at the right represents the results of the biochemist's analysis of the sample.

Element	Number of Atoms per Molecule
C	12
H	22
O	11
S	0
N	0
P	0

Based on these results, to which class of organic compounds did this sample belong?
 1 lipid 3 salt
 2 protein 4 carbohydrate

3 Which list of molecules is arranged in order of increasing size?
 1 oxygen, starch, glucose, sucrose
 2 sucrose, oxygen, starch, glucose
 3 oxygen, glucose, sucrose, starch
 4 starch, glucose, sucrose, oxygen

4 Which types of compounds are not classified as carbohydrates?
 1 lipids 2 sugars 3 starches 4 polysaccharides

5 Which is an example of an inorganic compound?
 1 glucose 2 maltose 3 water 4 starch

6 The process in which two monosaccharide molecules combine under certain conditions is an example of
 1 dehydration synthesis 3 enzyme deactivation
 2 protein synthesis 4 alcoholic fermentation

7 In most carbohydrate molecules, the ratio of hydrogen atoms to oxygen atoms is
 1 1:2 2 3:1 3 2:1 4 1:3

8 All carbohydrates are compounds that contain the elements
 1 iron, magnesium, and carbon
 2 oxygen, hydrogen, and nitrogen
 3 carbon, oxygen, and nitrogen
 4 carbon, hydrogen, and oxygen

9 Which organic compound is a building block of a triglyceride, containing three carbon atoms, five hydrogen atoms, and three hydroxyl groups?
1 glycerol
2 amino acid
3 saturated fatty acid
4 unsaturated fatty acid

10 Which is an organic compound found in most cells?
1 glucose
2 water
3 sodium chloride
4 oxygen gas

11 The organic chemical with the molecular formula $C_{18}H_{36}O_2$ is an example of
1 an amino acid
2 an element
3 a monosaccharide
4 a fatty acid

12 A carbon to nitrogen bond may also be called a (an)
1 peptide bond
2 hydrogen bond
3 ionic bond
4 polymer

13 Which of the following may be used as a building block for proteins?
1 monosaccharide
2 fatty acid
3 glycerol
4 amino acid

14 Which process produces peptide bonds?
1 digestion
2 hydrolysis
3 dehydration synthesis
4 enzyme deactivation

15 The reactions involving most chemical compounds in living systems depend upon the presence of
1 sulfur as an enzyme
2 salt as a substrate
3 water as a solvent
4 nitrogen as an energy carrier

16 Small soluble food molecules are converted to larger, insoluble molecules by the process of
1 hydrolysis
2 respiration
3 dehydration synthesis
4 fermentation

17 Which organic compound is correctly matched with the subunit that composes it?
1 maltose-amino acid
2 starch-glucose
3 protein-fatty acid
4 lipid-sucrose

18 Which substance is an inorganic compound that is necessary for most of the chemical reactions to take place in living cells?
1 glucose
2 water
3 starch
4 amino acid

19 If a specific carbohydrate molecule contains ten hydrogen atoms, the same molecule would most probably contain
1 one nitrogen atom
2 ten nitrogen atoms
3 five oxygen atoms
4 twenty oxygen atoms

20 A specific organic compound contains only the elements carbon, hydrogen, and oxygen in the ratio of 1:2:1 This compound is likely a
1 nucleic acid
2 monosaccharide
3 protein
4 lipid

3. Chemical Control

Living matter is in a state of dynamic chemical activity. Perhaps the most significant distinction between living and nonliving matter is the continuous and controlled chemical activity present in living systems.

Role of Enzymes

Enzymes may be referred to as *organic catalysts* since they are the principal regulators of most chemical activity in living systems. Enzymes are not changed during the reaction and therefore can be reused.

- *Each chemical reaction requires a specific enzyme.*
- *Enzymes modify the rate of reactions.*

Structure of Enzymes

Enzymes are large complex proteins consisting of one or more polypeptide chains whose names end in "**—ase.**" Enzymes are named for the substrate (chemical being acted upon). For example, Malt<u>ose</u> is hydrolyzed by Malt<u>ase</u>.

- **Protein Nature**. All enzymes are either exclusively proteins or are proteins with non-protein parts known as **coenzymes**. Often, coenzymes are vitamins.

- **Active Site**. Usually enzyme molecules are much larger than the molecules with which they interact. Only a small area of the enzyme is actually involved in the reaction. The specific way in which these chains fold results in the formation of pockets into which reacting molecules fit. This specificity of the enzyme is dependent on the **active site**.

Function of Enzymes

Evidence has accumulated to permit biochemists to develop a model of enzyme activity (catalytic action) which is useful in visualizing the nature of its function.

Enzyme Substrate Complex

It is thought that, for an enzyme to affect the rate of a reaction, the following events take place:

- The enzyme must form a temporary association with the substance or substances whose reaction rate it affects. These substances are known as substrates.

- This association between enzyme and substrate is thought to involve a close physical association between the molecules and is called the **enzyme-substrate complex**.

- While the enzyme-substrate complex is formed, the enzyme action takes place.

- Upon completion of the reaction, the enzyme and product(s) separate. The enzyme molecule is now available to form additional complexes.

Although enzymes may be reused in cells, they eventually are destroyed and new ones must be synthesized.

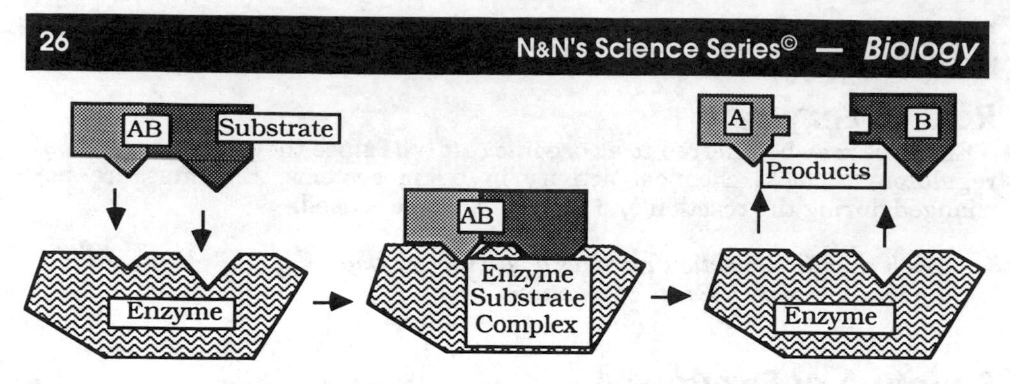

Lock and Key Model for Enzyme Activity
[Note that the enzyme and substrate join together in the enzyme-substrate complex, yet the enzyme is unaffected at the end of the reaction.] This illustration represents activity during hydrolysis.

"Lock and Key" Model

The suggestion that a particular enzyme molecule will only interact with a single type of substrate has given rise to the "lock and key" model of enzyme specificity. Like a key that will open only a particular lock, a particular enzyme will usually only form a complex with one particular type of substrate.

Factors Influencing Action of Enzymes

The rate of enzyme action is not fixed, but varies according to the environmental conditions of the reacting substances.

Such factors as pH, temperature, and relative amounts of enzyme substrate can determine the rate of enzyme action.

Extended Area: ***Biochemistry***

1. Temperature

In general, as temperature increases, the rate of enzyme action increases; that is, only to a point. The temperature at which an enzyme is most efficient is the optimum temperature.

At relatively high temperatures, however, the shape of enzyme molecules tends to be altered, thus making the enzyme ineffective. This distortion of enzyme molecules at high temperatures is **enzyme denaturation**. For many enzymes in the human body, denaturation begins to occur at around 40°C.

The response to changing temperature for many enzymes of humans is illustrated in the graph at the right.

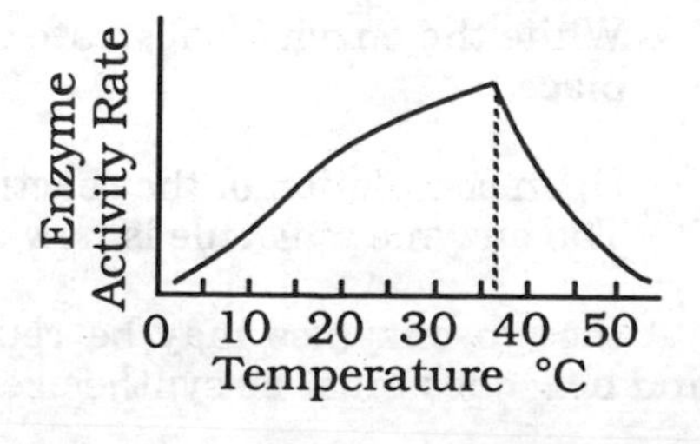

2. **Relative Amounts of Enzyme and Substrate**

The rate of enzyme action also varies according to the amount of available substrate molecules. When an excess of substrate is added to a system with a fixed concentration of enzymes, the rate of enzyme action tends to increase to a point and then remain fixed as long as the enzyme concentration remains constant.

The graph at the right illustrates the pattern of enzyme action rates when an excess of substrate is added to a system with a fixed enzyme concentration.

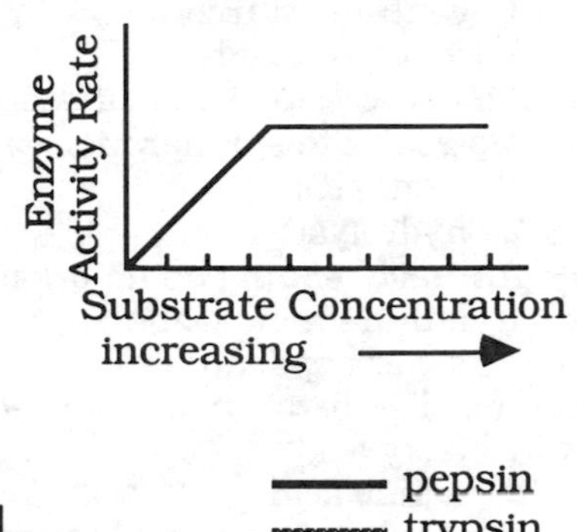

3. **pH**

The rate at which enzyme-regulated reactions occur varies according to the pH of the environment.

For many enzyme-controlled reactions, a pH of 7 provides the optimum environment. Other enzymes work best within different pH ranges. Specific pH ranges of enzymes are illustrated at the right.

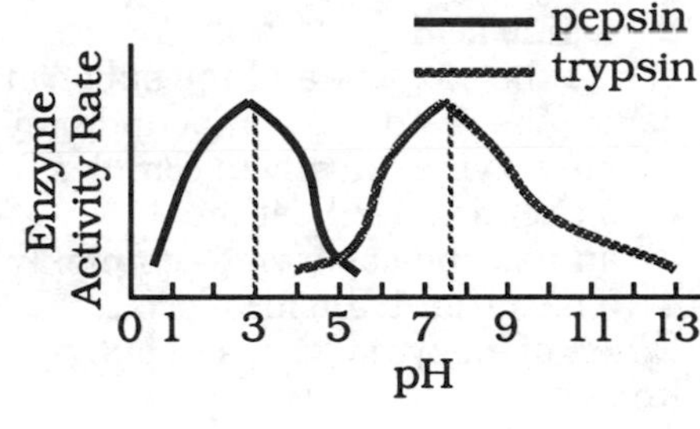

Questions

1 An enzyme-substrate complex may result from the interaction of molecules of
   1 glucose and lipase
   2 fat and amylase
   3 sucrose and maltase
   4 protein and protease

2 Since an enzyme both becomes a part of the reaction and does not become destroyed during the reaction, an enzyme may also be referred to as a(an)
   1 vitamin
   2 substrate
   3 organic catalyst
   4 nucleotide

3 A certain species of bacteria carries out its life processes most efficiently in the acidic waters of a spruce bog. The pH range for most enzyme action in this species would probably be
   1 2-5 pH
   2 8-11 pH
   3 7-9 pH
   4 10-13 pH

4 Which of the following variables has the least direct affect on the rate of a hydrolytic reaction regulated by enzymes?
   1 temperature
   2 carbon dioxide concentrations
   3 pH
   4 enzyme concentrations

5 Enzymes are produced as a direct result of which process?
   1 protein synthesis
   2 photosynthesis
   3 respiration
   4 enzymatic hydrolysis

6 Most chemical reactions in organisms are regulated by organic catalysts known as
1 enzymes
2 polysaccharides
3 hormones
4 nucleotides

7 The enzyme salivary amylase will act on starch but not on protein. This action illustrates that salivary amylase
1 contains starch
2 is not reusable
3 is chemically specific
4 lacks protein

8 The process by which digestive enzymes catalyze the breakdown of larger molecules to smaller molecules with the addition of water is known as
1 synthesis
2 hydrolysis
3 pinocytosis
4 photosynthesis

9 An "-ase" ending on an organic compound indicates
1 a protein molecule
2 a carbohydrate
3 an enzyme
4 a triglyceride

10 Which substance is produced as a result of the action of lipase?
1 fatty acid
2 amino acid
3 protease
4 glucose

11 When the enzyme and the substrate become chemically linked to each other, the resulting compound may be called
1 an enzyme-substrate complex
2 a chemical replication
3 an enzyme-coenzyme complex
4 a bacterial transformation

12 When an enzyme works to put two reactants together, the process is known as
1 dehydration synthesis
2 digestion
3 hydrolysis
4 respiration

13 The "lock and key" model of enzyme action illustrates that a particular enzyme molecule
1 forms a permanent enzyme-substrate complex
2 may be destroyed and resynthesized several times
3 interacts with a specific type of substrate molecule
4 reacts at identical rates under all environmental conditions

14 The enzyme lipase will act on a fat but not on protein. This action illustrates that lipase
1 contains a dipeptide
2 is not reusable
3 is specific
4 lacks a fatty acid

15 In a controlled experiment, 95% of the spores in one group germinate, but only 50% of the spores in a second group germinate. The temperature of the first group of spores is found to be 35°C., and the temperature of the second group of spores is found to be 20°C. The most probable cause for the reduced germination in the second group of spores is as a result of
1 dehydration synthesis
2 enzyme deactivation
3 oxidation of glucose
4 diffusion of oxygen

SELF-HELP: Unit I "Core" Questions

[Answers for all SELF–HELP questions are found after the College Board section towards the back of this book - pages 262-264.]

1 A biologist could determine that a bean plant is more closely related to a pea plant than to a corn plant on the basis of the
1 plant's ability to carry on their life function of respiration
2 arrangement of vascular tissue in the plant's stem
3 amount of nitrogen in the soil in which they grow
4 types of animals that feed on them

2 All of the chemical reactions that take place within a plant or animal cell are known as
1 transport 2 respiration 3 synthesis 4 metabolism

3 It is known that laboratory rats do not require vitamin C in their diet. However, vitamin C is found in their tissues. This vitamin C is present due to the process known as
1 osmosis 3 excretion
2 locomotion 4 synthesis

4 Which activity of life includes the absorption and distribution of essential materials throughout an organism?
1 excretion 2 synthesis 3 locomotion 4 transport

5 In which life function is the potential energy of organic compounds converted to a form of stored energy which can be used by the cell?
1 transport 3 respiration
2 excretion 4 regulation

6 Hydras, earthworms, grasshoppers, and humans are classified in the same
1 genus 2 species 3 phylum 4 kingdom

7 In classification systems, which group is the largest subdivision of a kingdom?
1 genus 2 phylum 3 order 4 class

8 The mosquito, *Anopheles quadrimaculatus*, is most closely related in structure to
1 *Aedes sollicitans* 3 *Culex pipiens*
2 *Ades aegypti* 4 *Anopheles punctulatus*

9 The classification group which shows the greatest similarity among its members is the
1 phylum 2 genus 3 kingdom 4 species

10 The modern classification system is based on structural similarities and
1 evolutionary relationships 3 habitat similarities
2 geographic distribution 4 Mendelian principles

11 Most cell membranes are composed principally of
1 DNA and ATP 3 chitin and starch
2 proteins and lipids 4 nucleotides and amino acids

12 Which cell organelles are the sites of aerobic cellular respiration in both plant and animal cells?
1 mitochondria 3 centrosomes
2 chloroplasts 4 nuclei

13 Chloroplasts are cell structures that are located in the
1 endoplasmic reticulum 3 cytoplasm
2 cell wall 4 nucleus

14 The excretory organelles of some unicellular organisms are contractile vacuoles and
1 cell membranes
2 cell walls
3 ribosomes
4 centrioles

15 Which organelle is present in the cells of a mouse but not in the cells of a bean plant?
1 cell wall
2 chloroplast
3 plasma membrane
4 centriole

16 Starch and glycogen are examples of a group of compounds classified as
1 peptides
2 polypeptides
3 disaccharides
4 polysaccharides

17 Which formula represents an organic compound?
1 NH_3
2 NaCl
3 H_2O
4 $C_6H_{12}O_6$

18 Which substances may function as coenzymes?
1 vitamins
2 fats
3 carbohydrates
4 proteins

19 What are the end products of carbohydrate hydrolysis?
1 amino acids
2 simple sugars
3 hydrogen ions
4 fatty acids

20 Which element is present in all lipids and proteins?
1 iron
2 nitrogen
3 carbon
4 calcium

21 The process by which proteins are broken down into amino acids is known as
1 dehydration synthesis
2 ingestion
3 intracellular digestion
4 hydrolysis

22 A chemical reaction involves the production of a dipeptide from two amino acids. This is an example of
1 hydrolysis
2 digestion
3 dehydration synthesis
4 carbon fixation

23 An organic compound has both an amino group and a carboxyl group. It is most probably a (an)
1 amino acid
2 monosaccharide
3 fatty acid
4 glycerol

Base your answers to questions 24 and 25 on the chemical equation below and on your knowledge of biology.

$$C_6H_{12}O_6 + C_6H_{12}O_6 \xrightarrow{X} C_{12}H_{22}O_{11} + H_2O$$

24 The process represented by the equation is known as
1 fermentation
2 hydrolysis
3 aerobic oxidation
4 dehydration synthesis

25 The substance represented by X is most likely
1 vitamin B
2 maltase
3 RNA
4 DNA

26 Digestive enzymes that hydrolyze molecules of fat into fatty acid and glycerol molecules are known as
1 proteases
2 maltases
3 lipases
4 vitamins

27 Glucose-1-phosphate can be rapidly converted to glucose-6-phosphate by the addition of phosphoglucomutase. In this reaction, the phosphoglucomutase acts as
1 a lipid
2 an enzyme
3 a sugar
4 a substrate

28 The hydrolysis of maltose is catalyzed by
1 water
2 glucose
3 protease
4 maltase

29 If all of the active sites on the available enzymes within a mixture are involved in the reaction, the addition of more substrate would most likely
1 decrease the rate of the reaction
2 increase the rate of the reaction
3 produce no change in the rate of the reaction

30 In an enzyme-catalyzed reaction, a protein is broken down. The enzyme most likely involved is
1 sucrase 2 lipase 3 protease 4 maltase

SELF–HELP: Unit I *"Extended"* Questions

Base your answers to questions 1 through 3 on the structural formula at the right and on your knowledge of biology.

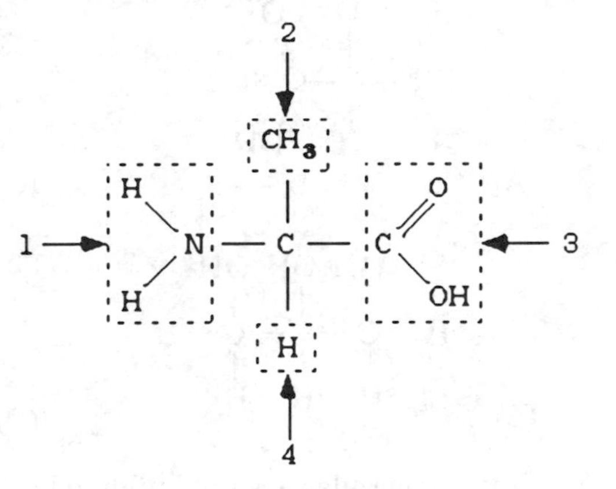

1 Which group of atoms varies from one type of amino acid to another?
1 1
2 2
3 3
4 4

2 A molecule of this substance may be formed as a result of the hydrolysis of a
1 lipid
2 disaccharide
3 nucleic acid
4 protein

3 Which groups are directly involved in the formation of peptide bonds?
1 1 and 2
2 2 and 3
3 2 and 4
4 1 and 3

4 The graph at the right represents the rate of enzyme action for gastric protease when the enzyme concentration is kept constant. The pattern of enzyme action shown in the graph most likely results from varying which experimental factor (*X* in the graph)?
1 hydrogen ion concentration
2 volume of enzyme
3 substrate concentration
4 intensity of light

Rate of Enzyme Action

X

Base your answers to questions 5 through 7 on the structural formula below and on your knowledge of biology.

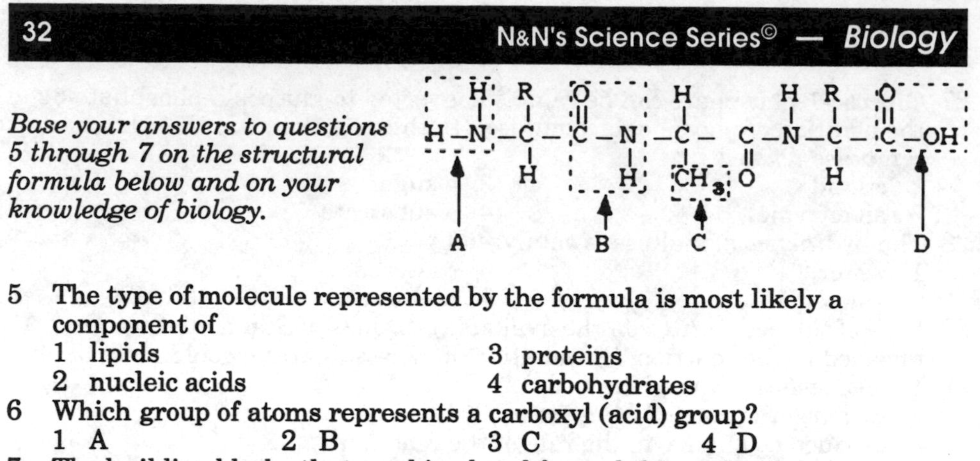

5 The type of molecule represented by the formula is most likely a component of
 1 lipids
 2 nucleic acids
 3 proteins
 4 carbohydrates

6 Which group of atoms represents a carboxyl (acid) group?
 1 A
 2 B
 3 C
 4 D

7 The building blocks that combined and formed this molecule are
 1 fatty acids
 2 amino acids
 3 inorganic acids
 4 nucleic acids

Base your answers to questions 8 through 10 on the structural formulas shown below.

(A)

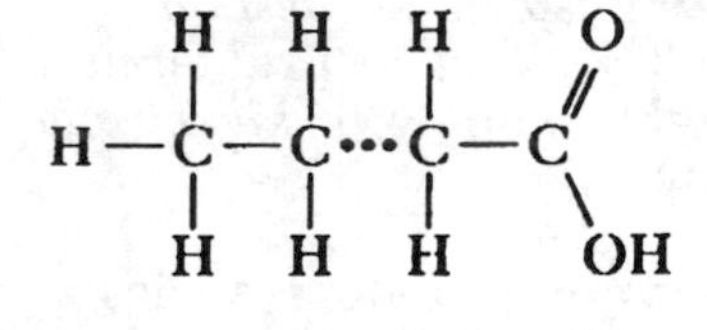

(C)

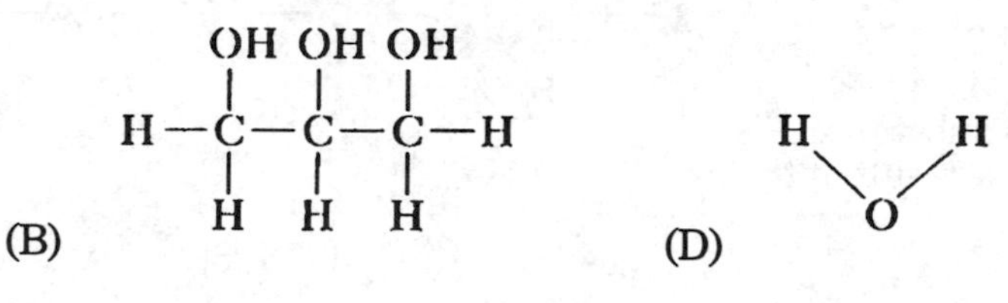

(B)

(D)

8 Which molecules are the building blocks of a lipid?
 1 A and D
 2 A and C
 3 B and C
 4 B and D

9 Which two molecules contain the carboxyl group?
 1 A and C
 2 B and A
 3 C and D
 4 A and D

10 Which molecule is inorganic?
 1 A
 2 B
 3 C
 4 D

11 Vegetable oil, such as corn oil, belong to which general class of organic substances?
 1 lipids
 2 proteins
 3 carbohydrates
 4 salts

12 Which organelle is the site of aerobic respiration in a maple tree?
 1 chloroplast
 2 mitochondrion
 3 vacuole
 4 ribosome

For each statement in questions 13 through 15, select the compound, chosen from the list below, which is most closely associated with that statement.

Compounds Found in Living Things
1) Carbohydrates
2) Lipids
3) Proteins
4) Vitamins
5) Water

13 Some of these molecules are made from three fatty acids bonded to a molecule of glycerol.

14 These molecules are added during the process of hydrolysis and are removed during dehydration synthesis.

15 These molecules often function as coenzymes.

16 Which graph below best illustrates the pattern of enzyme action rates when a specific substrate is slowly added to a system with a fixed enzyme concentration?

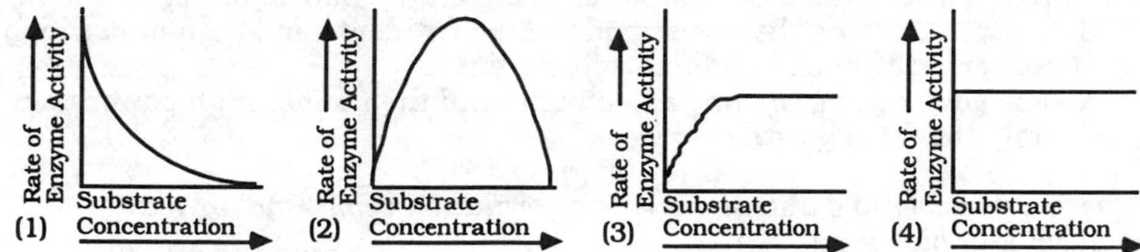

Base your answers to questions 17 through 20 on the diagram below which illustrates a biochemical reaction and on your knowledge of biology.

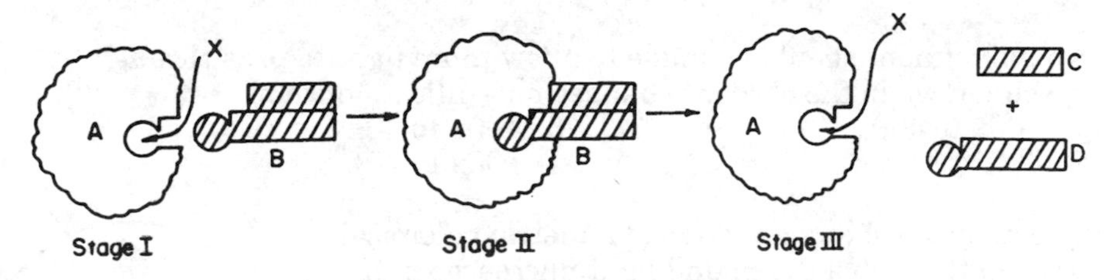

17 Which letter indicates a substrate molecule in this reaction?
1 A
2 B
3 C
4 D

18 Stage II in the diagram most correctly represents
1 an inorganic catalyst
2 a denatured enzyme
3 a vitamin
4 an enzyme-substrate complex

19 The area labeled X is known as
1 an atomic nucleus
2 an active site
3 a pH indicator
4 a temperature regulator

20 Which substance is needed in order for this biochemical reaction to occur?
1 water
2 iodine
3 table salt
4 Benedict's solution

SELF–HELP: Unit I *"Skill"* Questions

1 A compound light microscope has a 10x ocular, 10x low-power objective, and a 40x high-power objective. While observing Elodea under the high-power objective, a student noted that four cells end to end extended across the diameter of the field of view. If the microscope was switched to low power, the student should have been able to observe approximately how many cells across the diameter of his field of view?
1 1 2 8 3 16 4 4

2 To view cells under the high power of a compound microscope, a student places a slide of the cells on the stage and moves the stage clips over to secure the slide. The student then moves the high-power objective into place and focuses on the slide with the coarse adjustment. Two steps in this procedure are incorrect. For this procedure to be correct, the student should have focused under
1 low power using coarse and fine adjustments and then under high power using only the fine adjustment
2 high power first, then low power using on the fine adjustment
3 low power using the coarse and fine adjustments and then under high power using coarse and fine adjustments
4 low power using the fine adjustment and then under high power using only the fine adjustment

Base your answer to question 3 on the information below and on your knowledge of biology.

A student prepares a wet mount of an onion epidermis and observes it under three powers of magnification of a compound light microscope (40x, 100x, 400x).

3 An adjustment should be made to allow more light to pass through the specimen when the student changes magnification from
1 100x to 400x
2 400x to 100x
3 400x to 40x
4 100x to 40x

4 The diagram shows a section of a metric ruler scale as seen through a compound light microscope. If each division represents one millimeter (mm), what is the approximate width of the microscope's field of view in micrometers (μm)?
1 3,700 μm
2 4,200 μm
3 4,500 μm
4 5,000 μm

5 To best observe the nucleus of a cheek cell, a student should use
1 a compound light microscope and iodine solution
2 a compound light microscope and Benedict's solution
3 a dissecting microscope and bromthymol blue solution
4 a dissecting microscope and methylene blue solution

Base your answers to questions 6 and 7 on the enlarged photograph at the right which shows a piece of tissue stained with iodine solution, as viewed with a microscope under high power.

6 Which method would most likely be used to transfer structure A into another cell?
1 centrifugation
2 microdissection
3 fermentation
4 filtration

7 The microscope objective is changed from high power (40x) to low power (10x) and the tissue is brought into focus. Compared to the number of cells observed under high power, the number of cells observed under low power would be
1 less
2 greater
3 the same

8 The diagram at the right represents a compound light microscope. Choose one of the numbered parts. In a complete sentence, name the part selected and describe its function.

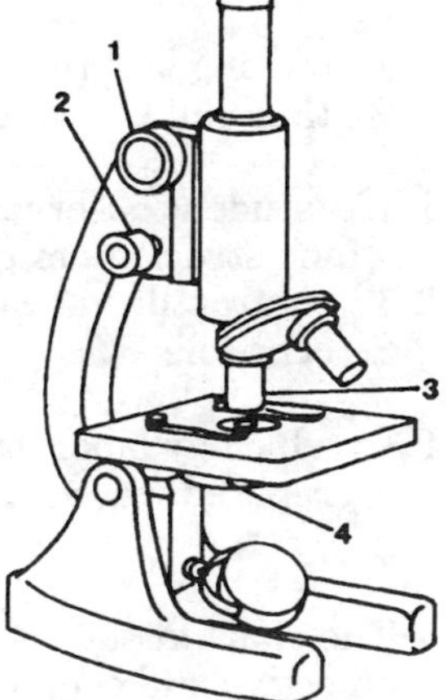

____

____

____

____

____

9 A student placed a slide, as shown in the diagram at the right, on the stage of the compound light microscope.

P

On the diagram at the right, draw in pencil or pen the position of the letter as it would be viewed when observed on the low power of the compound light microscope.

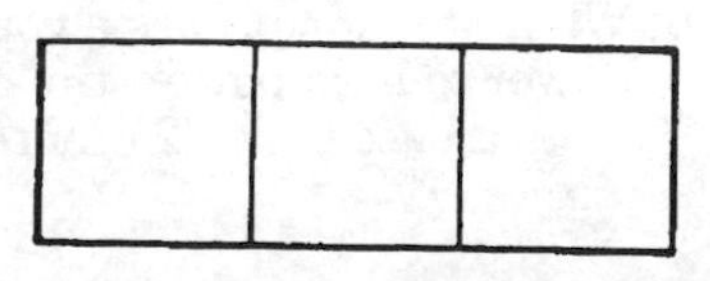

10 Diagram *A* represents the appearance of a wet mount of plant tissue as seen through a compound light microscope. Diagram *B* represents the appearance of the same field of view after the fine adjustment knob is turned. What is the best conclusion to be made from these observations?

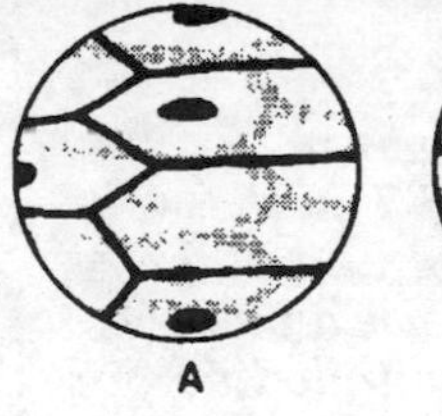

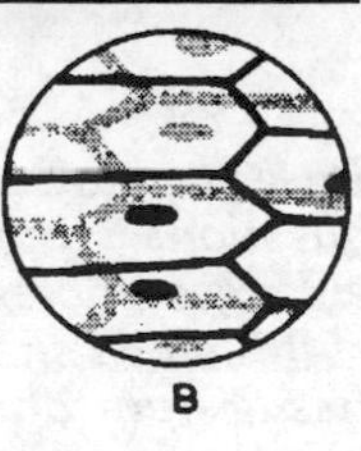

1 The tissue is composed of more than one layer of cells.
2 The tissue is composed of multi-nucleated cells.
3 The cells are undergoing mitotic cell division.
4 The cells are undergoing photosynthesis.

Base your answers to questions 11 and 12 on the information below and on your knowledge of biology.

A student prepares a wet mount of an onion epidermis and observes it under three powers of magnification of a compound light microscope (40x, 100x, 400x).

11 Iodine stain is added to the slide. Under 400x magnification, the student should be able to observe a
1 mitochondrion
2 nucleus
3 ribosome
4 centriole

12 A specimen that is suitable for observation under this microscope would be
1 moving and respiring
2 thin and transparent
3 alive and reproducing
4 stained with Benedict's

13 If a student observes a paramecium with a compound light microscope, which structure might be visible?
1 contractile vacuole
2 centriole
3 ribosome
4 sensory eye-spot

14 In a biology laboratory, four compound light microscopes were set up with a prepared wet mount slide of stained onion skin tissue on each stage.

The chart at the right shows the objective and ocular magnifications for each microscope.

Microscope	Ocular	Objective
1	5 x	10 x
2	5 x	20 x
3	10 x	40 x
4	10 x	50 x

In which microscope's field of view would the greatest number of stained onion cells most likely be observed?
1 1
2 2
3 3
4 4

15 When a student uses a compound microscope and switches from low power to high power, the diameter of the field of view will
1 decrease
2 increase
3 remain the same

Unit 2

Maintenance In Living Things

Introduction

Among living organisms there is a universality of the functions which maintain life. These include the obtaining, processing, and distribution of essential materials, the removal of metabolic wastes, and the regulation of all metabolic processes. While these functions are universal, organisms possess various structures and behavioral patterns which enable them to perform these functions efficiently in their environment. These structures and/or patterns are known as **adaptations**.

Objectives

The student should be able to:

- Identify and describe the basic functions necessary to maintain homeostasis.
- Identify and compare the adaptations of selected organisms for carrying out these life functions.
- Recognize that a unity of pattern underlies the diversity of living things.
- Correlate biochemical reactions with physiological functions.
- Observe and recognize that structure and function complement each other and culminate in an organism's successful adaptation to its environment.

Nutrition – Terms & Concepts

Autotrophic Nutrition
- Plants And Some Monera
- Chlorophyll - Chloroplasts
- Photosynthesis - Reactions
  - Photochemical (light)
  - Carbon-Fixation (dark)
- Leaves - Mesophyll
  - Palisade - Spongy
- Stomates - Guard Cells

Heterotrophic Nutrition
- Ingestion - Animals
- Digestion - Mech. - Chemical
  - Extracellular - Intracellular
- Egestion
- Fungi - Rhizoids
- Protozoans - Cilia, Pseudopods
  - Vacuoles - Lysosomes
- Hydra - Digestive Cavity
- Earthworm - Grasshopper - Human
  - One-way Digestive Tube

I. Nutrition

Nutrition includes those activities by which organisms *obtain* and *process* materials needed for energy, growth, repair, and regulation. The two types of nutrition are **autotrophic** (manufacturing of organic food from inorganic materials) and **heterotrophic** (obtaining pre-made organic food from other sources).

A. Autotrophic Nutrition

The ability of most green plants and certain Monerans and Protists to manufacture organic compounds from inorganic raw materials is autotrophic nutrition. These organisms are called **autotrophs** ("self-feeders"). The principle autotrophic process is **photosynthesis** ("light energy making"). A second mode of autotrophic nutrition is **chemosynthesis** ("chemical energy making").

1. Process of Photosynthesis

Photosynthesis, the process during which light energy is converted into the chemical energy of organic molecules, is one type of autotrophic nutrition.

Significance and Process of Photosynthesis

Most of the chemical energy available to organisms results from photosynthetic activity. In addition to food production, photosynthesis releases most of the oxygen that is in the air.

Most cells which carry on photosynthesis contain **chloroplasts**. These chloroplasts contain pigments which include **chlorophylls** ("green leaf"). Chlorophyll absorbs light energy. A technique called **chromatography** is used to separate and study the variety of pigments contained within the chloroplasts.

In the chloroplasts, carbon dioxide (CO_2) and water (H_2O) are used in the formation of simple sugar ($C_6H_{12}O_6$) molecules and oxygen (O_2).

On Earth, sunlight appears as "white" light. However, this light actually contains all of the colors of the spectrum (red, orange, yellow, green, blue, indigo, and violet).

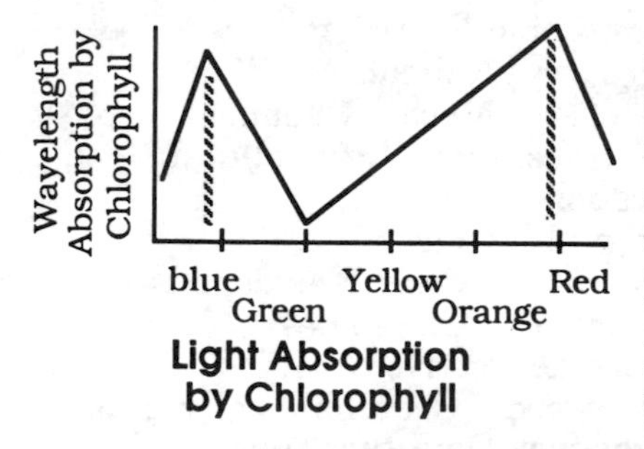

Light Absorption by Chlorophyll

During photosynthesis, light energy is trapped by chlorophyll and converted into the chemical energy of simple sugar molecules. *Red* and *blue* wavelengths of light are *most effective* for this energy conversion because of their relatively high degree of absorption by chlorophyll molecules. Wavelengths of green and yellow light are less effective than red or blue and are mostly reflected.

Simplified summary statements for photosynthesis are:

$$\text{Carbon Dioxide} + \text{Water} \xrightarrow[\text{Chlorophyll}]{\text{Light Energy}} \text{Glucose} + \text{Water} + \text{Oxygen}$$

$$6\,CO_2 + 12\,H_2O \xrightarrow{\text{Enzymes}} C_6H_{12}O_6 + 6\,H_2O + 6\,O_2$$

Extended Area: ***Biochemistry***

In photosynthesis, two major sets of reactions occur: light and dark reactions. The use of various isotopes (radioactive chemical tracers) has led to a greater understanding of these reactions.

Photochemical (*light*) Reactions

These "light needing" reactions occur within layered membranes inside the **chloroplasts**. Stacks of these membranes, the **grana**, contain the enzymes and the pigments necessary for the light reactions. Some of the energy absorbed by the chlorophyll pigments is used to "split" water molecules, producing hydrogen atoms and oxygen gas. This process is **photolysis**. The use of the isotope oxygen-18 has shown that *all* of the oxygen released during photosynthesis originates from water molecules.

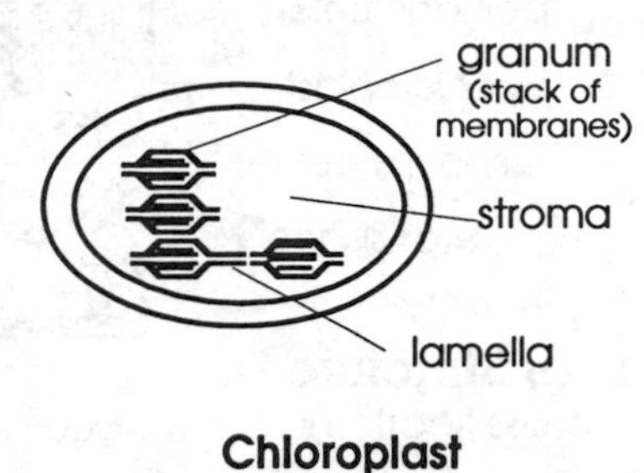

Chloroplast
(drawn from a photo made with an electron microscope)

In addition to the production of oxygen, **ATP**, an energy transfer compound, is produced within the chloroplasts during the light reactions. (ATP will be discussed in more detail in section III-Respiration.)

Carbon-Fixation (*dark*) Reactions

These reactions occur within the chloroplasts in the **stroma**. The stroma is the dense solution between the grana. Enzymes necessary for the dark reactions are found in the stroma. Hydrogen atoms from the light reaction and the carbon dioxide molecules participate in a series of chemical changes which produce a three-carbon sugar (**PGAL**) from which other molecules, including glucose, are synthesized. The isotope carbon-14 has been used to trace the pathways of carbon fixation.

Results of Photosynthesis

The glucose which is formed during photosynthesis may be:

- Used as an energy source in cellular respiration.
- Synthesized into other metabolic compounds.
- Converted into storage products by dehydration synthesis and other reactions.

Before these storage products can be used they are converted into simpler molecules by digestion within the cells (**intracellular digestion**), rather than in a specialized digestive system. Once digestion is completed, the end products can be used in the cell or they can be transported to other tissues.

2. Adaptations for Photosynthesis

In addition to the green plants, autotrophic algae (unicellular organisms) carry on photosynthesis. Since the algae are simple organisms, the raw materials necessary for photosynthesis are absorbed directly into the individual cells.

Algae are found in both the oceans (marine) and bodies of freshwater and are responsible for a large percentage of photosynthetic products found on the Earth.

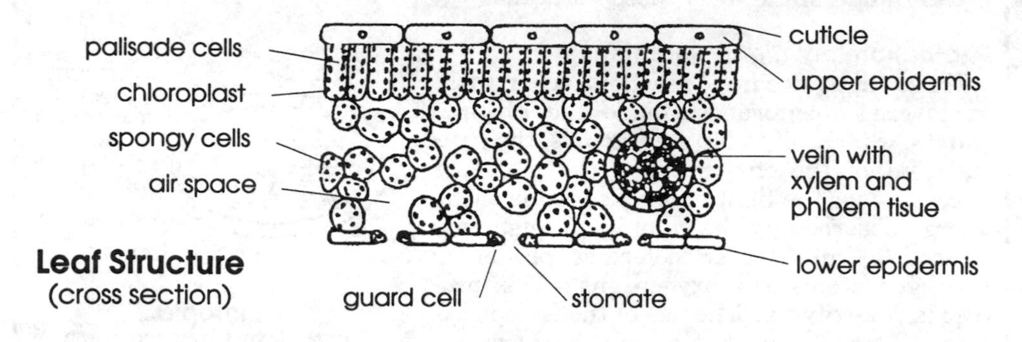

Leaf Structure
(cross section)

Most terrestrial (land) plants are complex organisms and have specialized organs adapted for photosynthesis. Although photosynthesis may occur in the stems of some plants, the organs which are most highly adapted for this process are the leaves.

- Most leaves provide a large surface area for the absorption of light energy.
- The chief functions of the outer cell layer of the leaf (**epidermis**) and the waxy covering of the epidermis (**cuticle**) are: the protection of the internal tissues of the leaf from excessive water loss, resistance to invasion by fungi, and protection from physical and mechanical injury.
- Openings in the cuticle and epidermis are called **stomates**. Their size is regulated by **guard cells**. Stomates allow the exchange of O_2, CO_2, and H_2O between the external environment and internal air spaces.
- Most photosynthesis occurring in leaves takes place in the **palisade layer** which is located under the upper epidermis.
- The **spongy layer** contains many interconnected air spaces which are surrounded by moist surfaces. The exchange and circulation of gases occur here. The spongy layer also carries on photosynthesis.
- In most leaves, **chloroplasts** are present in the palisade and spongy layers and in the guard cells.
- Conducting tissue, located in the **veins** of the leaves, carries water and minerals (xylem tissue) to the photosynthesizing cells and distributes food (phloem tissue) to other plant organs.

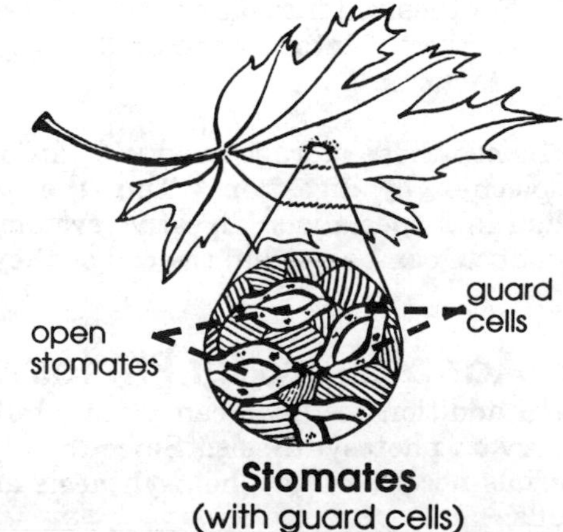

Stomates
(with guard cells)

Questions

1 The removal of the cuticle from a leaf would most likely result in an increase in the leaf's
1 surface area
2 food production
3 water loss
4 fungus resistance

2 Which process provides most of the oxygen found in Earth's atmosphere?
1 photosynthesis
2 aerobic respiration
3 dehydration synthesis
4 fermentation

3 The use of oxygen-18 (heavy oxygen) has aided biologists in determining that O_2 released in photosynthesis comes from the chemical breakdown of water. Which organisms did the biologists most likely use in this investigation?
1 fungi 2 algae 3 protozoa 4 arthropods

4 Four bean plants were cultivated under identical conditions, except that each was exposed to a different color of light: red, green, yellow, or blue. After 20 days the most probable observation is that
1 all four plants grew to the same size
2 the plant exposed to green light was smaller than the plant exposed to red light
3 the plant exposed to green light was larger than the plant exposed to yellow light
4 none of the plants grew without white light

Base your answers to questions 5 and 6 on the summary equation below of a metabolic process and on your knowledge of biology.

$$12\,H_2O + 6\,CO_2 \xrightarrow[\text{substance } X]{\text{light}} C_6H_{12}O_6 + 6\,H_2O + 6\,O_2$$

5 In the equation above, substance X is most likely
1 glucose 2 hemoglobin 3 urea 4 chlorophyll

6 This process occurs during daylight hours in
1 plants, but not animals
2 animals, but not plants
3 both in plants and animals
4 neither plants nor animals

7 A plant is illuminated with green light for 6 hours followed by illumination with red light of equal intensity for 6 hours. Which graph most likely indicates the relative amount of oxygen released by this plant?

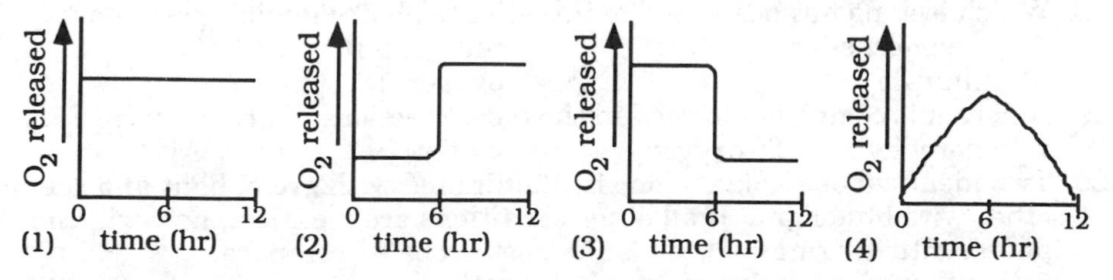

8 Starch molecules are synthesized from glucose within most cells of a leaf during the daylight hours. During the night, this starch may be changed back into glucose by the process of
1 hydrolysis
2 photosynthesis
3 dehydration synthesis
4 aerobic respiration

9 An important function of chlorophyll molecules during photosynthesis is
 1 absorbing and storing water in root cells
 2 converting water into carbon dioxide
 3 absorbing certain wavelengths of light energy
 4 converting chemical bond energy to light energy

10 The raw materials used by green plants for photosynthesis are
 1 oxygen and water
 2 oxygen and glucose
 3 carbon dioxide and water
 4 carbon dioxide and glucose

11 Light energy is absorbed by organic pigment molecules during
 1 anaerobic respiration
 2 aerobic respiration
 3 photochemical reactions of photosynthesis
 4 carbon-fixation reactions of photosynthesis

12 Which process is the source of most of the oxygen in the atmosphere?
 1 aerobic respiration
 2 fermentation
 3 transpiration
 4 photosynthesis

13 At optimum light intensity, which atmospheric gas most directly influences the rate of photosynthesis?
 1 nitrogen 2 oxygen 3 hydrogen 4 carbon dioxide

14 PGAL is synthesized during
 1 anaerobic respiration
 2 aerobic respiration
 3 photochemical reactions of photosynthesis
 4 carbon-fixation reactions of photosynthesis

15 Which word equation represents the process of photosynthesis?
 1 carbon dioxide + water → glucose + oxygen + water
 2 glucose → alcohol + carbon dioxide
 3 maltose + water → glucose + glucose
 4 glucose + oxygen → carbon dioxide + water

16 During photosynthesis, molecules of oxygen are released from the "splitting" of water molecules. This is a direct result of
 1 dark reactions
 2 light reactions
 3 formation of PGAL
 4 formation of carbon dioxide

17 Carbon dioxide, a product of plant metabolism, can be used by maple trees in the process of
 1 aerobic respiration
 2 photosynthesis
 3 fermentation
 4 transpiration

18 Which isotope has been used to investigate photochemical reactions?
 1 phosphorus-32
 2 sulfur-35
 3 nitrogen-15
 4 oxygen-18

19 An organic compound formed in the dark reactions of photosynthesis is
 1 chlorophyll 2 oxygen 3 water 4 PGAL

20 Two identical bean plants, one is illuminated with green light and the other with blue light. If all other conditions are identical, how will the photosynthetic rates of the plants most probably compare?
 1 Neither plant will carry on photosynthesis.
 2 Photosynthesis will occur at the same rate in both plants.
 3 The plant under green light will carry on photosynthesis at a greater rate than the one under blue light.
 4 The plant under blue light will carry on photosynthesis at a greater rate than the one under green light.

B. Heterotrophic Nutrition

Organisms unable to make organic molecules from inorganic raw materials are **heterotrophs**. Heterotrophic organisms must obtain and process *preformed* organic substances. Examples of groups of heterotrophic organisms include bacteria, fungi, protozoans, and animals.

1. Processes for Heterotrophic Nutrition

Heterotrophic organisms obtain preformed organic molecules from their environment.

Ingestion. Ingestion is the process of taking in food.

Digestion. Digestion is the conversion of large, insoluble molecules to smaller soluble molecules. Digestion can be **intracellular** (within the cell). Also, digestion can be **extracellular** (outside of the cell or organism) followed by the absorption of end-products.

- **Mechanical Breakdown**. Food is mechanically broken down by physical means, such as cutting, grinding, and tearing which *increases the surface area* of the foods prior to chemical digestion.

- **Chemical Digestion**. Large organic molecules are changed chemically to smaller organic molecules by enzymatic hydrolysis ("splitting with water").

Complete digestion of large molecules produces end products as indicated in the chart below. The end products of chemical digestion are the *same* in all organisms.

Chemical Digestion

Large Molecules	End Products
carbohydrates	simple sugars (glucose, galactose, fructose)
lipids (fats, waxes, oils)	fatty acids and glycerol
proteins	amino acids

Egestion. Egestion is the removal (elimination) of undigested or indigestible material by the heterotrophic organism. Heterotrophs may not be able to digest all food components either because they may lack specific enzymes, or because the time that their food is exposed to available enzymes may be too short.

2. Adaptations for Heterotrophic Nutrition

Fungi. Fungi live on or in their food supply. The filamentous body of a fungus, such as bread mold, contains special filaments called **rhizoids** which penetrate the food source and secrete digestive (hydrolytic) enzymes. This results in **extracellular digestion** and the subsequent **absorption** of the externally digested nutrients.

Bread Mold

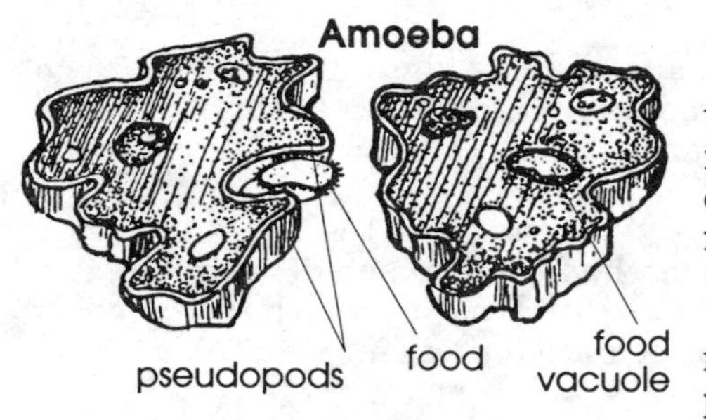

Protozoans. Since protozoans are unicellular, they do not have specialized organs. However, different groups of protozoans carry on the activity of nutrition in specialized ways.

In the **amoeba**, food is ingested by means of **pseudopods**. This engulfing process is **phagocytosis**.

In the **paramecium**, as a result of the action of **cilia**, food is ingested through a fixed opening located in the **oral groove**.

Intracellular digestion of this food occurs within a **food vacuole** after it merges with a lysosome. The end products of the digestion are then absorbed into the cytoplasm. In the paramecium, undigested materials are egested through a fixed opening called the **anal pore**.

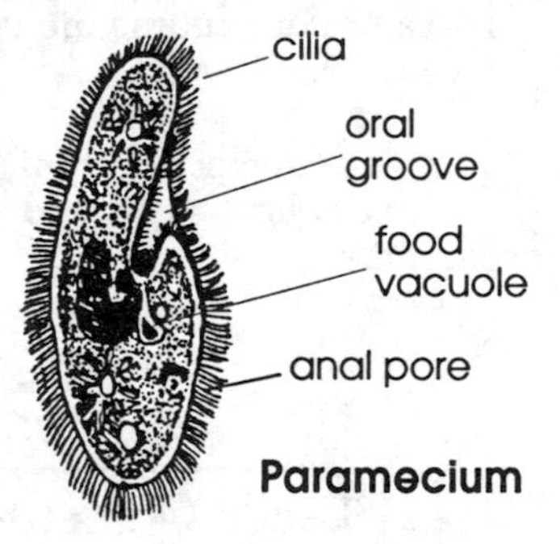

Paramecium

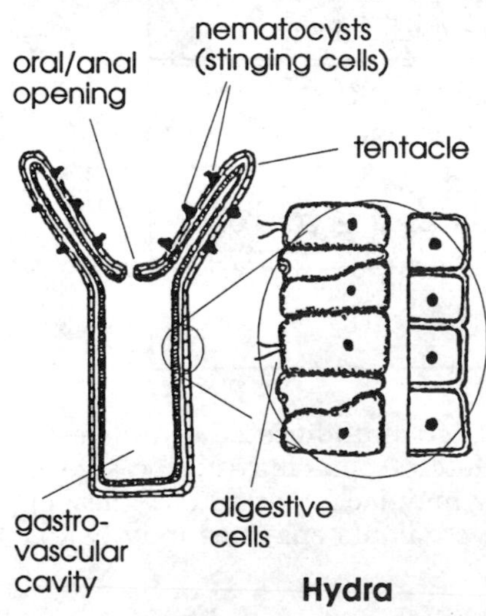

Hydra

Animals

Hydra. The hydra, a representative member of the Phylum Coelenterata, has a sac-like digestive cavity with a single opening. Food is ingested through this opening with the aid of tentacles and passes into the digestive cavity.

In the digestive cavity, **extracellular** digestion occurs as a result of enzymes which are secreted by specialized cells located in the lining of the cavity. Partially digested food is then engulfed by phagocytic cells lining the cavity and digestion is completed **intracellularly**.

Undigested material is egested through the same opening through which ingestion occurred; therefore, the hydra has a **"two-way"** digestive tube.

Earthworm. The earthworm is a segmented worm (Phylum Annelida) and has a tube-like digestive system with two openings. This **"one-way"** digestive tract allows food to be processed in an efficient manner as it passes through specialized organs.

A mixture of food and soil is ingested through the **mouth** and may be stored temporarily in the **pharynx** before processing. The mixture is then passed through the **esophagus** into the **crop** (a thin-walled organ) where it is stored. In the thick-walled, muscular **gizzard**, mechanical digestion increases the surface area of the food. From the gizzard, the food passes into the **intestine** where most chemical digestion takes place and where end products are absorbed into the circulatory system. Undigested material is egested through the **anus**.

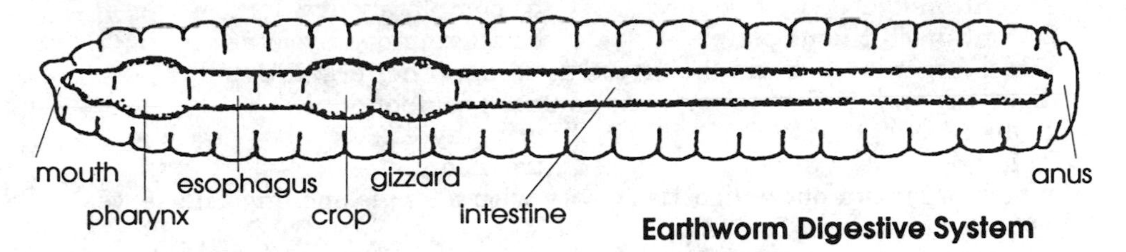

Earthworm Digestive System

Grasshopper. In the grasshopper, a member of the Phylum Arthropoda, the structure and function of the digestive system are essentially similar to that of the earthworm. Food moves in one direction from mouth to anus.

The grasshopper possesses highly specialized **mouth parts** for mechanical breakdown. **Salivary glands** and **gastric caeca** secrete hydrolytic enzymes into the digestive tract for chemical digestion. Food and metabolic wastes are stored in the rectum and eliminated through the anus.

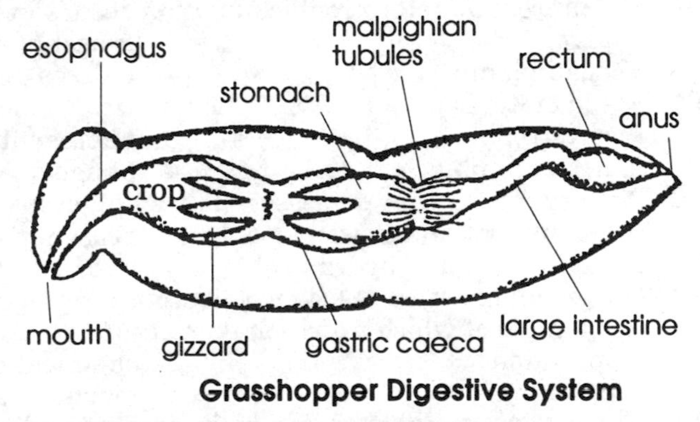

Grasshopper Digestive System

Human. The human digestive system is essentially similar to that of the grasshopper and earthworm in that it contains:

- A tube-like system with two openings — "one-way" digestive tube.
- Specialized organs and glands for mechanical breakdown and chemical digestion.

Note: Humans are included here as another representative organism. A detailed study of human physiology is found in Unit III.

Questions

1 Structures in the grasshopper that secrete enzymes into the digestive tract are the
1 gastric caeca and salivary glands
2 Malpighian tubules and gizzard
3 rectum and crop
4 tracheal tubes and spiracles

2 The presence of cilia, an oral groove and food vacuoles, and the absence of chloroplasts in a unicellular organism indicate that the organism carries on
1 sexual reproduction
2 autotrophic nutrition
3 extracellular digestion
4 heterotrophic nutrition

3 Which organism lacks the capability of autotrophic nutrition?
1 oak tree 2 mushroom 3 fern 4 maple tree

4 In order for a food vacuole within an ameba to carry on intracellular digestion effectively, the vacuole must
1 contain hydrolytic enzymes
2 contain chitinous plates
3 be completely dry
4 be surrounded by setae

5 Which organisms obtain their nutrition from other organisms by digesting and absorbing preformed organic compounds?
1 algae
2 fungi
3 bryophytes
4 tracheophytes

6 Which organism has a digestive cavity with a single opening that serves as both a mouth and an anus?
1 hydra
2 earthworm
3 paramecium
4 grasshopper

7 A cellular organelle in protozoans that preforms a function similar to that of the small intestine in humans is the
1 nucleus 2 cell wall 3 ribosome 4 food vacuole

8 The process of heterotrophic nutrition occurs in
1 algae
2 bean plants
3 grasshoppers
4 mosses

9 Based on their pattern of nutrition, most animals are classified as
1 autotrophic
2 heterotrophic
3 photosynthetic
4 phagocytotic

10 Two organisms that possess a one-way digestive tract are the
1 earthworm and grasshopper
2 earthworm and hydra
3 grasshopper and ameba
4 hydra and ameba

11 The grinding action of food which occurs in the gizzard of a grasshopper is an example of which nutritional process?
1 ingestion
2 egestion
3 chemical digestion
4 mechanical digestion

12 Which organism digests food by intracellular digestion only?
1 ameba
2 earthworm
3 grasshopper
4 human

13 Which organisms secrete enzymes that digest food outside their cells and then absorb the simple end products?
1 trees 2 fungi 3 grasses 4 algae

14 Nutrients are made soluble within the food vacuoles of the
1 grasshopper and earthworm
2 earthworm and human
3 Ameba and Paramecium
4 Hydra and grasshopper

15 A characteristic of animals that makes them similar to heterotrophic plants is that animals
1 obtain preformed organic molecules from other organisms
2 need to live in a sunny environment
3 are sessile for most of their lives
4 use energy to manufacture organic compounds

16 A fruit fly is classified as a heterotroph, rather than as an autotroph, because it is unable to
1 transport needed materials through the body
2 release energy from organic molecules
3 manufacture its own food
4 divide its cells mitotically

17 Which statement concerning the digestive processes in the Ameba and in the Hydra is true?
1 Digestion does not require water in either organism.
2 Digestion occurs by extracellular methods in both organisms.
3 Digestion occurs by intracellular methods in both organisms.
4 Digestion does not require enzymes in either organism.

18 The best description of the nutritional pattern of a protozoan such as the Paramecium is
1 heterotrophic and intracellular
2 heterotrophic and extracellular
3 autotrophic and intracellular
4 autotrophic and extracellular

19 In humans and earthworms, digested foods pass through the intestinal wall and are absorbed into the
1 stomach
2 Malpighian tubules
3 tracheal tubules
4 blood

20 Which organism lacks a one-way digestive system?
1 Hydra
2 earthworm
3 grasshopper
4 human

21 Which organism ingests food by engulfing it with pseudopods?
1 grasshopper
2 Paramecium
3 earthworm
4 Ameba

22 Animals can not synthesize nutrients from inorganic raw materials. Therefore, animals obtain their nutrients by
1 combining carbon dioxide with water
2 consuming preformed organic compounds
3 hydrolyzing large quantities of simple sugars
4 oxidizing inorganic molecules for energy

23 Which organism possesses a single opening which functions as a mouth for the ingestion of food and also as an anus for the elimination of undigested materials?
1 human
2 grasshopper
3 Hydra
4 earthworm

24 The process by which digestive enzymes catalyze the breakdown of larger molecules to smaller molecules with the addition of water is known as
1 synthesis
2 hydrolysis
3 pinocytosis
4 diffusion

25 An organism with a one-way digestive tube is the
1 Paramecium 2 earthworm 3 Ameba 4 Hydra

II. Transport

Transport – Terms & Concepts

Cell Membrane
- Fluid Mosaic Model - Singer

Transport - Diffusion
- Passive and Active
- Pinocytosis
- Phagocytosis

Plants - Vascular Tissue
- Root Hairs, Xylem and Phloem
- Stems, Leaves - Transpiration

Hydra - Gastrovascular Cavity
Earthworm, Human - Closed System
Grasshopper - Open System

The process of transport involves the absorption and circulation of materials throughout an organism.

A. Process of Transport

1. Absorption

Absorption is the process whereby the end products of digestion, as well as other dissolved solids and gases, enter the fluids and the cells of an organism through the plasma membrane. Absorption begins the process of transport.

Structure of the Plasma (Cell) Membrane

The plasma membrane selectively regulates the entry and exit of materials. This helps cells in maintaining **homeostasis**.

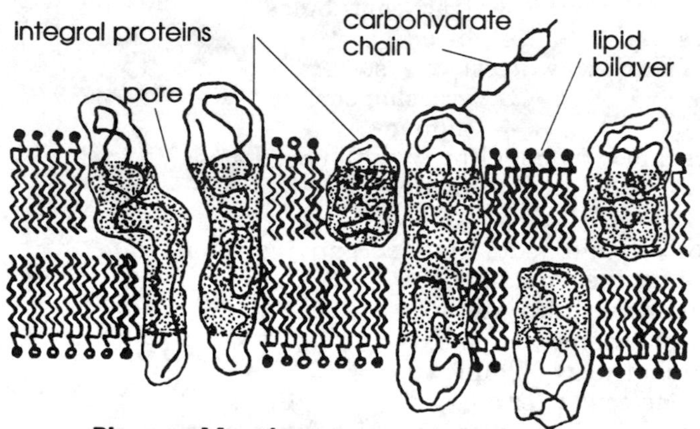

Plasma Membrane (Fluid Mosaic Model)

In 1972, S. J. Singer proposed a model for the plasma membrane that has become widely accepted, called the **fluid-mosaic model**. This model shows a double phospholipid layer in which large proteins are imbedded ("floating icebergs in a fatty sea").

Many small particles, such as amino acids and monosaccharides, diffuse through the membrane. Most larger molecules, such as proteins and starches, cannot diffuse into or out of cells unless they are first chemically digested. The passage of materials through the plasma membrane seems to be determined by both chemical factors and the size of the molecule.

Function of the Plasma (Cell) Membrane

The plasma membrane has both passive and active roles in transporting materials into and out of cells.

Passive Transport

In **passive transport** the movement of the materials through the plasma membrane is the result of the kinetic energy of the particles in motion. Since this movement does not use any cellular energy, it is known as "passive" transport.

Diffusion, a form of passive transport, is a process in which the net movement of ions or soluble molecules is from a region of higher concentration to a region of lower concentration. The diffusion of water through a semi-permeable membrane is **osmosis**.

Active Transport

Active transport is a process in which cellular energy is used to move ("pump") particles through a membrane. This movement is from a region of low concentration toward a region of high concentration. Carrier proteins embedded in the plasma membrane facilitate the transport of materials.

Pinocytosis. Pinocytosis ("cell drinking") is a process whereby vacuoles formed at the cell surface bring in large dissolved molecules. These large molecules can then be broken down by intracellular digestion.

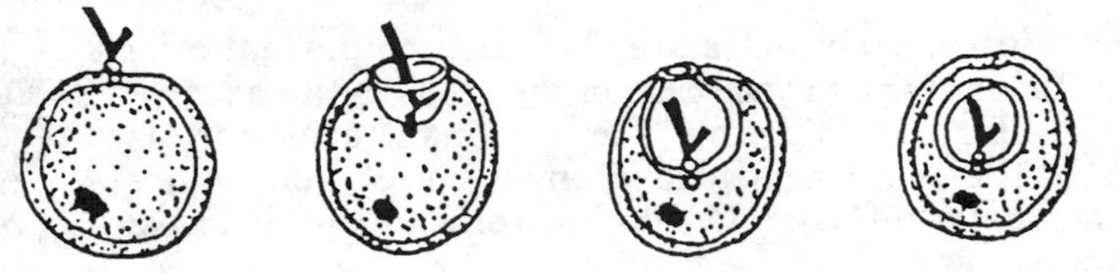

Phagocytosis. Phagocytosis ("cell eating") is a process in which a cell engulfs undissolved large particles by growing around them and enclosing them in a vacuole.

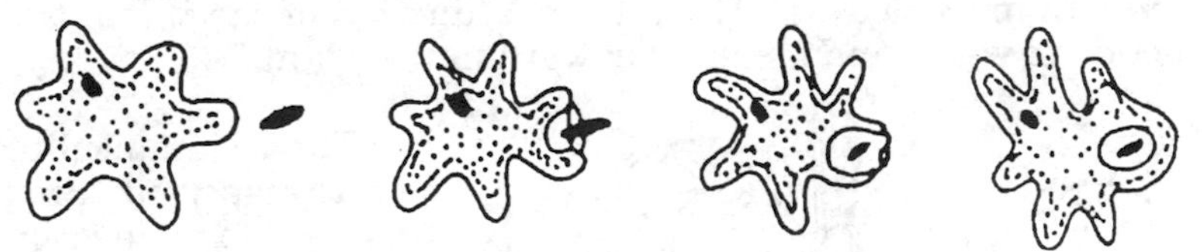

2. Circulation

The transport of materials within cells and/or throughout multicellular organisms is **circulation**, the second stage of transport.

- **Intra**cellular (within the cell) circulation is accomplished by diffusion, **cyclosis**, and possibly, by movement throughout the endoplasmic reticulum.
- **Inter**cellular (between the cells) circulation may be by diffusion or by transport through specialized conducting (vascular) tissue.

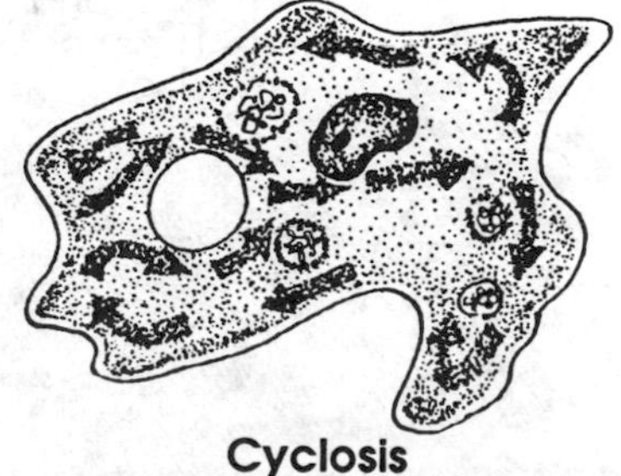

Cyclosis in Amoeba

B. Adaptations for Transport

Intracellular circulation occurs within *all* living cells. Intercellular circulation varies with the complexity of the organism.

Moss

1. Plants

Simple multicellular plants, **bryophytes** (such as moss), lack vascular tissue. They also lack "true" roots, stems, and leaves. Therefore, bryophytes accomplish intercellular transport by diffusion.

Higher plants, **tracheophytes** (such as grasses, shrubs, and trees), possess vascular tissue in their roots, stems, and leaves. Vascular tissues allow for intercellular transport.

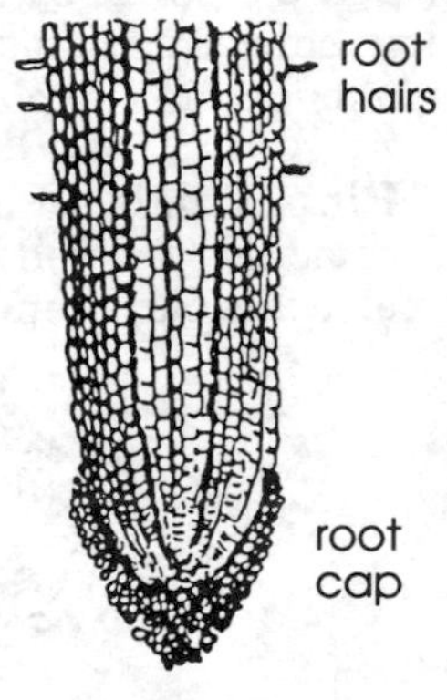

Root Tip

Roots

Roots are structures specialized for anchorage, nutrient storage, absorption of water and soluble salts, and for the conduction of materials to the stem.

Root Hairs. Root hairs are elongated epidermal cells which *increase the surface area* of the root for the absorption of water and minerals. The movement of materials through the semipermeable membrane of root hairs involves both diffusion, including osmosis, and active transport.

Xylem. The xylem is specialized transport (vascular) tissue extending from the roots to the leaves. The principal function of the xylem tissue is the conduction of water and minerals upward in the plant.

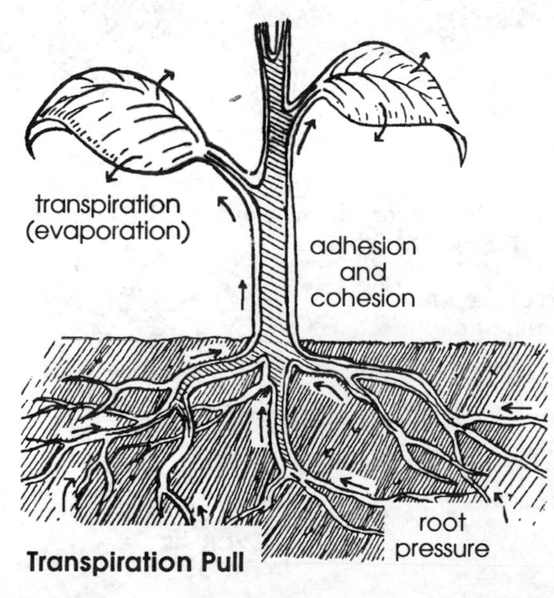

Transpiration Pull

The mechanism by which water is transported through the xylem is best explained by the hypothesis of **transpirational pull**. Transpirational pull involves the transpiration (evaporation) of water vapor through the stomates.

This exerts a pulling force on the column of water in the xylem. Because of the cohesive and adhesive properties of water, the water column does not break apart and is drawn up from the roots.

In small plants, when there is high humidity, root pressure may play a role in pushing water up through the xylem.

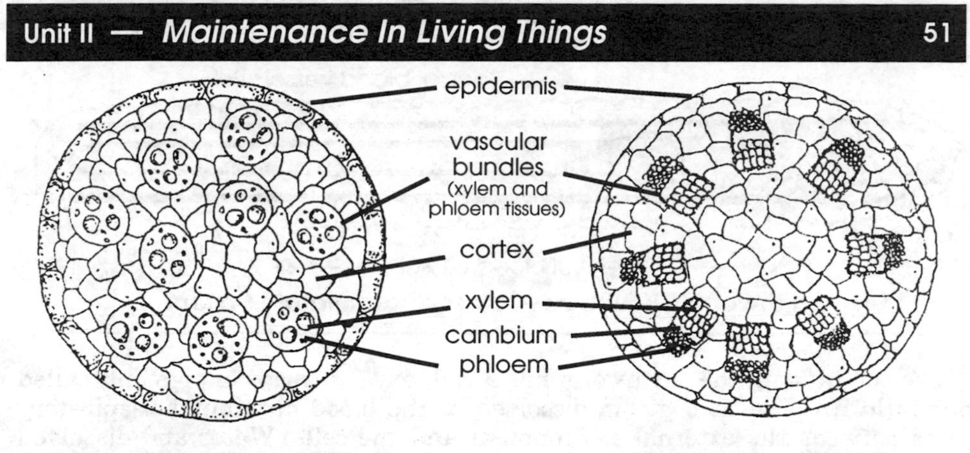

Herbaceous Monocot Stem **Herbaceous Dicot Stem**

Phloem. Phloem, a second type of vascular tissue, conducts organic food materials, such as sugar, both upward and downward to plant tissues for immediate use or storage.

Stems

Although the structure of the stem is different from that of the roots, the vascular tissues (xylem and phloem) are continuous and function in the same manner.

Leaves

Leaves, the photosynthetic organs of the plant, contain veins which are extensions of the conducting tissues in the stem. Water and dissolved minerals are conducted to the leaves through the xylem tissue of the roots and stems. The sugar produced during photosynthesis in the leaf is transported by the phloem to other parts of the plant for use and to the roots for storage.

bark (cork)
phloem
vascular cambium
annual growth rings
area of xylem
pith

Woody Dicot Stem

2. Animals

Hydra. Since most of the hydra's cells are in direct contact with a watery environment, it is able to survive without a special transport system. Flagellated cells aid in the circulation of materials throughout the gastrovascular cavity. Intercellular circulation is by diffusion.

Earthworm. The earthworm, a terrestrial organism, is a more complex organism than the hydra. Therefore, many cells of the earthworm are not in direct contact with the external environment. An internal, **closed circulatory system** (the blood is contained within vessels) transports materials throughout the organism.

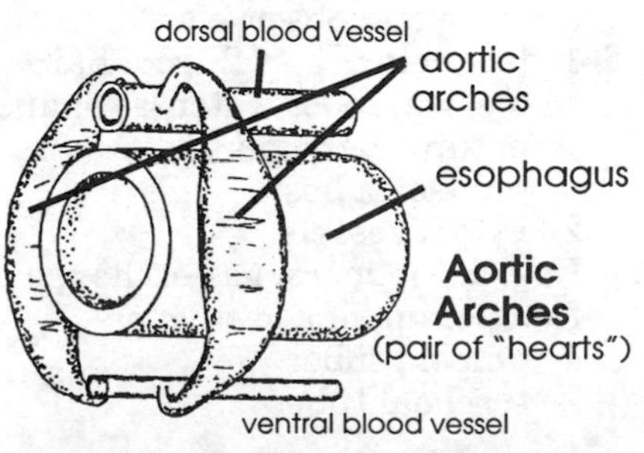

Aortic Arches
(pair of "hearts")

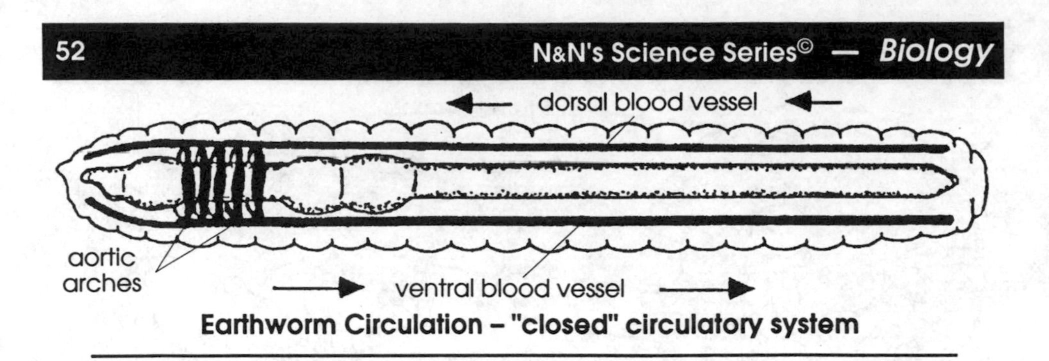

Earthworm Circulation – "closed" circulatory system

Like the human, the earthworm has a red, oxygen-carrying pigment, called **hemoglobin**. The hemoglobin dissolved in the blood distributes respiratory gases between the external environment and the cells. Water and dissolved materials, including the end products of digestion, are transported from the digestive system to the cells of the organism. **Aortic arches** pump the blood within the blood vessels. An infolding of the earthworm's digestive tube (*typhlosole*) is an adaptation which increases the absorptive surface where digestive end products enter the blood.

Grasshopper. The grasshopper has an internal, **open circulatory system** which brings materials in contact with all cells. In contrast to a closed circulatory system, in an open system, blood is distributed into sinuses by means of a pulsating blood vessel and primitive heart system.

The grasshopper's blood contains no hemoglobin and does not transport oxygen. As in the earthworm, the grasshopper has an infolded digestive tube which increases the absorption area.

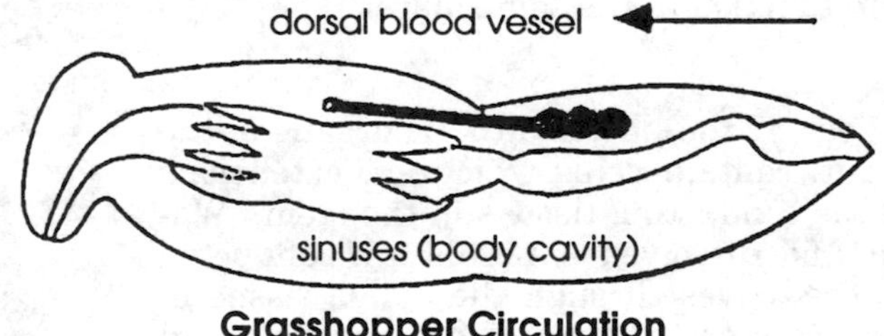

Grasshopper Circulation "open circulatory system"

Human. The human circulatory system is more like the earthworm's than the grasshopper's and is a closed system with hemoglobin for oxygen transport. There is a pumping structure called the heart. Infoldings of the digestive tube (villi) increase the absorption area for nutrients.

Questions

1 To study the structure of phloem and xylem tissue, a student could prepare and observe sections of a
  1 leaf vein 2 root hair 3 chloroplast 4 stomate

2 In a plant, most water is conducted from the roots to the leaves through structures known as
  1 phloem tubes 3 lenticels
  2 xylem vessels 4 capillaries

3 In grasshoppers, gas exchange and gas transport occur as a result of the functioning of a system of
  1 phloem tubes 3 ganglia
  2 tracheal tubes 4 setae

4 Which cells in corn plants have a function most similar to that of blood vessels in earthworms?
  1 epidermal cells and meristems
  2 guard cells and palisade cells
  3 spongy cells and root hairs
  4 xylem cells and phloem cells

5 The hydra survives without a special transport system since
  1 it possesses a nerve net
  2 it can ingest food with the aid of tentacles and stinging cells
  3 most of its cells are in contact with a watery environment
  4 all of its cells can live without using oxygen

6 The diagram at the right represents part of the lower surface of a bean leaf. With which process is area X most closely associated?
  1 digestion
  2 phototropism
  3 reproduction
  4 transpiration

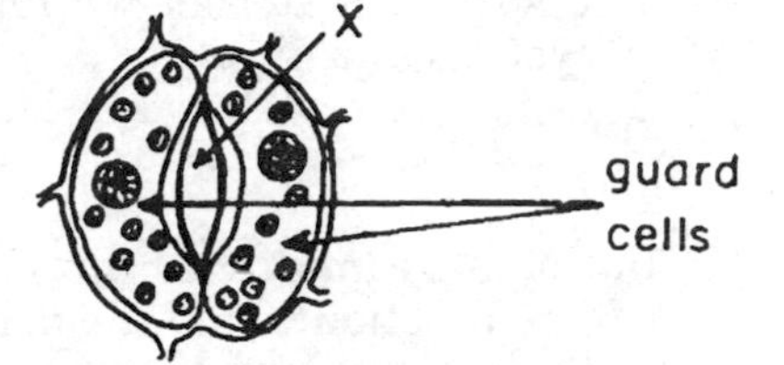

7 Oxygen from the atmosphere enters woody stems through
  1 capillaries 2 root hairs 3 lenticels 4 xylem cells

8 The passage of materials through the membrane of a plant cell is known as
  1 hydrolysis 3 circulation
  2 transport 4 transpiration

9 Molecules that are too large to pass through the pores of a plasma membrane may enter the cell by a process known as
  1 hydrolysis 3 cyclosis
  2 pinocytosis 4 synthesis

10 A biologist diluted a blood sample with distilled water. While observing the sample under a microscope, she noted that the red blood cells burst. This bursting is most likely the result of which process?
  1 staining 3 osmosis
  2 ingestion 4 active transport

11 Which cell part selectively regulates the entry and exit of substances as shown in the diagram at the right?
  1 plasma membrane
  3 nucleolus
  2 ribosome
  4 nuclear membrane

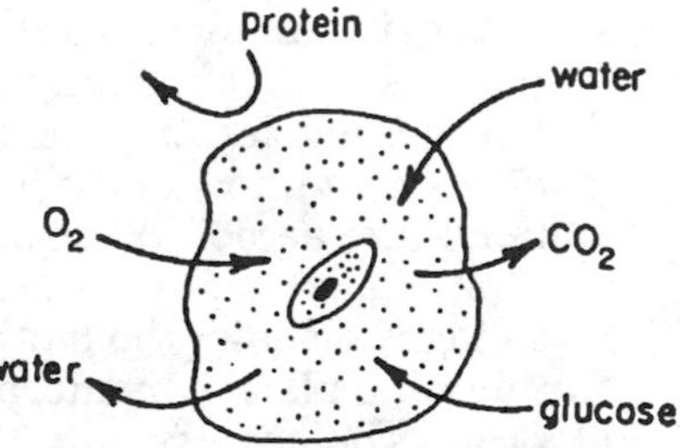

12 Which process directly aids in the upward transport of water and minerals in the xylem tubes?
  1 transpiration 3 hydrolysis
  2 digestion 4 phagocytosis

13 Which process requires cellular energy to move molecules across the cell membrane from a region of lower concentration to a region of higher concentration?
1 active transport
2 diffusion
3 osmosis
4 hydrolysis

14 Which organism has a closed circulatory system?
1 paramecium
2 hydra
3 earthworm
4 grasshopper

15 Which organism has a transport system most similar in structure to that of an earthworm?
1 hydra
2 grasshopper
3 human
4 paramecium

16 Transport of materials within protozoa occurs primarily by
1 cyclosis and diffusion
2 a closed circulatory system
3 pinocytosis and hydrolysis
4 an open circulatory system

17 In the diagram at the right of root cells, in which direction would the net flow of water be the greatest as a result of osmosis?
1 A to C
2 A to B
3 B to C
4 C to B

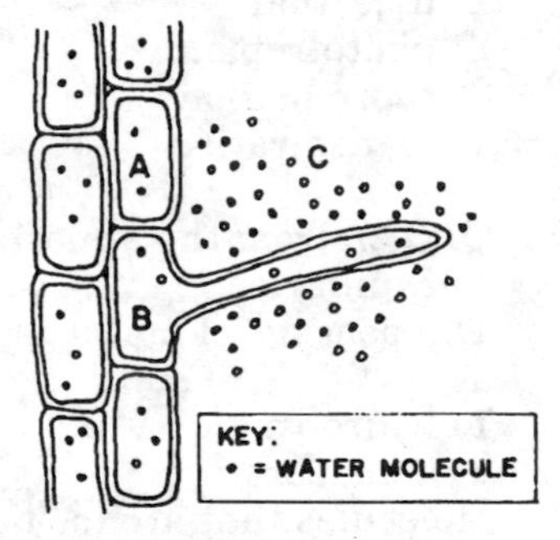

18 In vascular plants, the absorption of water from the soil into root hairs depends principally upon the presence of
1 phototropic response by the root hairs
2 geotrophic response by the conducting tissue
3 higher concentration of water in the soil than in the root hairs
4 higher concentration of water in the root hairs than in the soil

19 Ferns may grow several feet in height, whereas mosses seldom grow more than a few inches tall. What structural adaptation enables fern plants to grow taller than moss plants?
1 Ferns contain chlorophyll molecules, but mosses do not.
2 Ferns are heterotrophic, but mosses are not.
3 Ferns contain xylem and phloem, but mosses do not.
4 Ferns are aerobic, but mosses are not.

20 The diagram shows the method of entry of a molecule too large to diffuse through the plasma membrane of a cell. The process represented in the diagram is known as
1 homeostasis
2 pinocytosis
3 osmosis
4 cyclosis

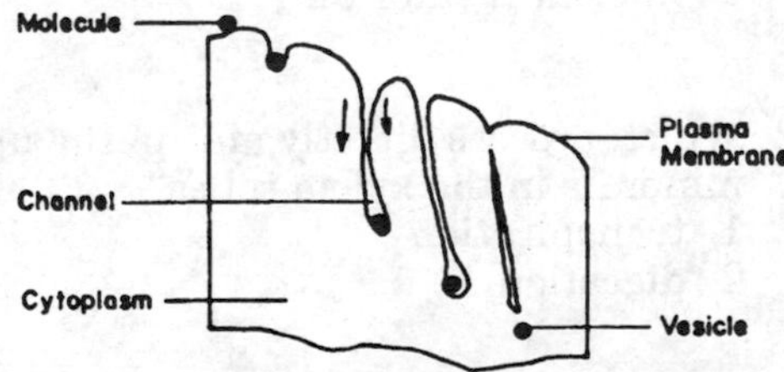

III. Respiration

Respiration – Terms & Concepts

Cellular Respiration
  ATP Production
Anaerobic Respiration
  Fermentation
    Lactic Acid or Alcohol/CO_2
Aerobic Respiration
  Prokaryotic Cells - 38 ATP gain
  Eukaryotic Cells - 36 ATP gain

Gas Exchange
  Diffusion - Cell Membranes
Plants - Stomates, Lenticels
Hydra - Diffusion
Earthworm - Moist, Thin Skin
  Hemoglobin Carries Oxygen
Grasshopper - Tracheal tubes,
  Spiracles - No Hemoglobin
Human - Lungs, Hemoglobin

Respiration is a process that occurs continuously in the cells of all living organisms. It involves the transfer of the stored chemical energy in food molecules to a form of energy readily usable by organisms.

A. Process

1. Cellular Respiration

Cellular respiration refers to those enzyme-controlled reactions in which the potential energy of organic molecules, such as glucose, is transferred to a more available form of energy. This available form of energy is stored in **adenosine triphosphate** molecules. When **ATP** is hydrolyzed (broken down), energy is released and ADP and phosphate are formed. This reaction is reversible. ATP can be used for metabolic activities which require energy.

$$H_2O + ATP \underset{\longrightarrow}{\overset{\longleftarrow}{\text{enzymes}}} ADP + P + \textit{Energy}$$

ATP Hydrolysis (to the right) **ATP Synthesis** (to the left)

If the energy transfer reactions involve the use of molecular oxygen, the process is **aerobic respiration**. The majority of organisms carry on aerobic respiration. If free oxygen is not used, the process is **anaerobic respiration**.

Anaerobic Respiration (Fermentation)

Some cells lack the enzymes necessary for aerobic respiration, and other cells revert to anaerobic respiration when oxygen is lacking. The enzymes for anaerobic respiration are located in the cytoplasm of cells.

Description. During most types of anaerobic respiration, glucose is gradually broken down in a series of enzyme-controlled reactions to either lactic acid or alcohol and CO_2. The end products vary depending on the type of organism. Lactic acid is produced in animals and is associated with muscle fatigue. Lactic acid is also produced by some bacteria and is important in the production of cheeses, buttermilk, and yogurt. Alcohol and CO_2 are usually produced by yeast and bacteria. These end products are useful in the baking and brewing industries. As a result of anaerobic respiration there is a net gain of 2 ATP.

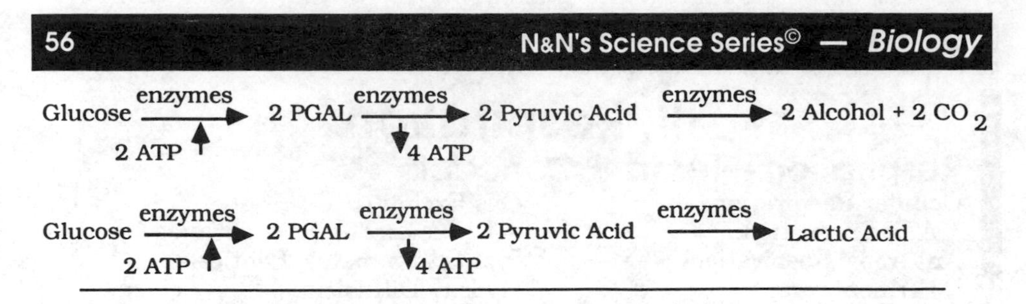

Because fermentation only partially uses the chemical bond energy of the glucose molecule, the end products, lactic acid and/or alcohol, still contain a great deal of potential energy.

Extended Area: ***Biochemistry***

Chemical Aspects of Anaerobic Respiration. The first series of reactions in anaerobic respiration involves the conversion of glucose to pyruvic acid and the production of 4 ATP's. This series of reactions requires specialized enzymes and activation energy (2 ATP's).

The conversion of pyruvic acid to either lactic acid or alcohol and CO_2 occurs without any further yield of energy. Therefore, there is a total net gain of only two molecules of ATP.

The reactions of anaerobic respiration may be summarized as follows (Note: The reaction pyruvic acid to lactic acid is reversible, but the conversion of pyruvic acid to alcohol and carbon dioxide is final.):

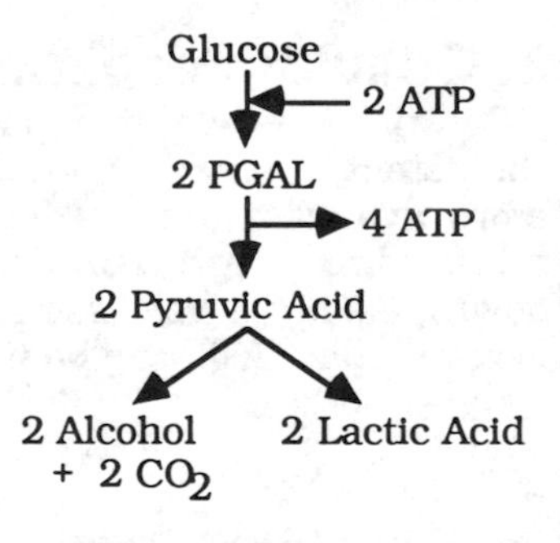

Aerobic Respiration

Many of the enzymes involved in aerobic cellular respiration are located in the **mitochondria**.

Description. During this process, and due to the presence of oxygen, the chemical energy of glucose is released gradually in a series of enzyme-controlled reactions.

$$\text{Glucose} + \text{Oxygen} \xrightarrow[\text{mitochondrion}]{\text{enzymes}} \text{ATP} + \text{Carbon Dioxide} + \text{Water}$$

$$C_6H_{12}O_6 + 6\,O_2 \xrightarrow[\text{mitochondrion}]{\text{enzymes}} 36\,\text{ATP} + 6\,CO_2 + 6\,H_2O$$

Since the glucose molecule is more completely broken down, aerobic respiration is more efficient than anaerobic. Authorities now believe that 38 ATP's are produced per molecule of glucose in all prokaryotic organisms, but that 36 ATP's are produced by respiration in eukaryotic organisms.

Extended Area: ***Biochemistry***

Chemical Aspects. Investigations of the chemical reactions occurring during Glucose oxidation indicate that, although there are many enzyme-catalyzed steps, there are two basic phases: the anaerobic phase and the aerobic phase.

In the **anaerobic phase**, glucose is oxidized to two molecules of pyruvic acid and the energy released is used to synthesize four molecules of ATP. Since the energy of two ATP molecules is required to activate this phase, there is a net gain of two ATP's.

$$\text{Glucose} + 2\text{ ATP} \xrightarrow[\text{(in cytoplasm)}]{\text{enzymes}} 2\text{ Pyruvic Acid} + 4\text{ ATP}$$

$$2\text{ Pyruvic Acid} + \text{Oxygen} \xrightarrow[\text{(in mitochondrion)}]{\text{enzymes}} \text{Carbon Dioxide} + \text{Water} + 34\text{ ATP}$$

In the **aerobic phase**, the pyruvic acid molecules are oxidized and more energy is released. Thirty-four ATP molecules are synthesized during this phase. Oxygen acts as a hydrogen acceptor, resulting in the formation of water. Carbon dioxide molecules are produced as a result of some intermediate reactions.

The net gain in the complete breakdown of one glucose molecule in aerobic respiration is therefore 36 ATP's.

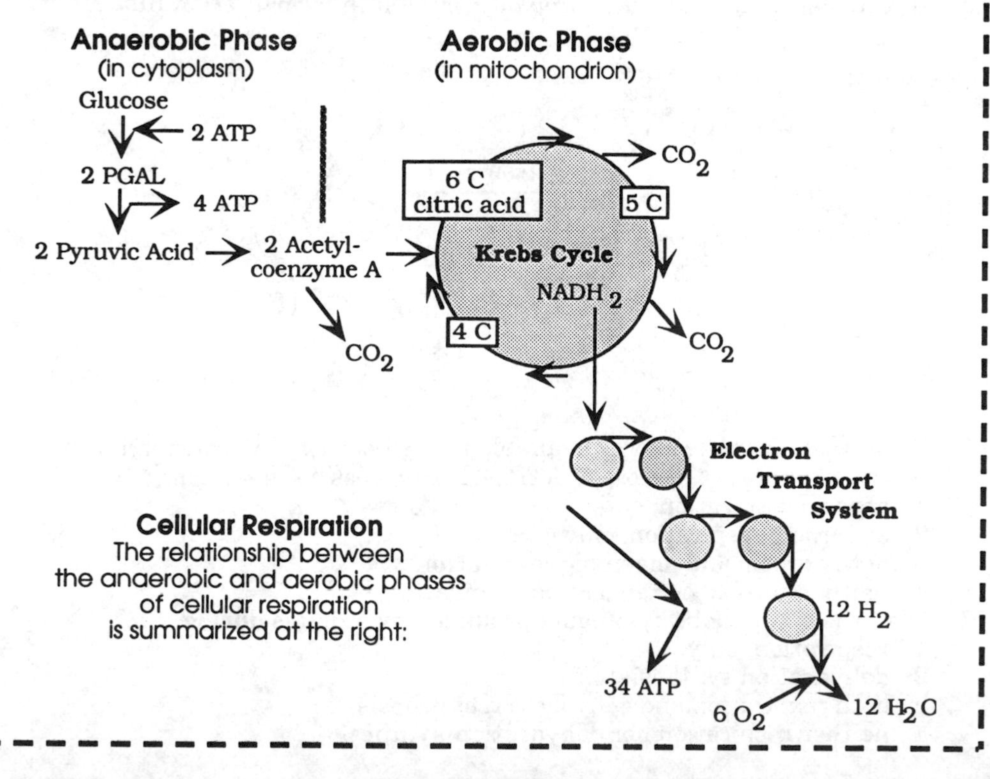

Cellular Respiration
The relationship between the anaerobic and aerobic phases of cellular respiration is summarized at the right:

Questions

1 Certain organisms break down glucose in a series of enzyme-controlled reactions. This series of reactions, which results in the production of alcohol and carbon dioxide, is an example of
1 dehydration synthesis
2 anaerobic respiration
3 hydrolysis
4 photosynthesis

2 A product of cellular respiration is
1 oxygen
2 ATP
3 PCB
4 glucose

3 During the process of aerobic respiration, energy stored in food is transferred to molecules of
1 ATP
2 DNA
3 glucose
4 enzymes

4 The energy released from the anaerobic respiration of a glucose molecule is less than that released from the aerobic respiration of a glucose molecule because
1 fewer bonds of the glucose molecule are broken in anaerobic respiration than in aerobic respiration
2 more enzymes are required for anaerobic respiration than for aerobic respiration
3 anaerobic respiration occurs 24 hours a day while aerobic respiration can only occur at night
4 anaerobic respiration requires oxygen but aerobic respiration does not require oxygen

5 In the diagram, what gas is probably present in fermentation tube *B*?

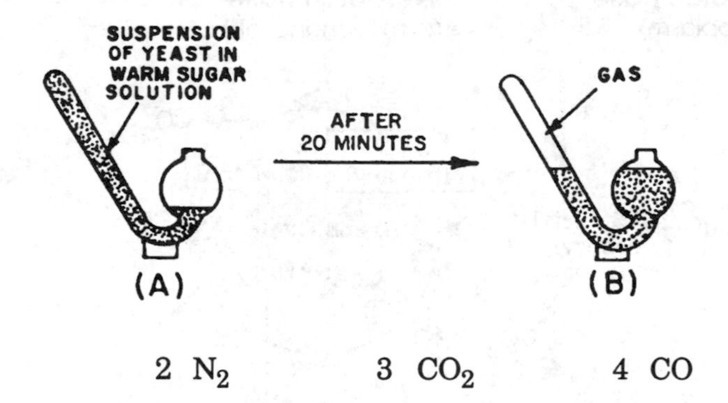

1 O_2
2 N_2
3 CO_2
4 CO

6 Most animals make energy available for cell activity by transferring the potential energy of glucose to ATP. This process occurs during
1 aerobic respiration, only
2 anaerobic respiration, only
3 both aerobic and anaerobic respiration
4 neither aerobic nor anaerobic respiration

7 Water is a metabolic by-product produced by animals during
1 respiration, only
2 dehydration synthesis, only
3 both respiration and dehydration synthesis
4 neither respiration nor dehydration synthesis

8 Which of the following processes releases the greatest amount of energy?
1 the oxidation of one glucose molecule to lactic acid molecules
2 the oxidation of one glucose molecule to carbon dioxide and water
3 the conversion of two glucose molecules to a maltose molecule
4 the conversion of one glucose molecule to alcohol and carbon dioxide

9 The site of aerobic cellular respiration is the
1 nucleus
2 ribosome
3 chromosome
4 mitochondrion

10 Compared to a cell that is carrying on anaerobic respiration, a cell that is carrying on aerobic respiration
1 uses less oxygen
2 produces more ATP
3 uses less carbon dioxide
4 produces more alcohol

11 Which process is illustrated by the summary equation below?

glucose + oxygen → water + carbon dioxide + 36 ATP

1 hydrolysis
2 photosynthesis
3 dehydration synthesis
4 aerobic respiration

12 Two end products of aerobic respiration are
1 oxygen and alcohol
2 carbon dioxide and water
3 oxygen and water
4 carbon dioxide and oxygen

13 The aerobic respiration of a molecule of glucose releases more energy than the anaerobic respiration of a molecule of glucose because in aerobic respiration
1 carbon dioxide is used.
2 oxygen is released.
3 more chemical bonds are broken.
4 lactic acid is formed.

Base your answers to questions 14 and 15 on the equation below concerning anaerobic cellular respiration.

$$\text{glucose} + 2X \xrightarrow{\text{enzymes}} 2 \text{ pyruvic acid} + 4 \text{ ATP}$$

14 The substance indicated by X is
1 O_2
2 H_2O
3 ATP
4 CO_2

15 In muscle cells, the pyruvic acid can be converted to
1 lactic acid
2 alcohol
3 oxygen
4 chlorophyll

2. Gas Exchange

Gas exchange involves the diffusion of gases between the organism and its environment. Oxygen diffuses inward, and carbon dioxide diffuses outward, both through the plasma membrane.

B. Adaptations

All living things must convert energy through the process of cellular respiration. While the chemical reactions of respiration are similar in most organisms, various adaptations for the exchange of respiratory gases are present in living things. A gas exchange surface must be *thin*, *moist*, and *in contact* with the external/internal environments.

1. Monera, Protista, and Fungi

These simple organisms are either single cells or have most of their cells close to their environment. Therefore, gas exchange occurs by diffusion through *thin and moist* membranes. This gas exchange surface is an external surface.

2. Plants

Plants are complex, multicellular organisms. Since many of their cells are not exposed to the environment, plants must have various methods for gas exchange.

Leaves. The outer covering is dry and impermeable and *not* a gas exchange surface. Gas exchange occurs across the membranes of internal cells. These cells are bounded by moist intercellular spaces. Access to these spaces is through the **stomates**.

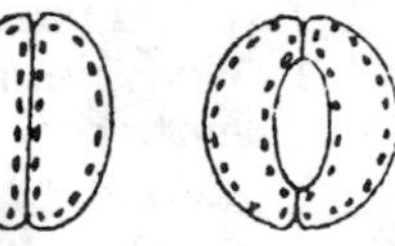

Stems. Stems of woody plants contain **lenticels** which permit the exchange of gases. Lenticels are tiny openings in "bark" of a stem.

Roots. Gas exchange occurs across the moist membranes of **root hairs** and other epidermal cells.

3. Animals

Hydra. Since each cell of the hydra is in contact with the watery environment, gas exchange occurs by simple diffusion.

Earthworm. As a terrestrial organism, the problem of the retention of a moist gas exchange surface is solved by the secretion of **mucus** by the skin. This maintains a moist surface which facilitates the diffusion of gases into and out of the blood. Hemoglobin aids in the transport of oxygen to the body cells.

Grasshopper. Since the grasshopper possesses an open circulatory system lacking hemoglobin, gas transport and gas exchange are accomplished by **tracheal tubes**. The tracheal tubes terminate internally in moist membranes where gases are exchanged. The external body surface is dry and impermeable. Access to the tracheal tubes is through **spiracles** (openings in the exoskeleton).

Human. Human skin is impermeable to respiratory gases. The lungs are a thin, moist, internal gas exchange surface. Hemoglobin aids in the transport of oxygen.

Questions

1 Which organism has tracheal tubes that serve as its primary respiratory structures?
1 ameba
2 grasshopper
3 earthworm
4 hydra

2 In which organisms does anaerobic respiration result in the production of ATP and alcohol?
1 algae
2 earthworms
3 ferns
4 yeast

3 Most plants are capable of exchanging gases with the environment because they possess
1 tracheal tubes and spiracles that function like lungs and gills
2 vacuoles that store more gases than animal vacuoles
3 moist intercellular spaces that open to the atmosphere
4 more efficient respiratory pigments than animals possess

4 Oxygen from the external environment enters a paramecium through the
1 contractile vacuole
2 plasma membrane
3 centriole
4 ribosome

5 When does the process of cellular respiration occur in bean plants?
1 in the daytime, only
2 at night, only
3 both in the daytime and at night
4 only when photosynthesis stops

6 Which processes most directly involves the exchange of gases through moist membranes?
1 digestion and locomotion
2 reproduction and growth
3 transport and regulation
4 respiration and excretion

7 Respiratory enzymes are present in
1 animal cells, but not in plant cells
2 plant cells, but not in animal cells
3 neither animal nor plant cells
4 both animal and plant cells

8 The respiratory system of an earthworm utilizes the skin as an external gas exchange surface. What additional system is used to carry gases to moist internal body tissues?
1 digestive
2 circulatory
3 nervous
4 endocrine

9 What do the tracheal tubes of the grasshopper and the air spaces of the geranium leaf have in common?
1 They regulate the flow of urea into and out of the organism.
2 They are the major sites for the ingestion of nutrients.
3 They have enzymes that convert light energy to chemical bond energy.
4 They are surrounded by moist internal surfaces for gas exchange.

10 During the winter in the Northeast U.S., deciduous trees carry on gas exchange through specialized stem structures known as
1 veins
2 lenticels
3 xylem vessels
4 phloem tubes

IV. Excretion

Excretion – Terms & Concepts

Metabolic Wastes
  Carbon Dioxide, Mineral Salts,
  Water, Nitrogenous Wastes
Plants - Recycle Wastes
  Toxic - Stored in Vacuoles
Hydra, Protists - Diffusion
Earthworm - Diffusion thru Skin
  Paired Nephridia
Grasshopper - Malpighian Tubes,
  Tracheal Tubes, Spiracles
Human - Lungs, Skin, Nephrons

A. Process

Excretion is the removal of cellular waste products of an organism. Cellular wastes are the results of metabolic activity.

1. Metabolic Waste Products

The metabolic activities of living organisms result in the production of waste materials. These products may be useful to other organisms.

- **Carbon dioxide** is the waste product of aerobic respiration.
- **Water** is the end product of dehydration synthesis and aerobic respiration.
- **Mineral salts** are produced during many metabolic processes.
- **Nitrogenous wastes** result from the metabolism of proteins and when excess amino acids are utilized in cellular respiration. Examples include: **ammonia** (very toxic), **urea** (less toxic), and **uric acid** (generally nontoxic).

2. Results of Excretion

Wastes may be either toxic or nontoxic. When toxic wastes are produced they are normally released, as in animals, or sealed off and stored, as in plants. Nontoxic wastes may either be retained, released, or recycled in other metabolic activities.

B. Adaptations

Organisms display various adaptations for excretion. Adaptations vary depending on the metabolic activities of the organism and the environment in which it lives.

1. Protista

Most unicellular organisms lack special excretory structures. Excretion is accomplished by diffusion through cell membranes.

Freshwater Protozoans. In the amoeba and paramecium, carbon dioxide, ammonia, and mineral salts diffuse (passive transport) through the plasma membrane directly into the watery environment.

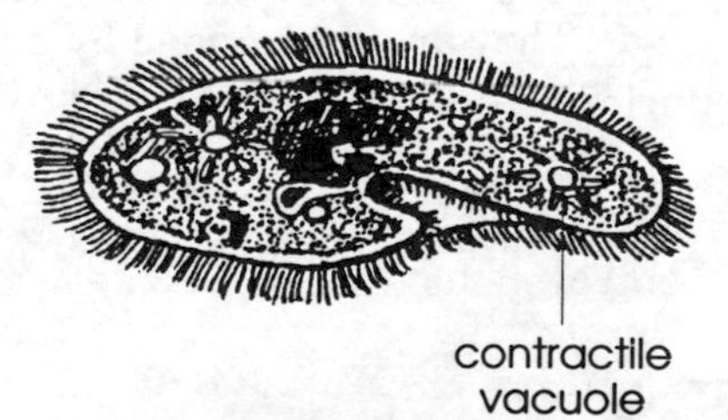

Paramecium

Large metabolic waste materials may be removed through active transport. In freshwater protozoans, the **contractile vacuole** functions in **osmoregulation** (water pressure control).

Algae. The waste product of respiration, carbon dioxide, can be recycled in the photosynthetic process. Also, some of the oxygen produced during photosynthesis is recycled for respiration.

2. Plants

Plants *recycle* the end products of photosynthesis (water vapor and oxygen) and respiration (carbon dioxide). In general, more oxygen is produced than can be utilized by the plant and is released to the environment. The excess gases leave the plant through the stomates, lenticels, and epidermal cells of the root.

Some waste products, such as organic acids which might be toxic, are stored in vacuoles where they cause no injury to the plant.

3. Animals

Hydra. Excretion in the hydra is essentially similar to that of the protozoans. Metabolic wastes are diffused through plasma membranes to the watery environment.

Earthworm. Carbon dioxide is excreted by diffusion through the moist skin of the earthworm. Pairs of excretory organs called **nephridia** (located in most body segments of the earthworm) excrete water, mineral salts, and urea through pores into the terrestrial environment.

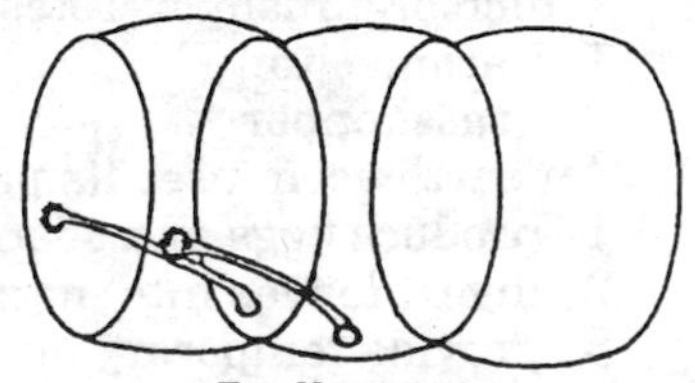

Earthworm
Paired Nephridia in Segments

Grasshopper. Carbon dioxide diffuses from the grasshopper's body fluids into **tracheal tubes** and is expelled through **spiracles**. Water, mineral salts, and insoluble uric acid crystals accumulate in the **Malpighian tubules** and are transported to the digestive tube where most of the water is reabsorbed. Minerals and uric acid are expelled with the fecal material.

The excretion of uric acid is a water-conserving mechanism of particular advantage to an egg-laying terrestrial organism. Because uric acid is insoluble, it can be stored in the egg without exerting harmful toxic or osmotic effects.

Human. Carbon dioxide and water are excreted through the respiratory system. Humans possess **nephrons**, excretory structures similar to the nephridia of earthworms, for the excretion of water, salt, and urea.

Questions

1 Nitrogenous wastes are removed from the blood of the earthworm through the
  1 Malpighian tubules
  2 nephridia
  3 nephrons
  4 tracheal tubes

2 Which metabolic waste is correctly matched with a process that produces it?
 1 urea - protein metabolism 3 mineral salts - respiration
 2 carbon dioxide - digestion 4 ammonia - digestion

3 Nitrogenous waste products are produced from complete metabolism of
 1 sugar 2 water 3 starch 4 protein

4 Excretory structures present in protists include
 1 cell membranes and contractile vacuoles
 2 cell walls and lungs
 3 nephridia and nephrons
 4 Malpighian tubules and spiracles

5 Which is a true statement concerning waste products in plants?
 1 Plants do not produce waste products.
 2 Malpighian tubules remove excess gaseous waste products from plants.
 3 The waste products of plants are often reused in other processes
 4 Contractile vacuoles in plant cells remove waste products.

6 Hydra excrete most of their nitrogenous wastes in the form of
 1 urea 2 uric acid 3 ammonia 4 nitrates

7 An organism containing Malpighian tubules would most likely possess
 1 a four-chambered heart 3 an open circulatory system
 2 an endoskeleton 4 a contractile vacuole

8 Which organism excretes most of its nitrogenous waste as ammonia?
 1 paramecium 3 human
 2 grasshopper 4 earthworm

9 An earthworm uses its paired nephridia to
 1 produce eggs and sperm
 2 digest leaves into inorganic nutrients
 3 provide temporary anchorage in soil
 4 excrete water, urea, and mineral salts

10 Which organism is correctly paired with its excretory structure?
 1 earthworm - nephridia 3 ameba - skin
 2 grasshopper - nephron 4 hydra - malpighian tubule

11 By which process do carbon dioxide molecules leave a plant and enter the atmosphere?
 1 digestion 2 osmosis 3 photosynthesis 4 diffusion

12 The leaf structure closely associated with both transpiration and excretion are
 1 'enticels 3 stomates
 2 elongated epidermal cells 4 waxy surfaces

13 Carbon dioxide released from the interior cells of a grasshopper is transported to the atmosphere through the
 1 Malpighian tubules 3 contractile vacuoles
 2 tracheae 4 lungs

14 A nitrogenous waste product resulting from the metabolism of protein molecules is
 1 carbon dioxide 3 mineral salts
 2 ammonia 4 water

15 Which organism produces uric acid as its principal nitrogenous waste product?
 1 Paramecium 2 Hydra 3 Ameba 4 Grasshopper

V. Regulation

Regulation – Terms & Concepts

Nerve Control
- Stimulus - Receptors
- Impulse - Transmission
- Response - Effectors

Neurons
- Dendrites - Cyton
- Axon - Terminal Branches

Synapse - Neurotransmitters
Hydra - Nerve Net
Earthworm - CNS
- Ventral Nerve Cord

Grasshopper - Sensory Organs
- Ventral Nerve Cord - Ganglia

Human - CNS - Highly Developed
- Chemical Control

Plant Hormones - Auxins
- Growth, Tropisms, Reproduction
- Meristematic Regions
- Flower, Seed, Fruit Formation

Animal - Endocrine (ductless glands)
- Metabolism, Metamorphosis, and Reproduction

Regulation is the coordination and control of the life activities. Coordination of all life functions depends on special control mechanisms. These mechanisms include nerve control (found in multicellular animals) and chemical control (common to all organisms).

A. Nerve Control

Nerve control depends mainly on the functioning of the tiny **neurons** (nerve cells) and their supporting cells.

1. Functional Definitions

Stimulus. Any *change* in the external or internal environment which initiates an impulse is a stimulus. Examples of stimuli include changes in chemicals, light, and sound.

Receptors are structures specialized to detect certain stimuli. Examples of specialized receptors are found in the skin, eyes, ears, etc.

Impulses are electro-chemical charges generated along a neuron. These charges are the result of changes in polarity along the membranes of neurons. Impulses transmit information to and from the central nervous system.

Response. The reaction to a specific stimulus, including glandular secretions and muscle movements, is a response.

Effectors are the organs of response such as muscles or glands.

Neuron. The nerve cell is a cell especially adapted for the transmission of impulses.

Neuron: Generalized Structure

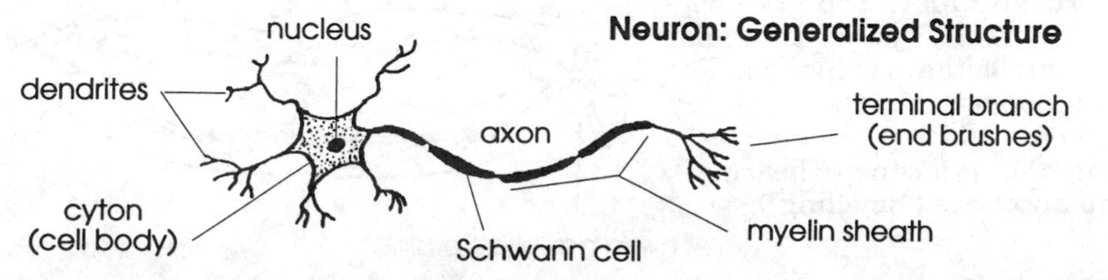

In general, the neuron is composed of:

- **Dendrites** are fibers which detect the stimulus and generate impulses toward the cyton.
- **Cyton** is the cell body containing the nucleus.
- **Axon** is a fiber that transmits the impulse away from the cyton toward the terminal branches.
- **Terminal Branches** are the ends of the axons which secrete neurotransmitters.

Synapse. Neurons and receptors and effectors do *not* directly contact each other. The synapse is a junction (gap or space) between adjacent neurons.

Neurotransmitters. Special chemicals, such as **acetylocholine**, are secreted from terminal branches and aid in the transmission across the synapse by stimulating the adjacent neuron or an effector.

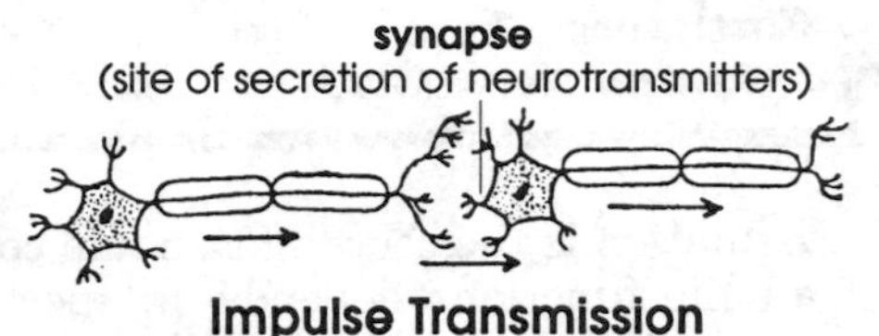

Impulse Transmission

2. Adaptations in Animals

Hydra. The Hydra possesses a **nerve net** composed of modified neurons. Responses are general since there is no central nervous system and impulses may travel in either direction over the neurons.

Hydra nerve net

Earthworm. The nervous system of the earthworm consists of a primitive "brain" composed of fused **ganglia**, a **ventral nerve cord**, and **peripheral nerves**. The presence of this central nervous system permits impulses to travel over definite pathways from receptors to effectors. This allows the earthworm to make a definite directional response to a specific stimulus.

Earthworm Nervous System

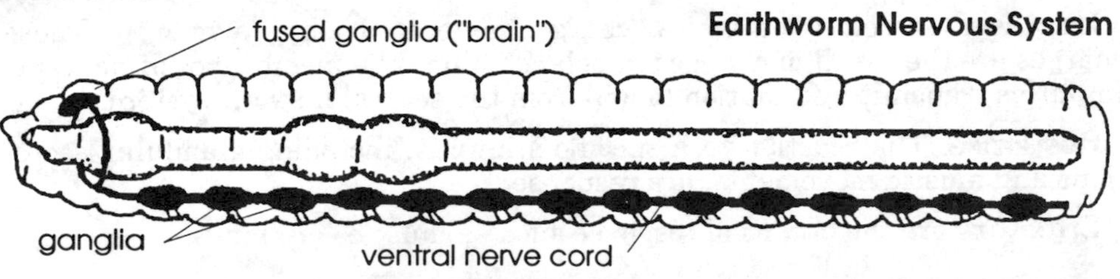

Grasshopper. The grasshopper's nervous system is similar to that of the earthworm. The grasshopper possesses sensory organs which include eyes ("sight"), tympana ("hearing"), and antennae ("smelling").

Grasshopper Nervous System

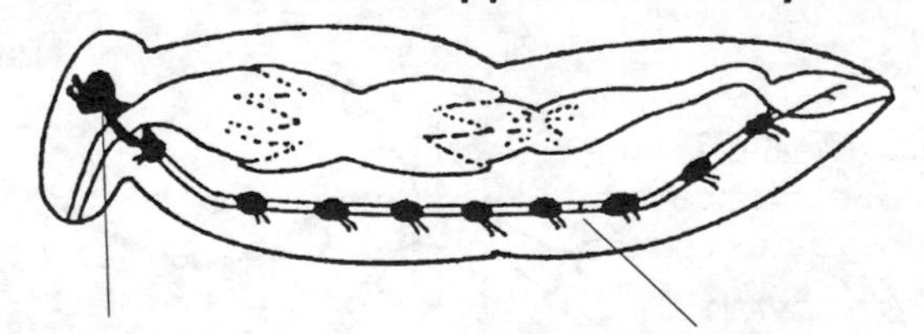

Human. The human central nervous system consists of a highly developed brain and dorsal nerve cord which permit impulses to travel over definite pathways. There are also many highly developed receptors.

Questions

1 A definite pathway for impulses from receptors to effectors is found in the
1 parmecium 2 hydra 3 earthworm 4 maple tree

2 In the diagram of the coelenterate at the right, the letter X indicates structures used to respond to stimuli. Which phrase best describes these structures?
1 ventral nerve cords with fused ganglia and connecting nerves
2 dorsal nerve cords with an anterior brain
3 neurons that transmit impulses in all directions
4 spinal cord connections to a simple anterior brain

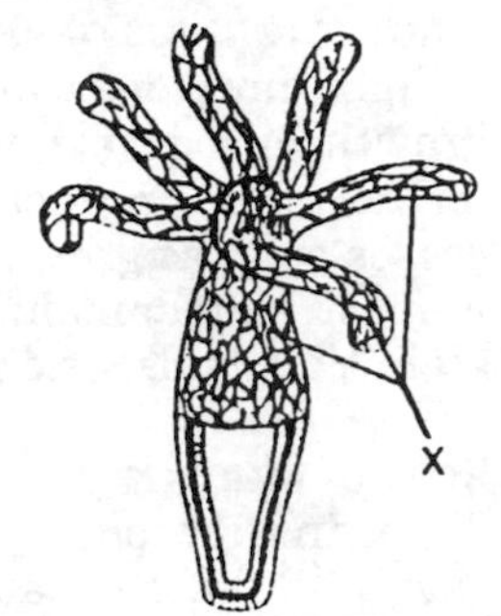

3 A student accidentally places her hand on a tack and quickly pulls her hand away. The tack represents
1 a stimulus 3 a response
2 an impulse 4 an effector

4 In the nervous system of an earthworm, fused ganglia make up
1 the primitive brain 3 the setae
2 the peripheral capillaries 4 the nephridia

5 The nervous system of the earthworm consists of
1 tympana, antennae, and a ventral nerve cord
2 tympana, antennae, and a dorsal nerve cord
3 fused ganglia and a ventral nerve cord
4 fused ganglia and a dorsal nerve cord

6 The space between two adjacent neurons is known as a
1 joint 2 linkage 3 synapse 4 dendrite

7 Impulses are transmitted from the cyton to the terminal branches of a neuron along the membranes of the
1 dendrite 2 axon 3 nucleus 4 mitochondrion

8 Neurotransmitters, such as acetylcholine, are initially detected by which part of a neuron?
1 dendrite 3 terminal branch
2 nucleus 4 mitochondrion

9 When a tentacle of a hydra is touched with a needle, the entire body responds as a result of impulses traveling to cells through
1 the central nervous system 3 a spinal cord
2 the posterior brain 4 a nerve net

10 During impulse transmission along neurons, a substance is added which interferes with the action of a neurotransmitter. The interfering substance would be added at which of the following points to have the greatest effect? At the
1 cyton 2 axon 3 nucleus 4 synapse

B. Chemical Control

Similar to nerve control, chemical control coordinates body processes by transmitting messages from one part of the organism to another part. Unlike nerve control, chemical control is achieved through the use of **hormone** action in both plants and animals.

1. Chemical Control in Plants

Characteristics of Plant Hormones

Plant hormones are chemicals produced by cells which affect the plant's growth and development. (There are *no* plant organs specialized exclusively for hormone production.) Plant hormones function in the coordination of processes such as growth, tropisms, and reproduction. Plant hormone production is most abundant in actively growing areas such as the cells at the root and stem tips, buds, and seeds.

Role of Plant Hormones

Auxins are one type of plant hormone. Auxins influence division, elongation, and differentiation of plant cells. Unequal distribution of auxins causes unequal growth responses called **tropisms**, which generally enhance the survival of the plant. Unequal auxin distribution can be caused by external stimuli such as light and gravity.

- **Phototropism** is the response to light (plant bends toward light source). Unequal distribution of auxins increase growth on side A, causing the plant to "bend" towards side B.

- **Geotropism** is the response to gravity (roots grow toward the earth).

- **Hydrotropism** is the response to water (roots grow toward a source of water).

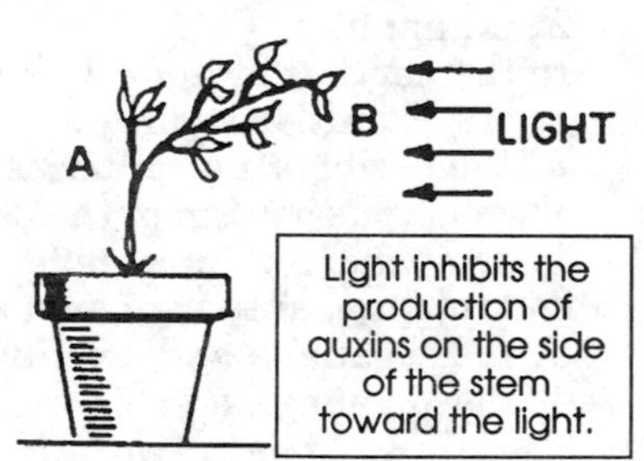
A
B
LIGHT
Light inhibits the production of auxins on the side of the stem toward the light.
Phototropism

Hormones promote other developmental changes including flowering, fruit formation, and seed development. Hormone responses depend upon the hormones involved, their concentrations, and the tissues affected.

2. Chemical Control in Animals

Chemical control in animals differs from that in plants since animals possess cells specialized solely for hormone production.

Endocrine Control

Endocrine glands synthesize and secrete hormones which control the activities of animals. **Hormones** are chemicals secreted in one area of the body which affect responses in other areas. The circulatory system (and blood) aids in the distribution of these hormones since the endocrine glands are ductless.

Role of Animal Hormones

Current research indicates that hormones exist in a wide variety of organisms and that the hormones themselves have wide distribution among animals. Hormones interact and exert control on metabolic activities such as metamorphosis and reproduction.

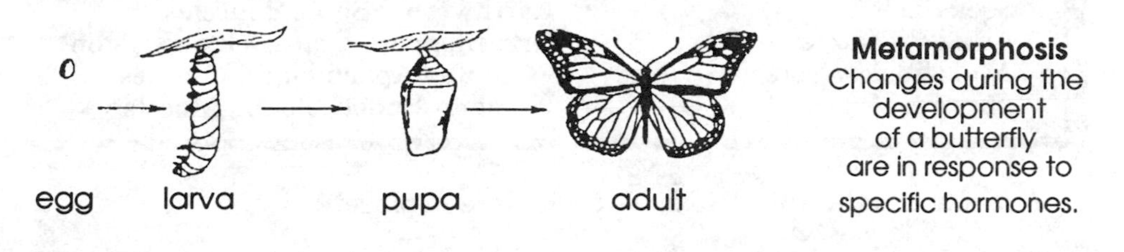

Metamorphosis
Changes during the development of a butterfly are in response to specific hormones.

Questions

1 Bean seeds were planted and put on a sunny windowsill. As the plants grew, their stems bent toward the window. This bending was most likely caused by an
   1 unequal distribution of auxin in the stem
   2 unequal distribution of a neurotransmitter in the stem
   3 equal distribution of auxin in the stem
   4 equal distribution of a neurotransmitter in the stem

2 An example of an animal response is
   1 a stone approaching the eye   3 the fragrance of a rose
   2 the buzz of a bee   4 the secretion of a hormone

3 A chemical growth regulator produced by plants is
   1 a neurotransmitter   3 a pigment
   2 an auxin   4 an isotope

4 The hormones released during insect metamorphosis serve as growth regulators. Which substance present in the stem of a bean plant has a similar function?
   1 DNA   2 auxin   3 ATP   4 ribose

5 Hormones are transported throughout many animals by the
   1 lysosomes   3 circulatory system
   2 synapses   4 excretory system

6 Geranium leaves grow in positions that permit the optimum use of light as a result of
   1 phototropic responses   3 transpiration pull
   2 capillary action   4 symbiotic relationships

7 Roots respond to gravity as a result of the unequal distribution of
   1 stomates   2 oxygen   3 lenticels   4 auxins

8 A chemical injected into a tadpole caused rapid metamorphosis into a frog. This chemical was most probably a(an)
   1 enzyme   2 lipid   3 hormone   4 blood protein

9 Secretions from ductless glands are known as
   1 enzymes   3 lachrymal fluids
   2 hormones   4 excretory fluids

10 The chief function of auxins in plants is to control the rate of
   1 growth   3 absorption
   2 photosynthesis   4 transpiration

VI. Locomotion

Locomotion – Terms & Concepts

Advantages
- Obtain Food
- Seek Shelter
- Avoid Predators
- Avoid Toxic Wastes
- Improve Mating

Protista - Flagella, Cilia, Pseudopods
Hydra - Sessile, Somersaulting
Earthworm - Setae, Muscles
Grasshopper - Chitin - Exoskeleton
Jointed Appendages, Muscles
Human - Endoskeleton, Muscles

Locomotion is the ability to move from place to place.

A. Advantages

Locomotion increases the probability of survival among animals and many protista. Some of the many advantages of locomotion (motility) include:

- increased opportunities to obtain food
- increased ability to seek shelter
- increased ability to avoid predators
- increased ability to move away from toxic wastes
- increased opportunities to find a mate

B. Adaptations

1. Protista

Algae. Some algae (and protists) move by means of flagella which are long whiplike projections.

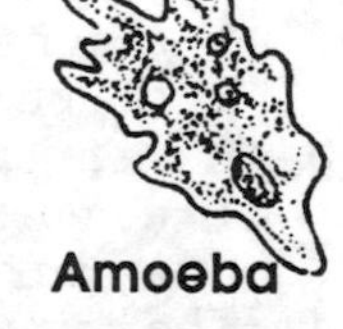

Euglena Amoeba

Protozoans. The locomotive structures utilized by paramecia are **cilia** (short hairlike projections). Locomotion in amoeba is by **pseudopods** (cytoplasmic extensions).

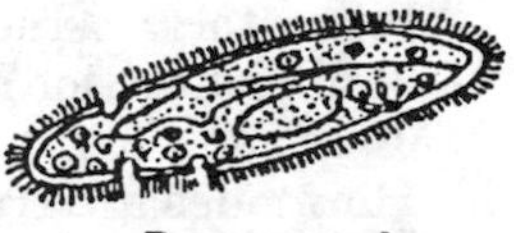

Paramecium

2. Animals

Hydra. The hydra is essentially a **sessile** (non-moving) organism. The presence of contractile fibers permits some motion, including a type of somersaulting.

Earthworm. Locomotion is accomplished through the interactions of **muscles** and **setae**. The setae (short bristles) permit temporary anchorage in the soil, and the movement is produced by alternate muscle contractions of the earthworms body.

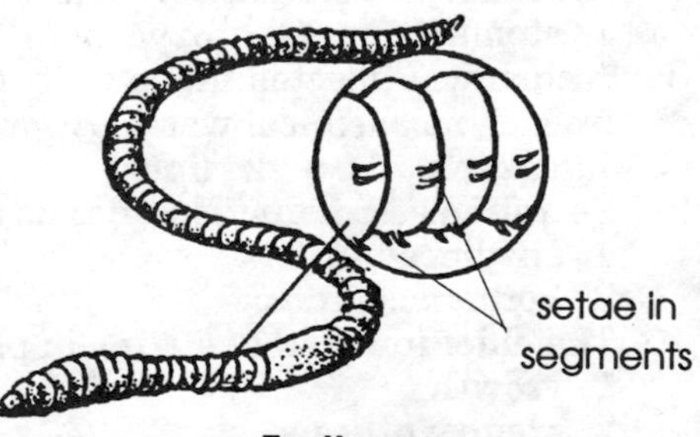

Earthworm

Grasshopper. The grasshopper has an **exoskeleton** made of **chitin.** The exoskeleton provides protection for internal organs, support, and points of attachment for muscles. Locomotion is accomplished by interactions of muscles with jointed, chitinous appendages.

Human. Humans have an **endoskeleton.** Locomotion is accomplished by the interaction of muscles and jointed appendages.

Questions

1 Locomotive structures found in some protists include
1 muscles
2 flagella
3 tentacles
4 contractile vacuoles

2 Organisms are classified as sessile or motile based on the life activity known as
1 locomotion
2 reproduction
3 synthesis
4 transport

3 Pseudopodia, chitinous appendages, and setae are structures that assist various animals in
1 the transmission of impulses
2 avoiding predators
3 the synthesis of protein
4 aerobic respiration

4 Locomotion in the earthworm is accomplished by the interaction of muscles and
1 bones
2 pseudopods
3 jointed appendages
4 paired setae

5 Which organ of the grasshopper is correctly matched with its function?
1 chitinous appendage - locomotion
2 Malpighian tubule - respiration
3 trachea - digestion
4 gonad - excretion

6 Locomotion does not increase an animal's opportunity to
1 obtain food
2 reproduce
3 escape from predators
4 transmit impulses

7 The ability of a paramecium to move from place to place is dependent upon the presence of
1 flagella
2 pseudopods
3 muscles
4 cilia

8 Which correctly matches an organism with the structures it uses for locomotion?
1 grasshopper - chitinous appendages and setae
2 human - muscles and bone
3 hydra - flagella and pseudopodia
4 paramecium - pseudopodia and muscles

9 Which two organisms are able to move due to the interaction of muscular and skeletal systems?
1 earthworm and human
2 hydra and earthworm
3 grasshopper and hydra
4 grasshopper and human

10 If a chemical introduced into the body of an earthworm interferes with the function of its setae, which life process would immediately be affected
1 respiration
2 reproduction
3 circulation
4 locomotion

SELF–HELP: Unit II *"Core"* Questions

1 The complete hydrolysis of a protein would result in the formation of
 1 fatty acids 3 amino acids
 2 glycerol 4 polysaccharides

2 An isotope is used to trace the chemical reactions of photosynthesis in a green plant. If the isotope is ultimately found in the starch stored in the green plant, the isotope would be
 1 nitrogen-14 3 sulfur-32
 2 carbon-14 4 carbon-10

3 A bean plant is an autotroph because it
 1 uses enzymes
 2 uses oxygen
 3 converts inorganic materials into organic nutrients
 4 absorbs organic nutrients from the soil

4 Large numbers of motile aerobic bacteria accumulate along portions of a filament of green algae exposed to blue and red light. Few bacteria accumulate in areas exposed to green and yellow light. The bacteria form this pattern because
 1 bacteria need red and blue light for fermentation
 2 more oxygen is released by algae cells in areas exposed to red and blue light
 3 more glucose diffuses out of the algae cells exposed to green and yellow light
 4 water is warmer in areas exposed to green and yellow light

5 An organism which makes its own food without the direct need for any light energy is known as a
 1 chemosynthetic heterotroph
 2 photosynthetic heterotroph
 3 chemosynthetic autotroph
 4 photosynthetic autotroph

6 In a rabbit, which process would be most directly affected by a poison that destroys some of the hydrolytic enzymes?
 1 photosynthesis 3 selective absorption
 2 ingestion 4 chemical digestion

7 Which occurs as a result of the action of hydrolytic enzymes?
 1 Inorganic substances are converted directly to organic substances.
 2 Complex organic molecules are made more soluble.
 3 Glucose molecules are converted to starches.
 4 Glucose molecules are converted to maltase molecules.

8 A common characteristic of animals and fungi is their ability to carry on
 1 heterotrophic nutrition
 2 auxin transport
 3 alcoholic fermentation
 4 transport through vascular tissue

9 What are the end products of carbohydrate hydrolysis?
 1 amino acids 3 glycerol
 2 simple sugars 4 fatty acids

10 Compared to the ingested food molecules, the end product molecules of digestion are usually
 1 smaller and more soluble 3 larger and more soluble
 2 smaller and less soluble 4 larger and less soluble

Base your answers to questions 11 and 12 on your knowledge of biology and on the diagram which illustrates a process by which protein molecules may enter a cell.

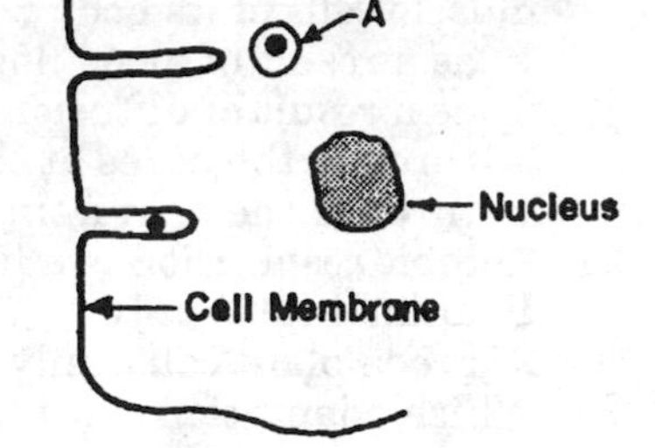

11 Which process is illustrated in this diagram?
1 pinocytosis
2 diffusion
3 osmosis
4 passive transport

12 Structure A is most likely a
1 ribosome
2 mitochondrion
3 nucleolus
4 vacuole

13 Active transport is different from passive transport in that active transport involves
1 the movement of molecules from high concentration to low concentration
2 an expenditure of energy
3 the use of ribosomes
4 a process which occurs only in the cells of simple plants and animals

14 The process of osmosis is best illustrated by the movement of
1 water into root hair cells
2 oxygen into red blood cells
3 carbon dioxide through stomates
4 glucose through phloem

15 Which organisms have internal, closed circulatory systems that can transport both nutrients and oxygen?
1 earthworms and hydras
2 earthworms and humans
3 grasshoppers and hydras
4 grasshoppers and humans

16 Carbohydrate molecules A and B come in contact with the cell membrane of the same cell. Molecule A passes through the membrane readily, but molecule B does not. It is most likely that molecule A is
1 a protein, and B is a lipid
2 a polysaccharide, and B is a monosaccharide
3 an amino acid, and B is a monosaccharide
4 a monosaccharide, and B is a polysaccharide

17 The movement of chloroplasts within a plant cell in a circular motion is most likely the result of
1 cyclosis 2 pinocytosis 3 ingestion 4 transpiration

18 The primary function of root hairs in a plant is to
1 prevent excessive water loss
2 provide increased surface area for absorption
3 conduct water and minerals upward
4 conduct organic food materials upward and downward

19 The tissue which conducts organic food throughout a vascular plant is composed of
1 cambium cells
2 xylem cells
3 phloem cells
4 epidermal cells

20 How are nutrients transported from the blood of an earthworm to the muscle cells of its body wall?
   1 as a result of blood flowing directly into muscle cells
   2 as a result of diffusion through capillary walls
   3 through the pores at the ends of nephridia
   4 through the skin from the outside environment

21 Aerobic respiration is a life function that occurs in
   1 animal cells, only
   2 green plant cells, only
   3 nongreen plant cells, only
   4 most animal and plant cells

22 Which adaptations for the exchange of gases do hydras, earthworms, and grasshoppers have in common?
   1 thin, moist membranes for absorption
   2 closed circulatory systems
   3 capillaries that distribute blood
   4 hearts that pump blood

23 Which is a true statement concerning aerobic respiration in humans and bean plants?
   1 Molecular oxygen is used by humans and by bean plants 24 hours per day.
   2 Molecular oxygen is used by humans 24 hours per day and by bean plants only in the sunlight.
   3 Molecular oxygen is used by humans 24 hours per day and by bean plants during darkness.
   4 Molecular oxygen is used by bean plants 24 hours per day and by humans only during darkness.

24 Which structures aid in the exchange of respiratory gases in grasshoppers?
   1 moist appendages
   2 tracheal tubules
   3 moist lungs
   4 Malpighian tubules

25 In animal cells, energy to convert ADP to ATP comes directly from
   1 hormones
   2 inorganic molecules
   3 organic molecules
   4 sunlight

26 Organisms make energy readily available by transferring the chemical bond energy of organic molecules to
   1 mineral salts
   2 light energy
   3 adenosine triphosphate
   4 nitrogenous wastes

27 The principal structure for excretion in the protozoans is the
   1 plasma membrane
   2 food vacuole
   3 lysosome
   4 nucleus

28 Metabolic wastes of animals most likely include
   1 water, carbon dioxide, oxygen, and salts
   2 carbon dioxide, nitrogenous compounds, water, and salts
   3 hormones, water salts, and carbon dioxide
   4 glucose, carbon dioxide, nitrogenous compounds, and water

29 Which statement best describes the excretion of nitrogenous wastes from Paramecia?
   1 Urea is excreted by nephrons.
   2 Uric acid is excreted by nephridia.
   3 Ammonia is excreted through cell membranes.
   4 Urea and uric acid are excreted through Malpighian tubules.

30 Which animal excretes wastes by the action of nephridia?
1 Paramecium 2 Hydra 3 earthworm 4 grasshopper

31 The bending of a stem toward light is the result of
1 transpiration pull
2 chemosynthetic nutrition
3 cambium stimulation
4 unequal auxin distribution

32 A ventral nerve cord and a closed circulatory system are characteristics of
1 a hydra
2 an earthworm
3 a human
4 a grasshopper

33 Plants respond to stimuli such as light and gravity. These responses are chiefly due to
1 the transmission of nerve impulses
2 the distribution of auxins
3 transpiration pull
4 capillary action

34 Metamorphosis in insects, allowing development of the egg, larva, pupa, and adult, is most directly regulated by
1 auxins 2 hormones 3 vitamins 4 minerals

35 Two systems that directly control homeostasis in most animals are the
1 nervous and endocrine
2 endocrine and excretory
3 nervous and locomotive
4 excretory and locomotive

36 Chemicals produced by the ends of neurons and secreted into synaptic spaces are
1 hormones
2 neurotransmitters
3 auxins
4 toxins

37 Which organism has a nervous system consisting of a simple anterior "brain" and a ventral nerve cord?
1 earthworm 2 hydra 3 ameba 4 human

38 Setae are structures for locomotion used by the
1 paramecium 2 earthworm 3 ameba 4 hydra

39 Locomotion is accomplished by the interaction of muscles and chitinous appendages in the
1 hydra 2 paramecium 3 grasshopper 4 human

40 An animal has increased opportunities to obtain food as a direct result of
1 circulation 2 locomotion 3 synthesis 4 excretion

SELF–HELP: Unit II *"Extended"* Questions

Base your answers to questions 1 through 3 on the chemical equation below which represents a metabolic activity.

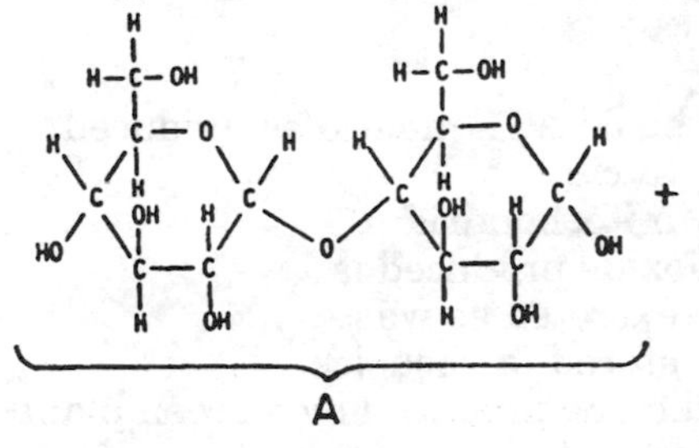

+ H_2O

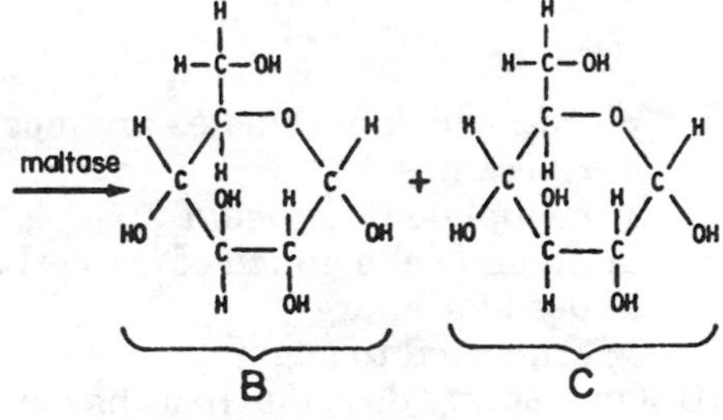

1 This metabolic activity is known as
1 glucose oxidation
2 enzymatic hydrolysis
3 polysaccharide formation
4 alcoholic fermentation

2 The chemical reaction represented can be performed by
1 vertebrate animals, only
2 multicellular plants, only
3 protists, only
4 plants, animals, and protists

3 When substance B and C combine chemically and produce substance A and water, this chemical process is known as
1 anaerobic respiration
2 digestion
3 dehydration synthesis
4 enzymatic hydrolysis

Base your answers to questions 4 through 6 on the chemical reactions below and on your knowledge of biology.

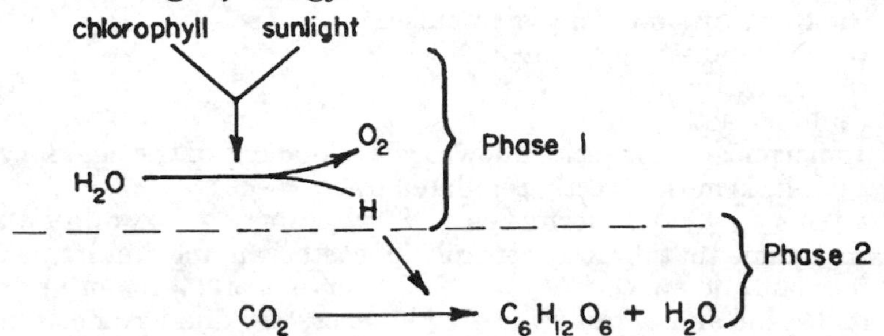

4 The process represented by the entire diagram is known as
1 digestion
2 photosynthesis
3 respiration
4 regulation

5 In which part of a chloroplast does phase 1 occur?
1 grana 2 ribosome 3 stroma 4 DNA

6 Phase 2 is also known as the
1 photochemical reaction
2 aerobic phase of respiration
3 carbon-fixation reaction
4 anaerobic phase of respiration

7 Enzymes for both photochemical and carbon-fixation reactions are found within organelles known as
1 chloroplasts
2 Golgi complexes
3 lysosomes
4 ribosomes

Base your answers to questions 8 through 10 on the equation below which represents a process that occurs in both plants and animals and on your knowledge of biology.

$$\text{glucose} + \text{oxygen} \xrightarrow{\text{enzymes}} \text{water} + \text{carbon dioxide} + 36\ \text{ATP}$$

8 Within which organelles are most of the 36 ATP molecules produced?
1 ribosomes
2 endoplasmic reticula
3 nuclei
4 mitochondria

9 In animal cells, much of the carbon dioxide produced is
1 used for energy
2 converted to sugar
3 excreted as waste
4 stored in vacuoles

10 On a sunny day, much of the carbon dioxide produced by a green plant may be
1 used for fermentation
2 used for photosynthesis
3 stored in vacuoles
4 converted to oxygen gas

11 An experiment is set up involving groups of autotrophic plants of the same species. Each group of plants is illuminated by a different color of light of the same intensity. In a given period of time, in which light would the plants probably release the *least* amount of oxygen gas?
1 red 2 orange 3 green 4 blue

12 Given the summary equation:

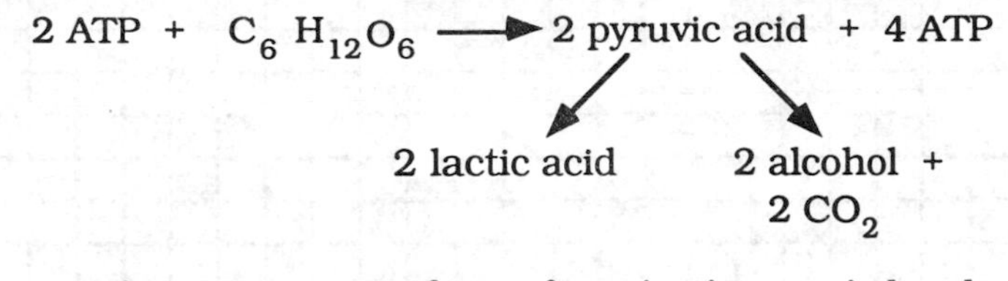

This equation represents a form of respiration carried on by some types of
1 algae and maple trees
2 paramecium and hydra
3 humans and grasshoppers
4 bacteria and yeast

13 In plants, glucose may be used in all of the following ways except for the
1 synthesis of starch
2 absorption of light energy
3 synthesis of fat
4 production of ATP

14 In the respiratory pathway of some organisms, when oxygen is unavailable, pyruvic acid is converted to
1 peptide chains
2 hydrolytic enzymes
3 lactic acid
4 nucleic acid

15 A similarity in the aerobic respiratory activity of animals is the
1 gases used and produced
2 net gain of four ATP molecules
3 type of alcohol produced
4 temperature of the respiratory organs

SELF–HELP: Unit II *"Skill"* Questions

Base your answers to questions 1 through 5 on the information below and on your knowledge of biology.

In a laboratory investigation, a bacteriologist counted the number of anaerobic bacteria present in a nutrient broth culture. He kept the culture at a constant temperature of 37°C. The table at the right includes the data the scientist recorded from this investigation.

Data Table

Hour	Number of Organisms per Milliliter of Broth
0	10
2	15
4	20
6	40
8	85
10	170
12	320
14	640
16	655
18	660

Using the information in the data table, construct a line graph on the next page, following the directions below.

1 Mark an appropriate scale on the axis labeled "Hour."
2 Mark an appropriate scale on the axis labeled "Number of Organisms per Milliliter of Broth."
3 Plot the data and connect the points to produce a line graph.

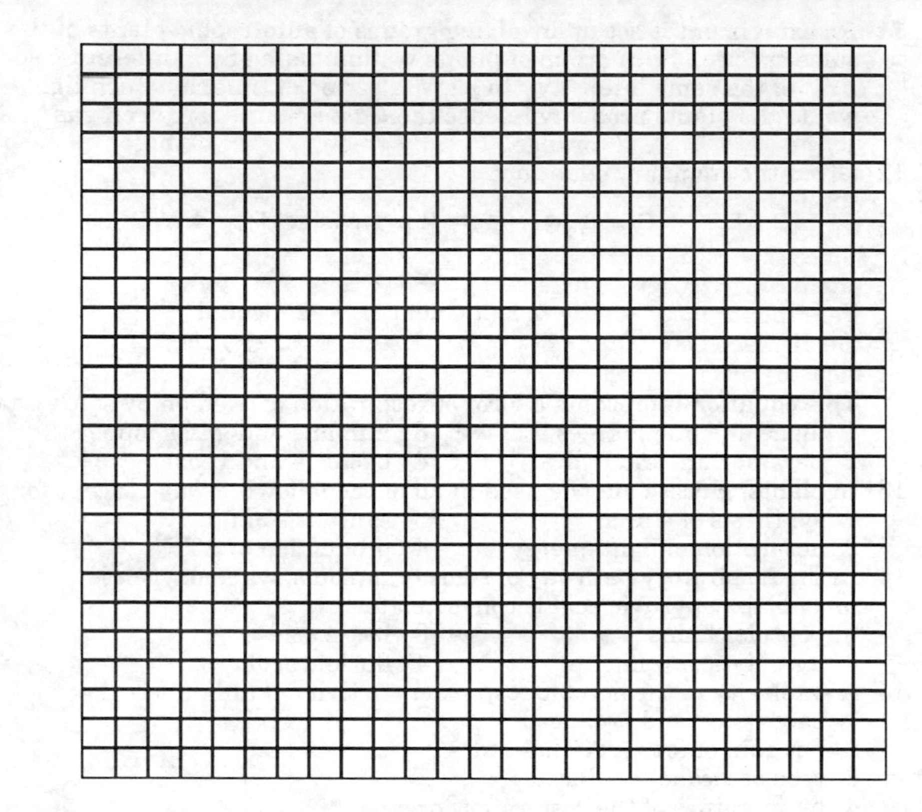

4 Which is an appropriate title for the graph?
 1 The Rate of Photosynthesis in Bacteria
 2 Oxygen Consumption by Bacteria
 3 The Rate of Enzymatic Hydrolysis in Bacteria
 4 A Growth Curve of Bacteria

5 How many organisms per milliliter would most likely be present in this nutrient broth at hour 20?
 1 850 2 660 3 1,000 4 1,320

Base you answers to questions 6 through 10 on the information and diagrams below and on your knowledge of biology.

An investigation was performed to determine the ability of yeast to metabolize different carbohydrates. Four experimental tubes, A, B, C, and D, each containing 0.5 gram of yeast, were filled with water as shown in the diagram at the right. Glucose was then added to tube A, maltose to tube C, and lactose to tube D in equal amounts. The tubes were then sealed. At the end of 45 minutes at 37°C, the displacement of liquid in each tube was measured to determine the amount of gas collected. The results are shown in the diagrams below.

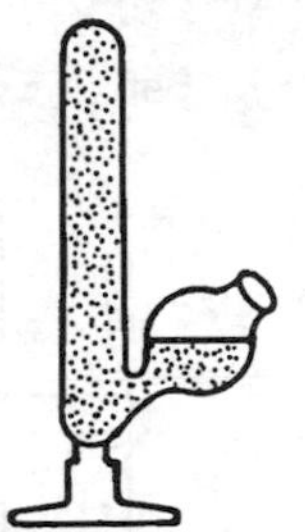

All Experimental Tubes at Start of Investigation

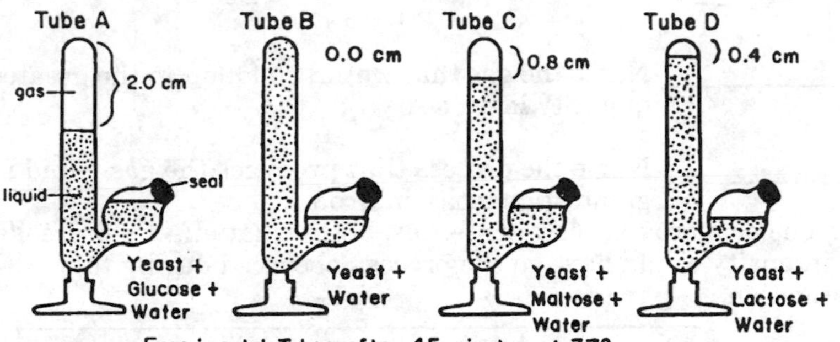

Experimental Tubes after 45 minutes at 37°

6 The gas collected and measured in this investigation was most likely
1 hydrogen
2 oxygen
3 nitrogen
4 carbon dioxide

7 Based on the results of this investigation, which substrate did the yeast metabolize most efficiently?
1 glucose
2 water
3 lactose
4 maltose

8 The metabolic process that produced the gas in the experimental tubes was most likely
1 photosynthesis
2 dehydration synthesis
3 alcoholic fermentation
4 lactic acid fermentation

9 Identify one variable in this investigation. ______________

____

10 Assume that the enzyme maltase had been added to the other ingredients in tube C, at the start of the investigation. Using a complete sentence, predict the effect of this addition on the amount of gas collected.

____

____

____

Base your answers to questions 11 through 13 on the diagrams below which represent a laboratory investigation.

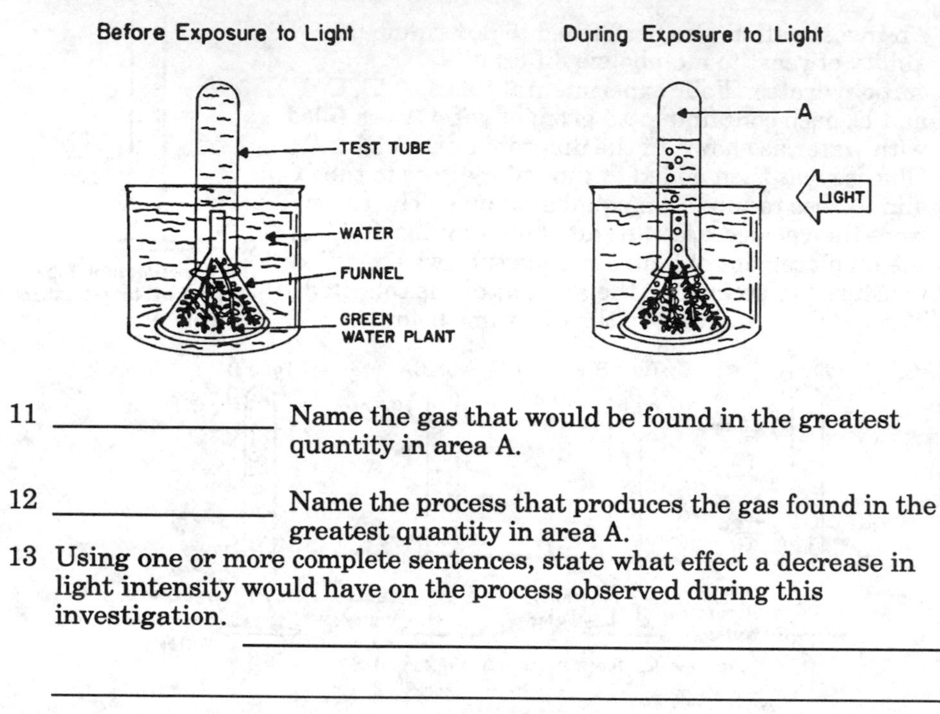

11 ________________ Name the gas that would be found in the greatest quantity in area A.

12 ________________ Name the process that produces the gas found in the greatest quantity in area A.

13 Using one or more complete sentences, state what effect a decrease in light intensity would have on the process observed during this investigation. ____

____

____

Base your answers to questions 14 through 15 on the information and the graph below and on your knowledge of biology.

Four pieces of apple were cut so that all were the same mass and shape, The pieces were placed in four different concentrations of sugar water. After 24 hours, the pieces were removed and their masses determined. The graph at the right indicates the change in the mass of each piece.

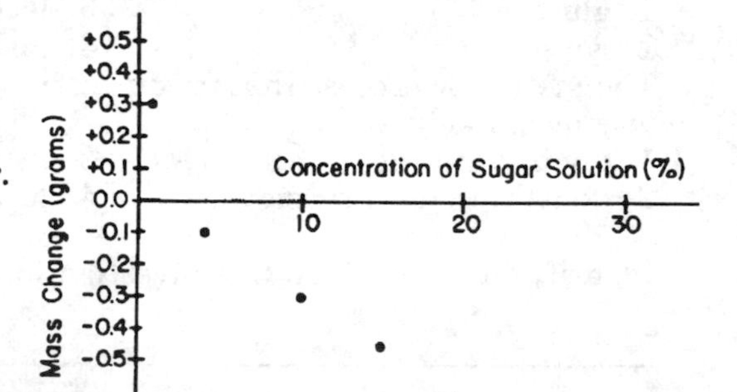

14 What was the change in mass of the apple piece in the 10% sugar solution?
1 a decrease of 0.45 gram
2 an increase of 0.30 gram
3 a decrease of 0.30 gram
4 an increase of 0.10 gram

15 At approximately what sugar concentration should the pieces neither lose nor gain weight?
1 6%
2 10%
3 3%
4 20%

Human Physiology

Objectives

The student should be able to:

- Recognize that humans are not unique in their performance of the functions necessary to maintain life.
- Apply scientific information to food choice decisions.
- Identify the major structures and functions of the human body and their role in the maintenance of homeostasis.
- Describe the interrelationships among the human body systems.
- Describe the structure and function of the major organs of the human body.

Nutrition – Terms & Concepts

Nutrition - Heterotrophic
Food - Nutrients and Roughage
  Carbohydrates and Lipids
  Proteins - Enzymes/Vitamins
  Minerals and Water
Digestion (one-way)
Gastro-intestinal Tract (GI)
  Oral Cavity - Salivary Glands
  Esophagus - Peristaltic Action
  Stomach - HCl / Gastric Glands
  Small Intestine
    Gall Bladder and Pancreas
Absorption - Villi
Large Intestine
  Water Absorption and Feces
Chemical Digestion
  Hydrolysis - Use of Enzymes
    Carbohydrate *to* Monosaccharide
    Protein *to* Amino Acid
    Lipid *to* Fatty Acid and Glycerol
Malfunctions
  Constipation and Diarrhea
  Ulcers and Appendicitis
  Gall Stones

I. Nutrition

Nutrition includes those activities by which organisms *obtain* and *process* nutrients needed for energy, growth, repair, and regulation. Humans are heterotrophic and therefore, must ingest food. Food includes **nutrients** and **roughage**. Nutrients include usable carbohydrates, proteins, lipids, minerals, vitamins, and water. Vitamins, minerals, and water are small molecules and can be absorbed without digestion. Carbohydrates, lipids, and proteins require digestion. Nutritional requirements vary with an individual's age, sex, and activities.

A. Functional Organization

The human digestive system consists of a continuous "one-way" gastrointestinal tract and the accessory organs which function in conjunction with the tract. Food is moved through the GI tract by slow, rhythmic muscular contractions called **peristalsis**.

1. Oral Cavity

Ingestion of food occurs through the **mouth (oral) cavity** which contains the teeth, tongue, and the openings from the salivary glands. The teeth function in the mechanical breakdown of food which serves to *increase the surface area* of the food for enzyme action.

The chemical digestion of carbohydrates begins here. The **salivary glands** secrete **saliva** which contains the enzyme amylase which digests starch into disaccharides.

Extended Area: ***Human Physiology***

Carbohydrates, which should constitute 50 percent of a balanced diet, are a primary source of energy for the body. Complex carbohydrates are found in fresh fruits and vegetables as well as whole grains.

In addition to serving as an energy source, complex carbohydrates provide nondigestible materials which increase the amount of roughage. Roughage is necessary for proper and regular egestion. In recent years roughage (fiber) has been shown be be effective in preventing some disorders of the GI tract.

2. Esophagus

As a result of swallowing, food moves into the esophagus. This begins the peristaltic action of the esophagus which moves the food to the stomach and through the GI tract.

3. Stomach

The stomach is a muscular organ in which food is temporarily stored, liquefied to chyme, and where protein digestion begins. Its lining contains **gastric glands** which secrete enzymes and hydrochloric acid. Hydrochloric acid provides an optimum pH for the hydrolytic activity of gastric protease. Under the influence of this enzyme, proteins are digested into polypeptides and dipeptides.

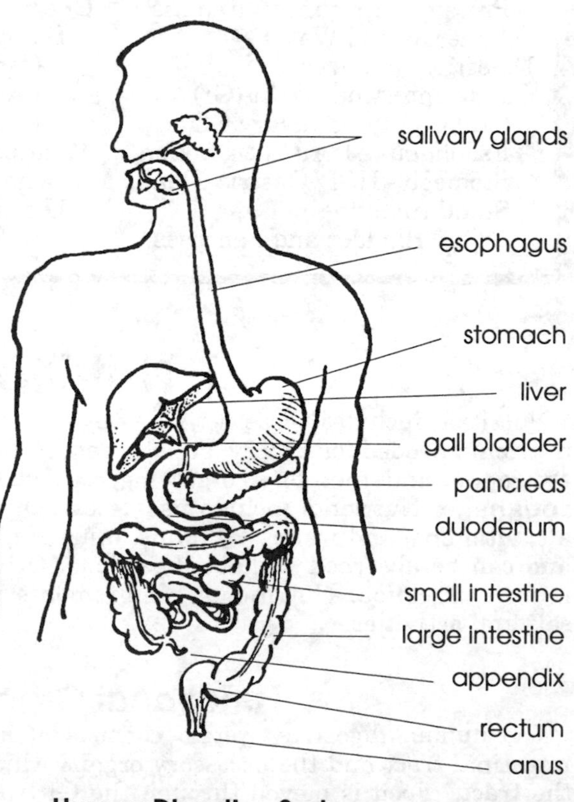

Human Digestive System

4. Small Intestine

The small intestine is a long, convoluted tube in which the major portion of food is digested. Partially digested, liquefied food enters the small intestine.

Accessory structures, the **gall bladder** and **pancreas**, empty their secretions into the small intestine. The secretion from the gall bladder is **bile**. Bile is produced in the liver and stored in the gall bladder. Bile is not an enzyme, but a chemical that emulsifies (physically break apart) fat which serves to *increase the surface area* of fats for subsequent chemical action.

Chemical Digestion

The pancreas secretes several enzymes including intestinal protease, lipase, and amylase. Intestinal glands that line the intestinal wall secrete protease, lipase, and disaccharidases, such as maltase.

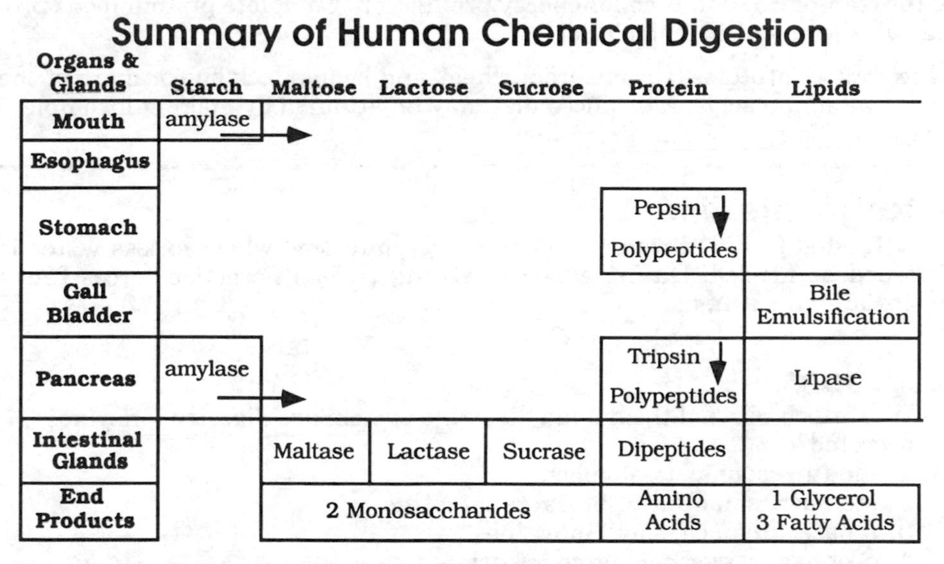

Summary of Human Chemical Digestion

Organs & Glands	Starch	Maltose	Lactose	Sucrose	Protein	Lipids
Mouth	amylase →					
Esophagus						
Stomach					Pepsin ↓ Polypeptides	
Gall Bladder						Bile Emulsification
Pancreas	amylase →				Tripsin ↓ Polypeptides	Lipase
Intestinal Glands		Maltase	Lactase	Sucrase	Dipeptides	
End Products		2 Monosaccharides			Amino Acids	1 Glycerol 3 Fatty Acids

The chemical digestion of proteins, lipids, and carbohydrates is completed in the small intestine.

Absorption in the Small Intestine

The lining of the small intestine contains numerous **villi** (small fingerlike projections) which *increase the surface area* of the small intestine to improve absorption. Capillaries and small lymphatic vessels, **lacteals**, extend into the villi.

Fatty acids and glycerol are absorbed through the villi into the lacteals and are transported in the lymph. Monosaccharides and amino acids are absorbed through the villi and enter the capillaries to be transported to the liver where they are temporarily stored. From there, they are available for distribution by the blood.

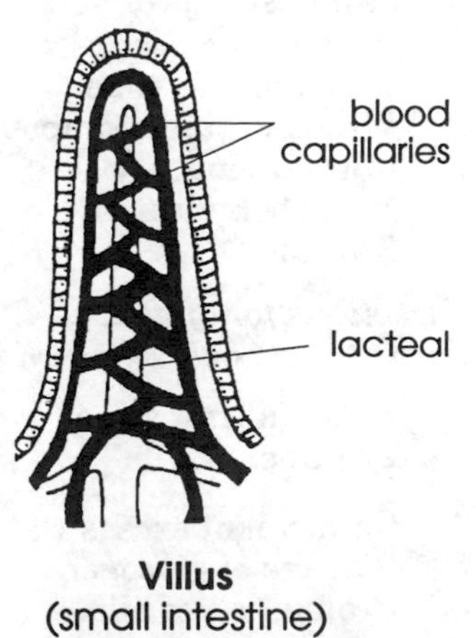

Villus
(small intestine)

Extended Area: ***Human Physiology***

Glucose is temporarily stored as the polysaccharide **glycogen** in the liver. Under the control of specific hormones, the breakdown of glycogen releases glucose for transport.

Amino acids are temporarily stored and distributed to the cells as needed for protein synthesis. Twenty amino acids are necessary for body cells to synthesize the required proteins needed to maintain and repair body tissues. Although humans can convert one amino acid into another, eight cannot be synthesized and must be consumed as part of the diet. These are called **essential amino acids**.

In order to synthesize new proteins, all the necessary amino acids must be present at the same time. If there is an insufficient supply of any one amino acid, protein synthesis will be limited, and the amino acids will be deaminated (broken-down) and used for energy production. Complete protein foods provide all the essential amino acids.

Incomplete proteins (such as from wheat and beans) lack one or more of the essential amino acids. A balanced diet may be attained by making incomplete protein foods complement each other.

5. Large Intestine

Undigested food and water enter the large intestine where excess water is absorbed as needed. During egestion, strong peristaltic action forces **feces** out through the **anus**.

Questions

1 Into which parts of the human digestive system are digestive enzymes secreted?
   1 mouth, esophagus, stomach
   2 stomach, small intestine, large intestine
   3 mouth, stomach, small intestine
   4 esophagus, stomach, large intestine

2 After a person's stomach was surgically removed, the chemical digestion of ingested protein would probably begin in the
   1 mouth
   2 small intestine
   3 large intestine
   4 liver

3 Which organic compounds undergo partial chemical digestion in the human mouth?
   1 carbohydrates
   2 fats
   3 proteins
   4 amino acids

Directions for questions 4 to 10: For each function , select the organ, chosen from the list below, which is most closely associated with that function.

1) Small Intestine
2) Pancreas
3) Large Intestine
4) Liver
5) Stomach
6) Gall Bladder

4 Removal of excess water from the GI tract.
5 Storage of glycogen.
6 Production of bile.

7 Completion of all chemical digestion.
8 Provides enzymes for chemical digestion in the small intestine.
9 Begins protein digestion.
10 Contains villi for the absorption of all nutrients.

B. Mechanism for Chemical Digestion

Hydrolysis is the splitting of large, insoluble molecules into small, soluble molecules with the addition of water. In organisms, this process is regulated by hydrolytic enzymes and is illustrated by the following:

maltose + water $\xrightarrow{\textit{maltase}}$ **glucose + glucose**

proteins + water $\xrightarrow{\textit{protease}}$ **amino acids**

lipids + water $\xrightarrow{\textit{lipase}}$ **3 fatty acids + 1 glycerol**

Polysaccharides (carbohydrates), such as starch, are completely hydrolyzed to simple sugars. In the presence of water and protease, the peptide bonds of proteins are broken resulting in the production of amino acids. In similar fashion, lipids are hydrolyzed to one glycerol and three fatty acid molecules.

Extended Area: ***Human Physiology***

Review Unit I, pages 20 through 22 for illustrations of the hydrolysis of carbohydrates, lipids, and proteins.

Fats contain a high quantity of potential energy and are necessary for the synthesis of cell membranes. However, increased fat consumption (such as cholesterol) represents a potentially dangerous change in an individual's dietary pattern. It is important to be aware of not only how much but also what kind of fat is eaten. Fats are described as **saturated** or **unsaturated**.

Saturated fats are solid at room temperature. There is evidence that increased intake of saturated fats is one of many factors that predisposes humans to cardiovascular disease.

Saturated Fatty Acid

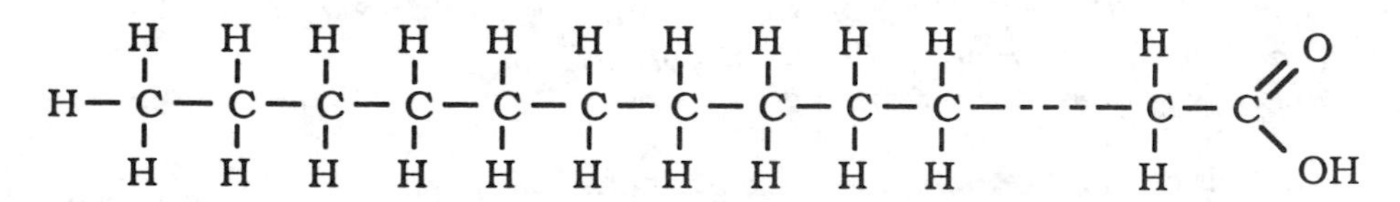

Polyunsaturated fats (oils) are liquid at room temperature and do not appear to be linked to cardiovascular diseases.

Unsaturated Fatty Acid

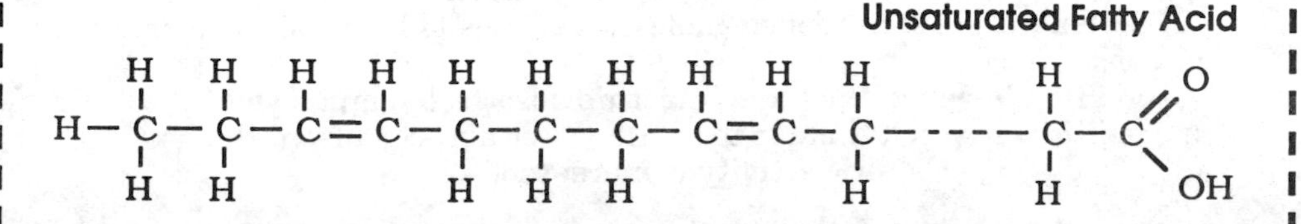

C. Some Malfunctions of the Digestive System

Ulcers. An ulcer is the erosion of the surface of the alimentary canal generally associated with some kind of irritant.

Constipation is a condition in which the large intestine is emptied with difficulty. Too much water is reabsorbed and the solid waste hardens.

Diarrhea is a gastrointestinal disturbance characterized by decreased water absorption and increased peristaltic activity of the large intestine. This results in increased, multiple, watery feces. This condition may result in severe dehydration especially in infants.

Appendicitis is an inflammation of the appendix due to infection.

Gallstones are an accumulation of hardened cholesterol and/or calcium deposits in the gall bladder.

Questions

For questions 1 through 6 select the organ, chosen from the diagram at the right, which is most closely related to the statement.

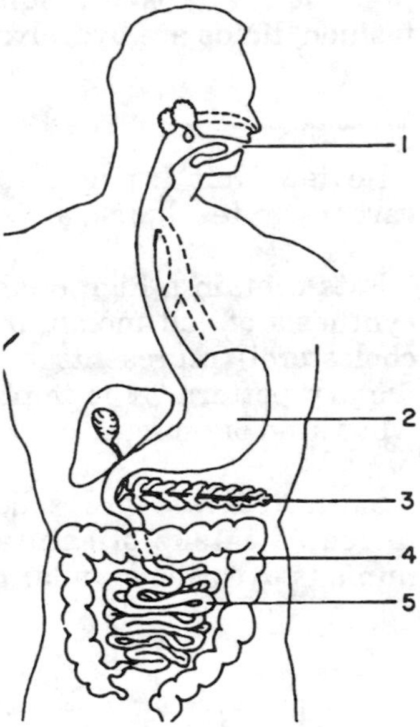

1 The chemical digestion of proteins begins in this organ.
2 This organ produces and secretes hydrochloric acid.
3 This organ contains many villi.
4 The chemical hydrolysis of carbohydrates begins in this organ.
5 Water reabsorption occurs in this organ.
6 Which substance is formed as a result of the chemical breakdown of glycogen?
 1 starch 3 cellulose
 2 glucose 4 glucagon

1
2
3
4
5

For questions 7 through 10, select the malfunction, chosen from the list below, which is most closely related to the statement.

1) diarrhea
2) ulcer
3) appendicitis
4) constipation
5) gallstones

7 An erosion of the lining of the alimentary canal.
8 An accumulation of hardened cholesterol deposits in the storage organ for bile.
9 A condition in which the large intestine absorbs too much water.
10 A condition in which there is too little water absorbed from the large intestine and peristaltic activity is increased.

II. Transport

Transport – Terms & Concepts

Circulatory System - Closed
Blood - Fluid Tissue
  Plasma - Liquid
  Red Blood Cells - Oxygen
  Platelets - Clotting
  White Blood Cells - Disease
    Lymphocytes
    Immunity - Active/Passive
    Allergies - Infection
Blood Types
  A, B, AB, O (blood groups)
  Antigens - Antibodies
Organ Transplants
  Matching and Rejection
ICF and Lymph
Arteries and Veins
Capillaries and Lymph Vessels
Heart - 4 Chambers
  Atria, Ventricles - Valves
Aorta and Vena Cava
Blood Pressure - Systole/Diastole
Pulmonary Circulation - Lungs
Coronary Circulation - Heart
Lymphatic Circulation - Body
  Cardiovascular Disease
  High Blood Pressure
  Coronary Thrombosis
  Angina Pectoris
Blood Conditions
  Anemia, Leukemia, and HIV

The process of transport involves the absorption and circulation (distribution) of materials throughout an organism.

A. Functional Organization

A function of the human circulatory system is the transport of dissolved and suspended materials throughout the body.

1. Transport Media

Blood

Blood is a fluid tissue composed of **plasma** in which red blood cells, white blood cells, and platelets are suspended. Blood functions as a transport medium helping to maintain **homeostasis** for all body cells.

- **Plasma**. The plasma, which is made up mostly of water, contains dissolved inorganic ions, wastes, hormones, nutrients, and a variety of proteins including antibodies, clotting factors, and enzymes.

- **Red Blood Cells**. Red blood cells are small (diameter of 7-9 micrometers) disk-shaped cells which lack nuclei when mature and are not able to reproduce. RBC's contain hemoglobin which combines with oxygen to form oxyhemoglobin. This is used to distribute oxygen to all cells.

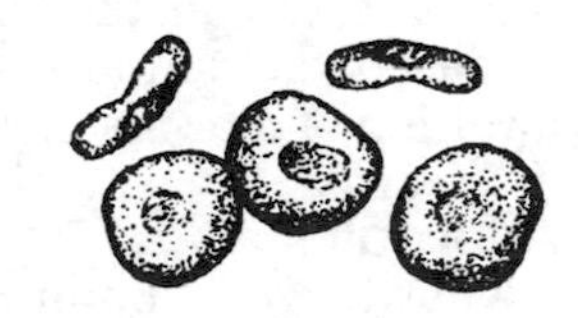

- **Platelets**. Platelets are noncellular (cell fragments) of the blood. They are smaller than either red or white blood cells and play a key role in blood clot formation.

Extended Area: ***Human Physiology***

Blood Clotting

Blood clotting involves a series of enzyme-controlled reactions resulting in the formation of protein fibers that trap blood cells and form a clot. Although all reactants are present in the blood, the rupturing of the platelets and the release of an enzyme appear to initiate the process. When the clot forms at a break in a blood vessel, the clot prevents further loss of blood and protects the area of injury.

- **White Blood Cells.** Several types of white blood cells exist.
- **Phagocytic White Blood Cells.** Phagocytic white blood cells engulf and destroy bacteria at the site of infection by the process of phagocytosis. This is a normal defense against infection.
- **Lymphocytes.** Lymphocytes are another type of white blood cell that are associated with the immune response. These white blood cells produce specific antibodies which act against foreign protein molecules known as **antigens.**

Extended Area: ***Human Physiology***

Immunity

Immunity involves the accumulation of specific **antibodies** in the plasma of the blood enabling the individual to resist specific diseases. These diseases are caused by invading foreign organisms and the products they make and are called **antigens.**

Immunity can be acquired in the following ways:

- **Active Immunity.** The antigen - antibody reaction occurs within the body in response to either contact with the disease-causing organisms or their products or by receiving a **vaccination.** A vaccination involves an inoculation with a weak or dead disease organism and is generally long lasting.
- **Passive Immunity.** A temporary form of immunity can be produced by the introduction of antibodies into the body. These preformed antibodies (medicines) have the same effect on antigens and are made from humans or animals.

Allergies

Many people are allergic to various substances — dust, pollen, insect bites, foods, drugs, and more. The body responds to these substances as if they were antigens and produces antibodies. These antibodies may cause inflammations and/or result in the release of a chemical substance called **histamine.** The released histamine causes an allergic response such as the irritation and swelling of mucous membranes.

Applications

Knowledge of immunity has resulted in the ability to type blood and transplant organs.

- **Blood Typing.** More than 50 antigens can be found in human blood. Of these, the most commonly tested antigens are of the ABO blood group.

Blood typing in the ABO blood group is based on the presence or absence of antigens on the surface of red blood cells. Two types of antigens are known: "A" and "B." In addition, plasma may contain the antibodies, anti-A and/or anti-B.

Antigens on red blood cells	**Antibodies in plasma**	**Blood Type**
A	Anti-B	A
B	Anti-A	B
A and B	Neither Anti-A nor Anti-B	AB
Neither A nor B	Anti-A and Anti B	O

- **Organ Transplants.** In order for a blood transfusion and an organ transplant to be successful, the antigens of the recipient and donor must be the *same or very close.* Rejection of organ transplants occurs when the recipient's body produces antibodies in response to the antigens present in the donor's organ.

ICF and Lymph. Intercellular fluid (**ICF**) is derived from blood plasma and surrounds all living cells of the body. This fluid is rich in the materials needed to help maintain the homeostasis of cells. When ICF passes into the lymph vessels it is called **lymph.**

2. Transport Vessels

Arteries. Arteries are relatively thick-walled, muscular blood vessels which transport blood away from the heart to all parts of the body. Contraction of their muscular walls (**pulse**) aids in the flow of blood.

Capillaries. Found at the "end" of small arteries and at the "beginning" of small veins, capillaries are tiny blood vessels with walls only one cell thick. They readily exchange dissolved materials by diffusion between the blood and the intercellular fluid surrounding all body cells.

Veins. Veins are relatively thin-walled blood vessels possessing valves which prevent the backflow of blood. The veins return blood to the heart.

Lymph Vessels. Lymph vessels include extremely small tubes with walls only one cell thick. These vessels branch through all the body tissues. Major lymph vessels have lymph nodes which contain phagocytic cells which filter bacteria and dead cells from the lymph. Some lymph vessels have valves, similar to those found in veins that aid in the movement of the lymph.

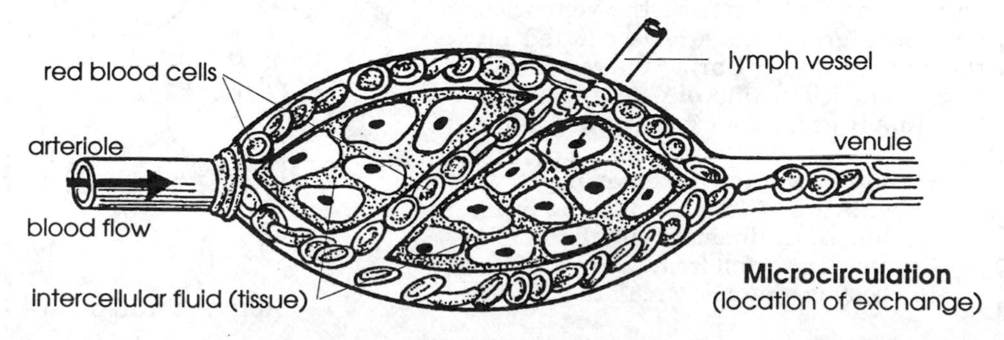

Microcirculation
(location of exchange)

3. Transport Mechanisms

Structure and Function. The muscular heart is a four-chambered double pump composed of two **atria** and two **ventricles**. The two atria are found at the top of the heart and receive blood from the body (right atria) and the lungs (left atria). The ventricles have thicker walls than the atria and pump blood to the lungs (right ventricle) and the body (left ventricle). The heart pumps blood through the arteries creating a **blood pressure** throughout the body.

Extended Area: ***Human Physiology***

Circulation Through the Heart. The **right atrium** receives deoxygenated (low in oxygen) blood from the body through the **vena cava**. The **left atrium** receives oxygenated (high in oxygen) blood from the lungs through the **pulmonary vein**. Blood passes through valves from the atria to the **ventricles**. The valves prevent the backflow of blood into the atria. The right ventricle pumps deoxygenated blood to the lungs through the **pulmonary artery**, and the left ventricle pumps oxygenated blood to the rest of the body through the **aorta**. Valves between these arteries and the ventricles prevent the backflow of blood into the ventricles.

pulmonary artery (to lungs)
superior vena cava (from upper body)
right atrium
pulmonary valve
right AV valve
right ventricle
inferior vena cava (from lower body)
aorta (to body)
pulmonary veins (from lungs)
left atrium
left AV valve
aortic valve
left ventricle
septum

Human Heart
(Major Structures)

Blood pressure refers to the pressure exerted on the walls of the arteries during the pumping action of the heart. During the contraction of the ventricles (**systole**), great pressure is exerted on the arterial walls. During the relaxation of the ventricles (**diastole**), less pressure is normally exerted on the arterial walls.

- **Pulmonary and Systematic Circulation**. Circulation to and from the lungs is pulmonary circulation. Circulation to and from the rest of the body is systemic circulation.

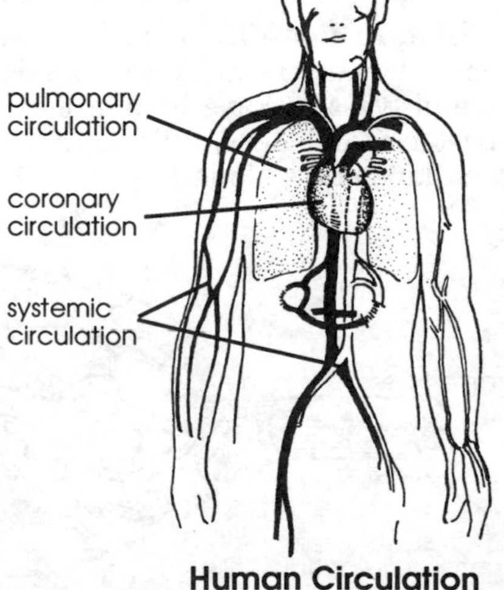

Human Circulation

- **Coronary Circulation.** The muscle tissue of the heart is supplied with blood through a system of coronary blood vessels (branches of the aorta).

- **Lymphatic Circulation.** Tissue fluid may be drained through lymph vessels. These lymph vessels join larger vessels which ultimately form two main trunks that empty lymph into veins of the circulatory system in the shoulder region.

B. Some Malfunctions of the Transport system

Cardiovascular Diseases

Cardiovascular diseases are malfunctions involving the heart and the blood vessels.

High Blood Pressure. High blood pressure (hypertension) is the most common form of cardiovascular disease and is characterized by increased arterial pressure. This can be caused by a number of variables including stress, dietary factors, heredity, cigarette smoking, and aging. High blood pressure can lead to damage to the lining of the arteries and a weakening of the heart muscle.

"Heart Attack". Heart attacks, as they are called, include:

- **Coronary Thrombosis.** Coronary Thrombosis is a blockage in the coronary artery or its branches resulting in oxygen deprivation in the heart muscle. The deprived muscle usually becomes damaged.

- **Angina Pectoris.** Angina Pectoris is a narrowing of the coronary arteries causing an inadequate supply of oxygen to the heart muscle. Often, an intense pain radiating from chest to shoulder and arms is felt.

Blood Conditions

Blood conditions are abnormalities in the circulatory fluid.

- **Anemia.** Anemia is the impaired ability of the blood to transport sufficient amounts of oxygen. This can be due to reduced amounts of hemoglobin and/or red blood cells.

- **Leukemia.** Leukemia, a form of cancer, is a disease of the bone marrow characterized by uncontrolled production of non-functional white blood cells.

- **AIDS**, acquired immunodeficiency syndrome, is a disease caused by the Human Immunodeficiency Virus (HIV). The disease affects the infected person by preventing his/her immune system from fighting other diseases. The Human Immunodeficiency Virus is transferred between humans through the direct contact of body fluids. It has spread throughout the world, and because there is no known cure, it has become a major health problem for all nations.

Questions

1 In humans, a function of intercellular fluid is to
 1 produce red blood cells
 2 serve as a transport medium
 3 produce white blood cells
 4 serve as a filter for uric acid

2 Which is a characteristic of lymph nodes?
 1 They carry blood under great pressure.
 2 The move fluids by means of a muscular pump.
 3 They produce new red blood cells.
 4 They contain phagocytic cells.

3 Oxygenated blood leaves the heart and is transported to other parts of the body through
 1 lymph vessels
 2 veins
 3 capillaries
 4 arteries

Base your answers to questions 4 and 5 on the diagram at the right and on your knowledge of biology.

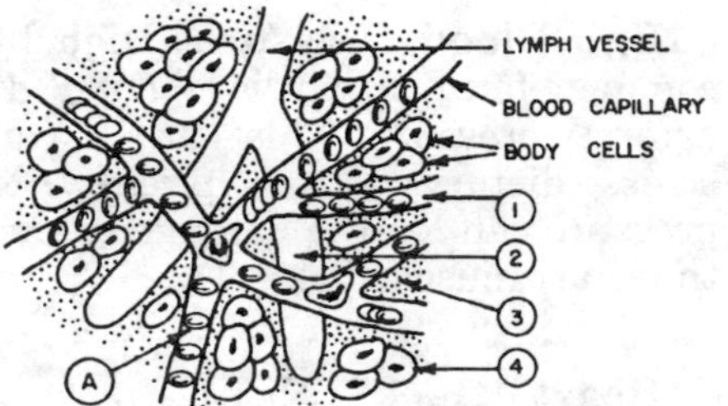

4 The major function of cell A is
 1 to initiate blood clot formation
 2 to produce antibodies
 3 to engulf invading bacteria
 4 to transport oxygen

5 Materials passing between body cells and blood must pass through intercellular fluid (ICF), which is located in area
 1 1
 2 2
 3 3
 4 4

6 Which part of the blood is correctly matched with its normal function?
 1 platelet - blood clot formation
 2 lymphocyte - carbon dioxide transport
 3 red blood cell - defense against infection
 4 phagocyte - oxygen transport

7 A biochemical compound that readily combines with oxygen and distributes it throughout the human body is
 1 urea
 2 water
 3 acetylcholine
 4 hemoglobin

8 Many bacteria that enter the circulatory system are engulfed and destroyed by
 1 phagocytic white blood cells
 2 pinocytic red blood cells
 3 plasma
 4 platelets

9 In humans, the concentration of carbon dioxide in the plasma
 1 causes increased production of hydrochloric acid
 2 regulates gastric acid production by forming carbonic acid
 3 regulates breathing rate by its effect on the medulla
 4 causes inflammation of the tissues of the bronchial tubes

10 In humans, which of the following is produced within certain bones?
 1 RBC's
 2 urine
 3 bile
 4 urea

11 In the human circulatory system, most of the end products of nutrition are carried in the
 1 white blood cells
 2 red blood cells
 3 platelets
 4 plasma

12 The thin-walled vessels of the circulatory system where most oxygen and carbon dioxide are exchanged are
1 alveoli
2 capillaries
3 veins
4 arteries

13 An injury to a blood vessel may result in the formation of a blood clot when
1 bone marrow cells decrease platelet production
2 kidney tubules synthesize clotting factors
3 ruptured platelets release enzyme molecules
4 white blood cells release antibodies

14 The right ventricle is the chamber of the heart which contains
1 deoxygenated blood and pumps this blood to the lungs
2 deoxygenated blood and pumps this blood to the brain
3 oxygenated blood and pumps this blood to the lungs
4 oxygenated blood and pumps this blood to the brain

15 Which heart chambers contain blood which has the highest concentration of oxygen?
1 right atrium and right ventricle
2 left atrium and left ventricle
3 right atrium and left ventricle
4 left atrium and right ventricle

16 Which vessel most likely contains the greatest amount of carbon dioxide?
1 pulmonary vein
2 artery
3 coronary artery
4 vein

Base your answers to questions 17 through 20 on the following list of heart structures.

1) right atrium
2) right ventricle
3) inferior vena cava
4) pulmonary artery
5) left atrium
6) left ventricle
7) aorta

17 Which number indicates the receiving chamber for blood returning from the body?
1 1 2 2 3 3 4 4

18 The last heart chamber the blood leaves before moving toward capillaries in the arm is indicated by number
1 1 2 5 3 6 4 4

19 Blood entering the inferior vena cava would be expected to have more
1 oxygen than in the right atrium
2 oxygen than in the aorta
3 carbon dioxide than in the right ventricle
4 carbon dioxide than in the aorta

20 Examination of the walls of the inferior vena cava and the aorta would reveal
1 more cardiac muscle in the inferior vena cava than in the aorta
2 less smooth muscle in the inferior vena cava than in the aorta
3 equal amounts of smooth muscle in both vessels
4 less voluntary muscle in the aorta than in the inferior vena cava

III. Respiration

Respiration/Excretion – Terms & Concepts

Cellular Respiration
Aerobic and Anaerobic
Lung - Gas Exchange
Nasal Cavity and Pharynx
Trachea
Bronchi and Bronchioles
Alveoli - Capillaries (exchange)
Breathing - Diaphragm
Malfunctions - Bronchitis
Asthma and Emphysema

Excretion - Metabolic Wastes
Lungs - Water, Carbon Dioxide, Heat
Liver - Deamination of Protein
Skin - Sweat Glands
Water, Salt, and Urea
Kidneys - Homeostasis
Nephron - Urine
Filtration and Reabsorption
Ureter, Urinary Bladder, and Urethra
Malfunctions- Kidney Disease, Gout

Respiration involves the processes of cellular respiration and gas exchange. It is the conversion of the chemical bond energy in foods to a usable form of energy (ATP) in human cells.

A. Cellular Respiration

In humans the process of cellular respiration is essentially the same as that of other aerobic organisms. However, under conditions of oxygen deprivation (insufficient oxygen), muscle cells respire in an anaerobic manner, and lactic acid is produced. (Review Unit II)

B. Gas Exchange

The function of the human respiratory system is to transport gases between the external environment and the internal thin, moist membrane surfaces to allow for gas exchange.

1. Functional Organization of Respiratory System

The respiratory system is composed of a network of passageways which permit air to flow from the external environment to the lungs.

Nasal Cavity. The nasal cavity is exposed to the air through nostrils. This cavity is lined with a ciliated mucous membrane which filters, warms, and moistens the air.

Pharynx. The pharynx is the area in which the oral cavity and nasal cavity meet. Food is prevented from entering the trachea by a flap of tissue called the **epiglottis**.

Trachea. The trachea is a tube that conducts air between the pharynx and the bronchi and is kept open by partial rings of cartilage. The ciliated mucous membrane which lines the trachea traps microscopic particles and sweeps them toward the pharynx where the particles can be expelled or swallowed. Deposits from cigarette smoke and other atmospheric pollutants may interfere with the action of cilia.

Bronchi. The two major subdivisions of the trachea are the bronchi. The bronchi are lined with mucous membrane and ringed with cartilage. Each bronchus extends into a lung where it subdivides many times, forming progressively smaller bronchioles.

Bronchioles. Bronchioles are lined with mucous membrane but lack cartilage rings. Tiny bronchioles terminate with the alveoli.

Alveoli. Located at the ends of the bronchioles, the alveoli are the functional units for gas exchange. They are thin, moist, and surrounded by capillaries.

Lung. The lungs contain the tiny "air sacs." Each lung includes one of the two bronchus with its bronchioles and alveoli.

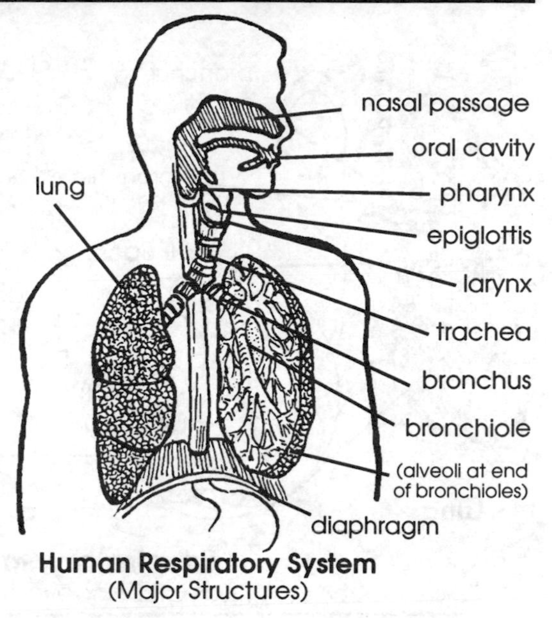

Human Respiratory System
(Major Structures)

2. Mechanisms for Gas Exchange

Breathing. The lungs are highly elastic and respond passively to the actions of the diaphragm and rib cage. Movements of the diaphragm and rib cage cause pressure changes in the chest cavity which move air into or out of the lungs. When the chest cavity expands, pressure is lowered within the chest cavity and air is *pulled* into the lungs. When the chest cavity decreases in size, greater pressure is exerted in the chest cavity, and the air is *pushed* out of the lungs. This process is known as **breathing** or ventilation.

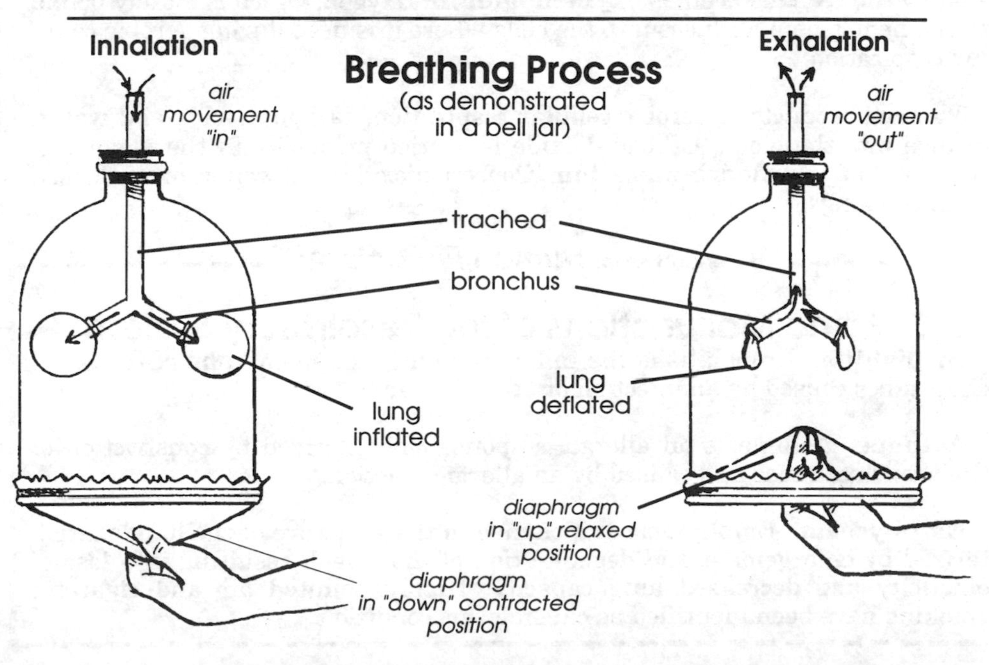

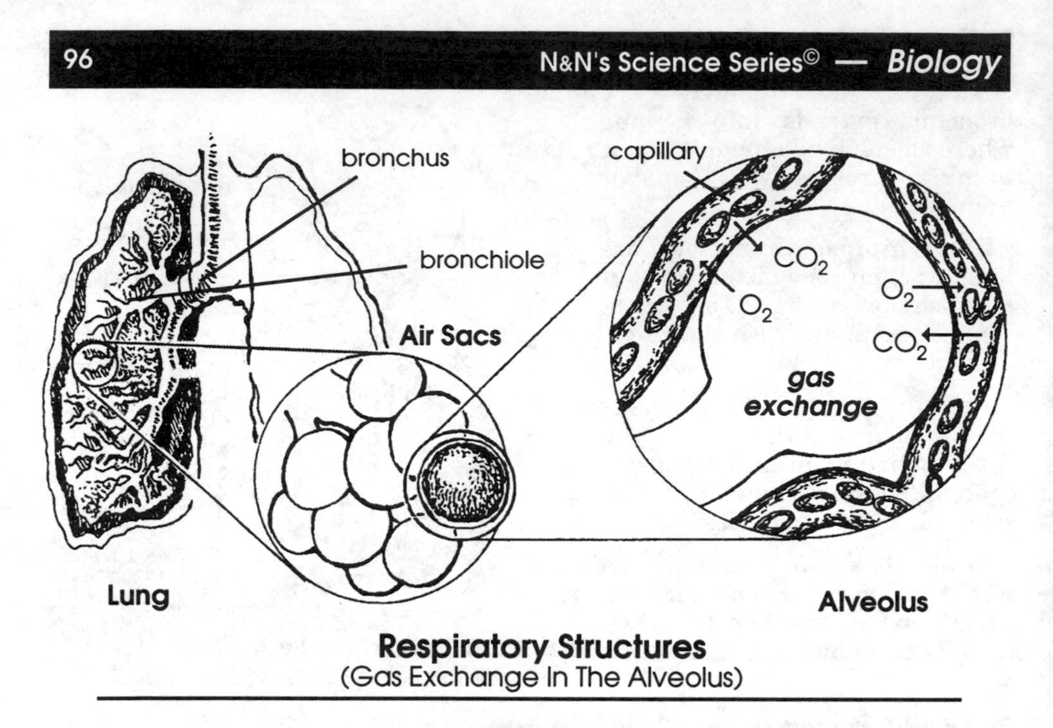

Respiratory Structures
(Gas Exchange In The Alveolus)

The breathing rate is regulated by the concentration of carbon dioxide in the blood (H_2CO_3 – lowers pH of blood) and its effect on the medulla of the brain. This is an example of a **feedback mechanism** which aids in the maintenance of homeostasis.

Gas Exchange. The capillaries which surround the air sacs are involved in gas exchange between the blood and the alveoli. In the blood, oxygen is carried by the red blood cell as **oxyhemoglobin**. Oxygen, which is loosely bound to the hemoglobin, diffuses into the cells where it is used during aerobic cellular respiration.

The end products of aerobic cellular respiration, carbon dioxide and water, diffuse into the blood. Carbon dioxide is carried primarily in the plasma in the form of the **bicarbonate ion**. Carbon dioxide and water are released from the lungs.

Extended Area: ***Human Physiology***

C. Some Malfunctions of the Respiratory System

Bronchitis. Bronchitis is the inflammation of the membrane of the bronchial tubes caused by an infection or other irritant.

Asthma. Asthma is an allergic response characterized by constriction of the bronchial tubes and caused by an allergic reaction.

Emphysema. Emphysema is a change in the structure of the lung characterized by enlargement and degeneration of the alveoli resulting in a loss of elasticity and decreased lung capacity. Highly polluted air and cigarette smoking have been identified as causing this condition.

Questions

1 The exchange of air between the human body and the environment is a result of the rhythmic contractions of the rib cage muscles and the
1 diaphragm 2 lung 3 trachea 4 heart

2 Which sequence most correctly represents the network of passageways which normally permits air to flow from the external environment to the human lungs?
1 nasal cavity, bronchioles, trachea, pharynx, bronchi
2 nasal cavity, pharynx, trachea, bronchi, bronchioles
3 nasal cavity, trachea, pharynx, bronchi, bronchioles
4 nasal cavity, bronchi, bronchioles, pharynx, trachea

3 Bronchitis, an inflammation or irritation of the bronchi in humans, may be expected to occur within a human system called the
1 respiratory system 3 circulatory system
2 nervous system 4 digestive system

4 The human trachea is prevented from collapsing due to the presence of
1 respiratory cilia 3 cartilage rings
2 smooth muscle 4 striated muscle

5 In order to function effectively in gas exchange, alveoli must be in close association with
1 arteries 3 capillaries
2 veins 4 lacteals

6 In which structure do deposits from cigarette smoke interfere with the action of cilia?
1 trachea 3 esophagus
2 larynx 4 pharynx

7 In humans, which substance is produced through anaerobic respiration during strenuous activity?
1 lactic acid 3 carbon dioxide
2 glycogen 4 alcohol

8 The breathing rate of humans is principally regulated by the concentration of
1 carbon dioxide in the blood 3 oxygen in the blood
2 platelets in the blood 4 white blood cells in the blood

9 Which function of a part of the human respiratory system is represented in the diagram at the right?
1 gas exchange
2 external cyclosis
3 cellular respiration
4 active transport

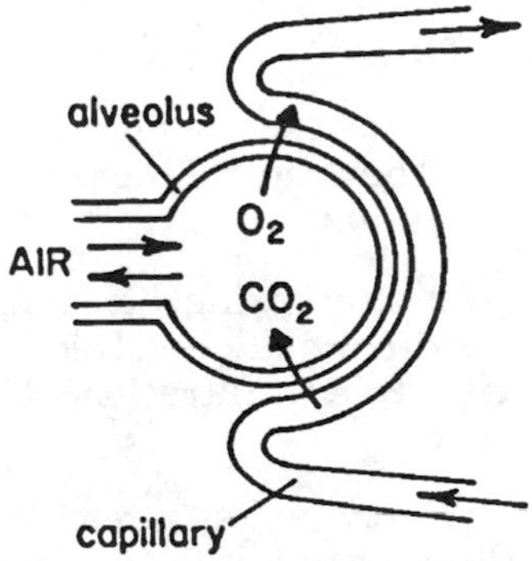

10 In humans, the amount of energy that is present in a lactic acid molecule is greater than the amount of energy in a molecule of
1 starch 2 glucose 3 sucrose 4 water

IV. Excretion

In humans, many organs are involved in the removal of metabolic wastes from the body cells to the environment.

A. Functional Organization

1. Lungs

Carbon dioxide and water, waste products of respiration, diffuse from the blood into the alveoli. These gases along with excess heat are removed from the body during exhalation.

2. Liver

The liver is a large, multi-purpose organ whose excretory functions include the breakdown of red blood cells, recycling of usable materials, and the production of urea following amino acid deamination.

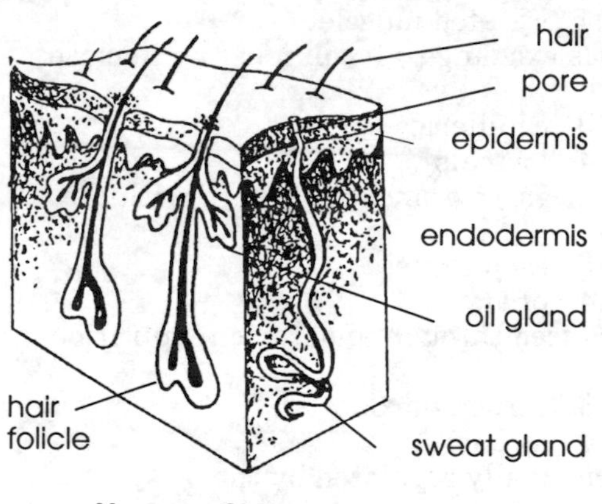

Human Skin

3. Skin/Sweat Glands

Water, salts, and some urea diffuse from the blood into the sweat glands of the skin and are subsequently excreted as perspiration. Perspiration is only incidentally excretory since its primary function is temperature regulation. Evaporation of the sweat (98% water and 2% salts and urea) occurs when heat is absorbed from skin cells. This absorption of heat lowers body temperature. Temperature regulation is an example of **homeostasis**.

4. Urinary System

Kidneys. The kidneys perform two major functions:

- They excrete most of the urea.
- They control the concentration of most of the constituents of the body fluids.

Arteries bring blood to the kidneys where microscopic structures called **nephrons** are involved in filtration and reabsorption. Blood pressure plays a vital role in filtration. If the pressure in the arteries is too low, filtration is poor.

Water, salts, urea, amino acids, and glucose are filtered from the **glomerulus** into the cuplike **Bowman's capsule**. As these materials move through the **tubule** of the nephron, water, minerals, and digestive end products are reabsorbed by mostly active transport into capillaries associated with the tubule. Veins carry blood away from the kidneys. After reabsorption, the fluid that remains in the tubule is urine.

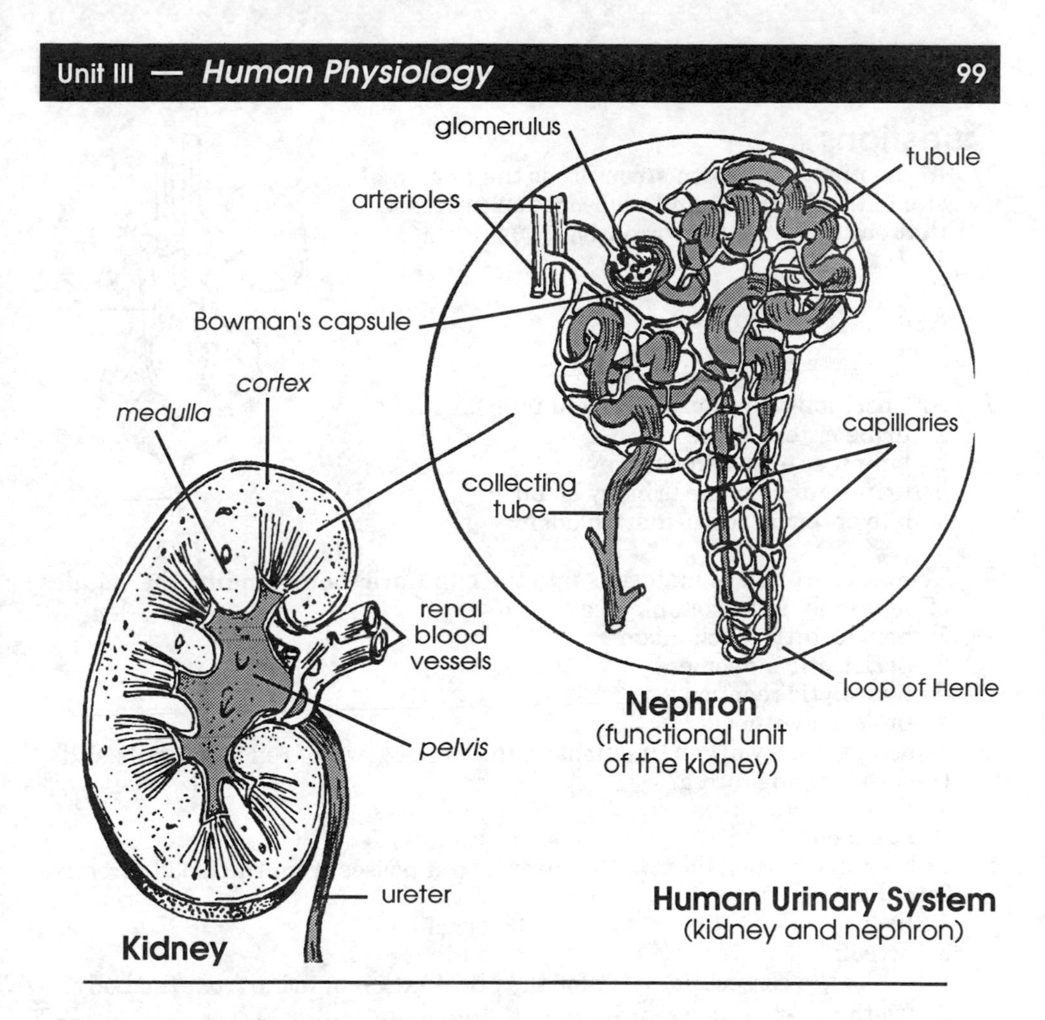

Human Urinary System
(kidney and nephron)

Ureter. Urine flows from the kidneys through the two ureters (one from each kidney) to the urinary bladder.

Urinary Bladder. The urinary bladder stores the urine until it is eliminated through the urethra.

Urethra. Periodically, urine is excreted from the urinary bladder through a small tube called the urethra.

Extended Area: ***Human Physiology***

B. Malfunctions

Kidney Diseases are associated with the malfunctioning of the kidney or the nephron of the kidney. Kidney failure must be considered serious, since without filtration toxic metabolic wastes build up in the body cells. Poor diet, high blood pressure, salt and chemical imbalance, and severe physical stress can bring about kidney failure.

Gout is a disease caused by inflammation in the joints associated with uric acid production and its deposition resulting in arthritic-like, painful attacks.

Questions

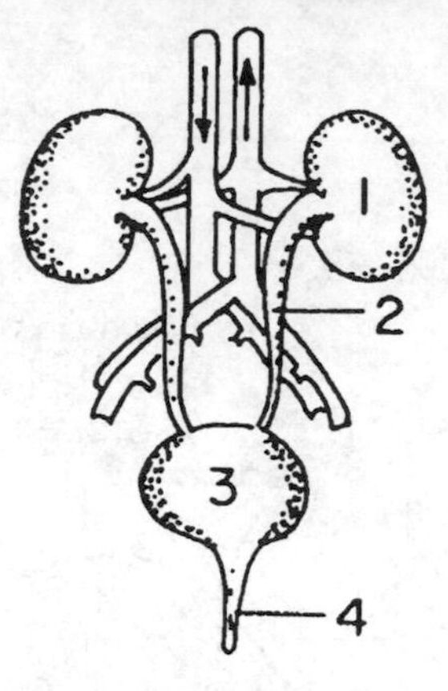

1 In the diagram, which structure is the principal site of the filtration and reabsorption processes that occur during the formation of urine?
 1 *1*
 2 *2*
 3 *3*
 4 *4*

2 An obstruction in a ureter would interfere with
 1 urine entering the kidney
 2 lymph entering the kidney
 3 urine entering the urinary bladder
 4 bile entering the urinary bladder

3 The reabsorption of materials into the capillaries surrounding the tubules of nephrons is an example of a
 1 homeostatic mechanism
 2 peristaltic movement
 3 hydrolytic reaction
 4 protein synthesis

4 Which excretory organ functions in the breakdown of red blood cells and the production of urea?
 1 liver
 2 stomach
 3 pancreas
 4 lung

5 In humans, carbon dioxide that is excreted passes from the blood directly into the
 1 liver
 2 alveoli
 3 trachea
 4 kidneys

6 Which is the correct pathway for the elimination of urine from the body?
 1 urethra → bladder → ureters → kidneys
 2 kidneys → bladder → urethra → ureters
 3 kidneys → ureters → bladder → urethra
 4 kidneys → urethra → ureters → bladder

For each function in questions 7 through 10, select the organ, chosen from the list below, which is most closely associated with the function.

1) Kidney 2) Lung 3) Liver 4) Skin

7 Removal of carbon dioxide from the blood
8 Storage of glycogen
9 Production of urea
10 Maintaining homeostasis by removing body heat, water, and other wastes

V. Regulation

In the human as in other animals, regulation is achieved by the integration of the nervous system and the endocrine system. The nervous and endocrine systems of humans show certain similarities and certain differences.

Similarities include:
- Both secrete chemical messengers. Nervous system secretes neurotransmitters and the endocrine system secretes hormones.
- Both play a major role in homeostasis.

Differences include:
- Nerve responses are more rapid than endocrine responses.
- Nerve responses are shorter in duration than endocrine responses.

A. Nervous System

Nervous System – Terms & Concepts

Nervous System - Neurons
  Sensory - Sense Organs
  Motor - Effectors
  Interneurons - Relays
Nerves - Impulse - Response
Malfunctions
  Cerebral Palsy - Polio
  Meningitis - Stroke

Central Nervous System (CNS)
  Brain - Control Center
    Cerebrum, Cerebellum, Medulla
  Spinal Cord - Reflexes
Peripheral Nervous System (PNS)
  Somatic - Voluntary
  Autonomic - Involuntary

1. Functional Organization

Neurons. The neuron is the basic cellular unit of the nervous system. The nervous system is composed of three structurally different types of nerve cells (neurons): sensory neurons, interneurons, and motor neurons.

- **Sensory neurons** transmit impulses from receptors to the central nervous system (brain and spinal cord). Sense organs are structures where sensory neurons are concentrated. These organs include the eyes, ears, tongue, nose, and skin.

- **Interneurons**, located mainly in the central nervous system, interpret and relay nerve impulses between sensory and motor neurons. The interneuron was formerly called an association neuron.

- **Motor neurons** transmit impulses from the central nervous system to effectors. These impulses cause muscles, effectors, and glands to respond.

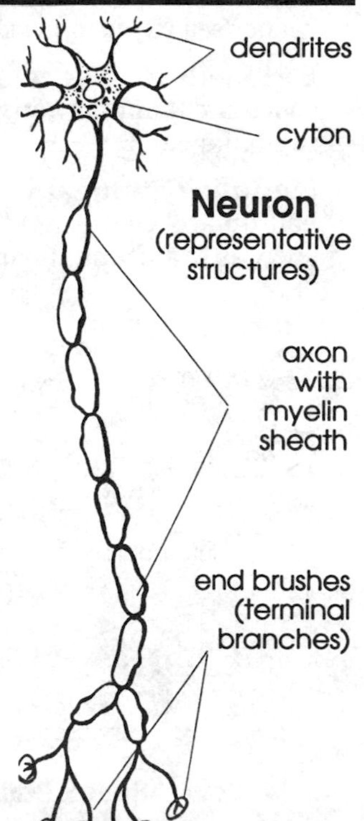

Neuron (representative structures)

Nerves. Nerves are bundles of neurons or parts of neurons specialized for long distance and high speed impulse transmission. They can be sensory nerves, motor nerves, or mixed nerves. Most nerves have fatty (myelin) sheaths for protection and to prevent contact with other nerves and tissues.

Central Nervous System (CNS)

Brain

The brain is a large mass of neurons located in the cranial cavity and is responsible for controlling and coordinating most of the activities of the human. The cerebrum, the cerebellum, and the medulla are the three major divisions of the brain, each having specialized functions.

- **Cerebrum**. The cerebrum is the center for voluntary activity. In specific areas, sensory impulses are interpreted, motor activities may be initiated, and memory, thinking, and reasoning occur.

  Habits, which are acquired by repetition, are examples of **conditioned behavior**. The repetition establishes pathways for nerve impulse transmission which permit rapid automatic responses to various stimuli.

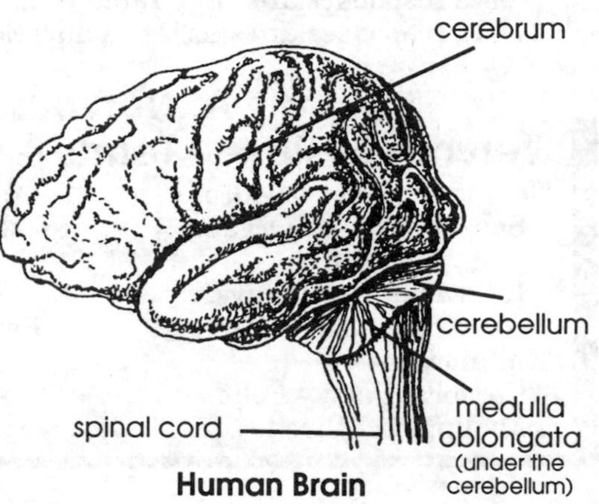

Human Brain

- **Cerebellum**. The cerebellum, located behind and under the cerebrum, coordinates motor activities (muscle movements) and helps in maintaining balance.
- **Medulla Oblongata**. The medulla oblongata, located beneath the cerebellum and connected to the spinal cord, controls involuntary activities such as breathing, heartbeat, blood pressure, and peristalsis.

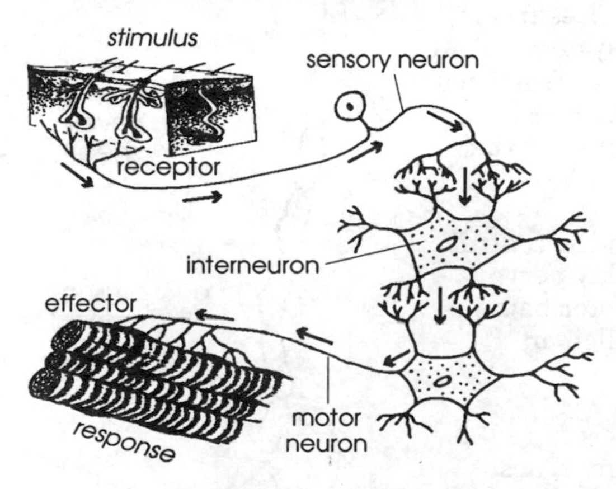

Spinal Reflex Pathway
"stimulus" — receptor, sensory neuron, interneuron (spinal cord) motor neuron, effector — "response"

Spinal Cord

The spinal cord lies within, and is protected by, the vertebrae of the spinal column. The spinal cord is continuous with the medulla of the brain. The spinal cord coordinates activities between the brain and other body structures. It is a center for reflex actions and does not require any activity of the brain to bring about a reflex response.

- Reflex actions are inborn, involuntary patterns of behavior.
- Reflex action involves a pathway (**reflex arc**) over which

impulses travel. In a spinal reflex there is a pathway from receptors to a sensory neuron to interneurons in the spinal cord to a motor neuron to an effector.

Peripheral Nervous System (PNS)

The peripheral nervous system is located outside the central nervous system and consists of nerves extending throughout the body.

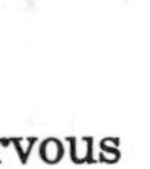

Extended Area: ***Human Physiology***

The peripheral nervous system is separated into the somatic and autonomic nervous systems.

- **Somatic Nervous System** consists of the nerves that control the voluntary muscles of the skeleton.
- **Autonomic Nervous System** consists of the nerves that control cardiac muscle, glands, and smooth muscle. It is generally considered to be an involuntary system.

Note: The distinction between "voluntary" and "involuntary" is not clear-cut. There is an overlap between their responsibilities. For example, through concentration (conscious thought) it is possible to "override" some involuntary responses with voluntary ones.

2. Some Malfunctions of the Nervous System

Cerebral Palsy is a group of congenital (to be born with) diseases characterized by a disturbance of motor functions. Although speech and muscle actions are distorted, the intelligence of a person with CP falls within the range of normal.

Meningitis is the inflammation of the membranes surrounding the brain and spinal cord.

Stroke. A stroke is caused from a cerebral hemorrhage (burst blood vessel) or a blood clot in the cerebral vessel and may result in brain damage.

Polio is a viral disease of the central nervous system which may result in paralysis and is preventable through immunization.

Questions

1 Neurons are to neurotransmitters as endocrine glands are to
   1 hormones   3 enzymes
   2 vitamins   4 nucleic acids

2 In the diagram at the right, which number indicates the region of the human brain that maintains balance and coordinates motor activities?
   1 1   3 3
   2 2   4 4

3 In humans, which structural unit transmits electrochemical impulses?
1 nephron
2 alveolus
3 sweat gland
4 neuron

4 The breathing rate of a human increases following rapid exercise. The part of the nervous system that controls this involuntary action is the
1 cerebrum
2 cerebellum
3 medulla
4 spinal cord

5 The correct sequence for the pathway of an impulse in a simple reflex arc is
1 effector → motor neuron → sensory neuron → interneuron → receptor
2 receptor → interneuron → sensory neuron → motor neuron → effector
3 receptor → sensory neuron → interneuron → motor neuron → effector
4 effector → sensory neuron → receptor → interneuron → motor neuron

6 An example of a human response is
1 a stone approaching the eye
2 the buzz of a bee
3 the fragrance of a rose
4 the secretion of a hormone

Base your answers to questions 7 and 8 on the diagram below which represents a simple nerve pathway in the human body and on your knowledge of biology.

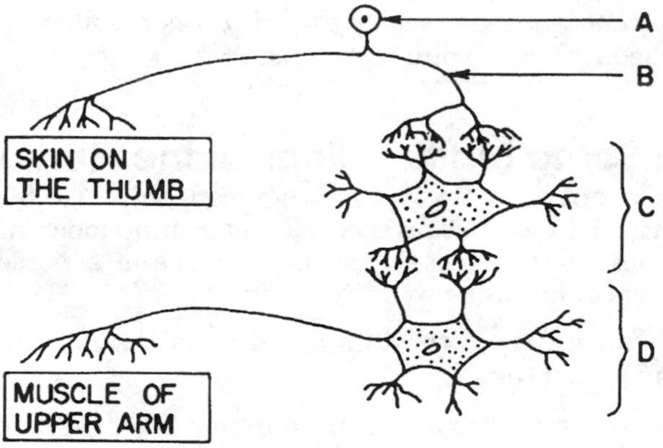

7 Which statement best describes the skin on the thumb as represented in the diagram?
1 It contains receptors that interpret stimuli.
2 It contains receptors that detect stimuli.
3 It is an effector that interprets stimuli
4 It is an effector that detects stimuli.

8 An interneuron is represented by
1 A
2 B
3 C
4 D

9 The chief function of the sensory neuron is to transmit impulses
1 to the central nervous system
2 away from the effectors
3 away from the central nervous system
4 to the receptors

10 Neurotransmitters are produced as a result of
1 excretion in nephrons
2 synthesis in nerve cells
3 respiration in alveoli
4 hydrolysis in muscle cells

B. Endocrine System

Endocrine – Terms & Concepts

Endocrine System - Feedback
Target Glands - Hormones
Hypothalamus
Pituitary Gland (Master Gland)
TSH, FSH, GSH
Thyroid Gland - Thyroxine
Adrenal Glands
Cortex - Steroids
Medulla - Adrenaline
Parathyroids - Parathormone
Islets of Langerhans (Pancreas)
Insulin and Glucagon
Gonads
Testes (male) - Testosterone
Ovaries (female) - Estrogen
Negative Feedback System
Malfunctions of Hormones
Goiter and Diabetes

The endocrine glands, located in various parts of the body, and their hormones, make up the endocrine system. Hormones are transported by the circulatory system and affect various tissues or "target" organs.

Extended Area: ***Human Physiology***

1. Functional Organization

Hypothalamus. The hypothalamus is a small region of the brain. Although part of the central nervous system, it has an endocrine function. It produces hormones which influence the pituitary gland.

Pituitary Gland. The pituitary gland, "master gland" of the body and located at the base of the brain, secretes numerous hormones. One of these hormones, a **Growth-stimulating Hormone**, has widespread effect on the body. The growth-stimulating hormone stimulates the elongation of long bones. Other pituitary hormones control specific endocrine glands.

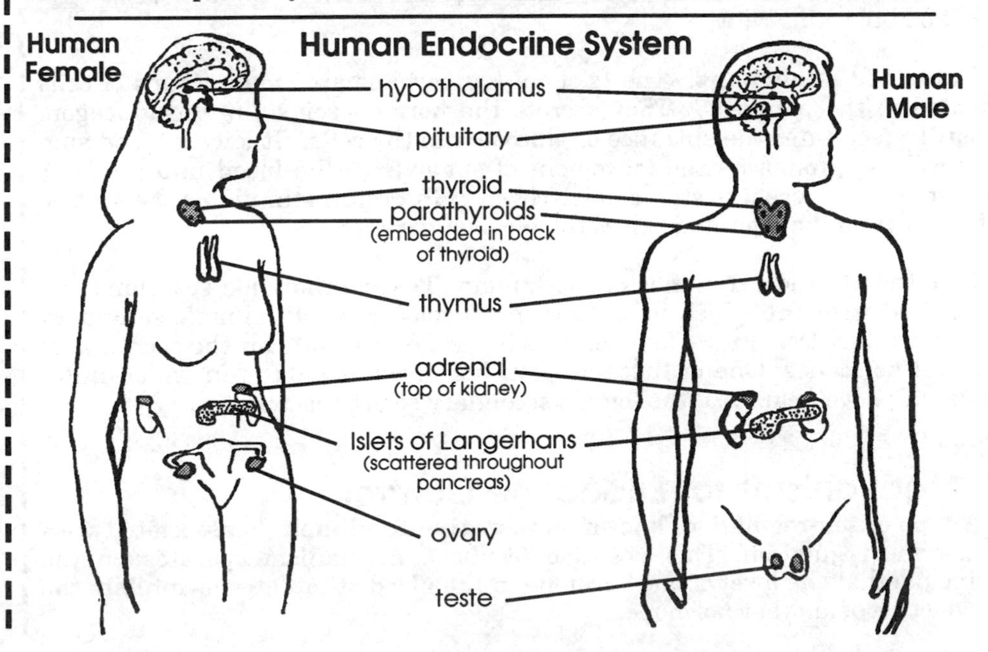

Two examples of pituitary gland hormones include:

- **Thyroid Stimulating Hormone (TSH)** — stimulates the thyroid gland to produce its hormone, **thyroxine.**

- **Follicle Stimulating Hormone (FSH)** — stimulates activity in the ovaries and testes.

Thyroid Gland. The thyroid gland, located in the neck, produces **thyroxine**, which contains iodine. Thyroxine regulates the rate of metabolism in the body and is essential for normal physical and mental development.

Parathyroid Glands. The parathyroid gland, patches of tissue embedded in the thyroid gland, produce and secrete the hormone **parathormone**. Parathormone controls the metabolism of calcium which is necessary for nerve function, blood clotting, and proper growth of teeth and bones.

Adrenal Glands. The adrenal glands are two small glands located on top of the kidneys, but, there are no direct connections with the kidneys. Each gland consists of two distinct regions. The outer portion is the adrenal cortex, and the inner mass is the adrenal medulla.

- The **adrenal cortex** secretes two types of steroid hormones. (1) One type promotes the conversion of body fat and protein into glucose. (2) The other type promotes the reabsorption into the blood stream of sodium and chloride ions by the kidney tubules. This affects the water balance and helps maintain blood pressure.

- The **adrenal medulla** secretes **adrenalin** in times of emergency. Adrenalin increases the blood sugar level and accelerates the heart and breathing rates.

Islets of Langerhans. The Islets of Langerhans are small groups of cells located in the pancreas. They secrete the hormones insulin and glucagon. **Insulin** facilitates the entrance of glucose into the cells. It lowers blood sugar levels by promoting the movement of sugar from the blood into the liver and muscles where it is stored as glycogen. **Glucagon** stimulates the release of sugar from the liver and raises the blood sugar level.

Gonads are the sex glands of the human. **Testes**, the male sex gland, secrete **testosterone** which influences the development of the male secondary sex characteristics. In the female, **ovaries** are responsible for the secretion of several hormones. One of these ovarian hormones is **estrogen** which influences the development of the female secondary sex characteristics.

2. Mechanisms for Endocrine Control

A type of self-regulation, known as **negative feedback**, is associated with endocrine regulation. This negative feedback mechanism operates on the principle that the level of one hormone in the blood stimulates or inhibits the production of another hormone.

The relationship between TSH and thyroxine is an illustration of the homeostatic feedback mechanism in the body. The pituitary gland secretes TSH which stimulates the thyroid gland to secrete thyroxine. As the level of thyroxine increases in the blood, TSH is reduced which causes a reduction in the stimulation of the thyroid, thus a reduction in the amount of thyroxine released.

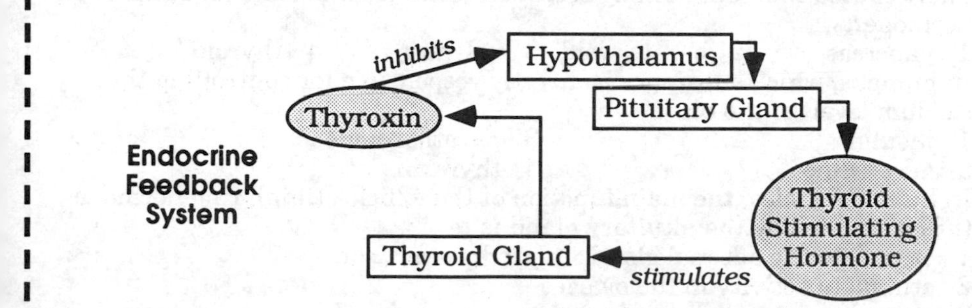

3. Malfunctions

Goiter. A goiter is an enlargement of the thyroid gland usually resulting from the gland's inability to manufacture thyroxine. This is often associated with an iodine deficiency in the diet.

Diabetes is a disorder characterized by an insulin deficiency which results in an elevated blood sugar level and an inability of the body to store the sugar as glycogen in the liver.

Questions

1 The maintenance of proper blood sugar level involves the storage of excess sugar in the
 1 salivary glands
 2 stomach
 3 pancreas
 4 liver

2 The body normally responds to low concentrations of sugar in the blood by secreting
 1 insulin 2 glucagon 3 estrogen 4 testosterone

3 Into which system do endocrine glands secrete hormones?
 1 circulatory 2 nervous 3 digestive 4 reproductive

4 Hormones that are secreted in one area of the human body affect responses in other areas. Hormones reach these different areas of the body by way of
 1 ducts
 2 nephrons
 3 respiratory tubules
 4 blood vessels

5 A difference between the human nervous system and the endocrine system is that
 1 nerve responses are more rapid than endocrine responses
 2 endocrine responses are of shorter duration than nerve responses
 3 nerve impulses travel by way of the transport system while hormones travel by way of neurons
 4 the endocrine system secretes neurotransmitters and the nervous system secretes hormones

6 Within the liver cells of animals, glycogen is most directly converted to glucose by the process of
 1 enzymatic hydrolysis
 2 cellular respiration
 3 active transport
 4 molecular adhesion

7 A person was admitted to the hospital with abnormally high blood sugar and an abnormally high sugar content in his urine. Which gland most likely caused this condition by secreting lower than normal amounts of its hormone?
 1 pancreas
 2 parathyroid
 3 salivary
 4 thyroid

8 In humans, which substance is directly responsible for controlling the calcium levels of the blood?
 1 insulin
 2 adrenaline
 3 parathormone
 4 thyroxine

9 In human females, the main function of the follicle stimulating hormone (FSH) secreted by the pituitary gland is to
 1 stimulate the adrenal glands to produce cortisone
 2 stimulate activity in the ovaries
 3 control the metabolism of calcium
 4 regulate the rate of oxidation in the body

10 Which structure secretes the substance that it produces directly into the bloodstream?
 1 gall bladder
 2 skin
 3 salivary gland
 4 adrenal gland

VI. Locomotion

Locomotion – Terms & Concepts

Endoskeleton (internal)
Bones - Marrow (blood/fat)
Connective Tissue
Cartilage
 Flexible, Elastic, Fibrous
Muscles
 Visceral - Involuntary/Smooth
 Cardiac - Involuntary/Striated
 Skeletal - Voluntary/Striated
Anaerobic Respiration
 Lactic Acid, Fatigue, Oxygen Debt
Tendons (bone to muscle)
 Tough, Inelastic, Fibrous
Ligaments (bone to bone)
 Tough and Fibrous
Joints - Extensors/Flexors
Malfunctions
 Arthritis and Tendonitis

Human locomotion involves the interaction of bones, cartilage, muscles, tendons, and ligaments. It allows humans to move from place to place.

A. Functional Organization

Bones. The human endoskeleton (internal) consists of bones of various shapes and sizes. The functions of the bones include: support and protection of body structures, anchorage sites for muscle action, leverage for body movement, and production of blood cells in the bone marrow.

Cartilage. Although the human skeleton consists primarily of bone, another type of connective tissue, cartilage, is also present. Cartilage is flexible, fibrous, and elastic.

The functions of cartilage include: pliable support, flexibility of joints, and cushioning effects in joints. Cartilage is found in both the embryo and the adult.

- **Embryo** — Cartilage makes up most of the embryo's skeleton. By adulthood most of this cartilage is replaced by bone.
- **Adult** — Cartilage is found at the end of ribs, between vertebrae, at the ends of bones, and in the nose, ears, bronchi, and trachea.

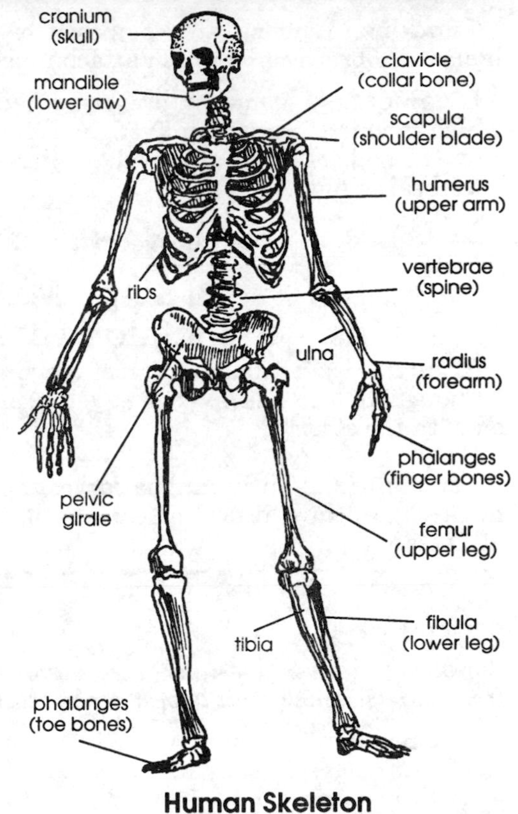

Human Skeleton

Muscles. There are three major types of muscles in the human body: **visceral** (smooth, involuntary) muscles which are involuntary in action and smooth in appearance; **cardiac** (heart) muscles which are involuntary in action and striated ("striped") in appearance; and **skeletal** (striated, voluntary) muscles which are voluntary in action and striated in appearance.

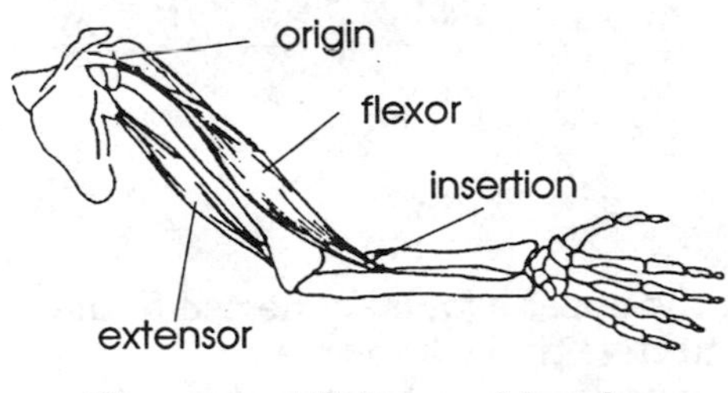

Flexor and Extensor Muscles

Skeletal muscles are controlled by the central nervous system. They serve to move the bones in a coordinated manner. Muscles usually operate in pairs which "pull" on the bones either side of a joint. They include **extensors** which extend limbs (straighten joints) and **flexors** which return the limbs (bend joints).

Vigorous activity of skeletal muscles may lead to an oxygen deficiency which can result in anaerobic respiration and a buildup of lactic acid. Lactic acid production is associated with fatigue. When sufficient oxygen is supplied to the muscle cells, lactic acid is converted back to pyruvic acid, and aerobic respiration is restored.

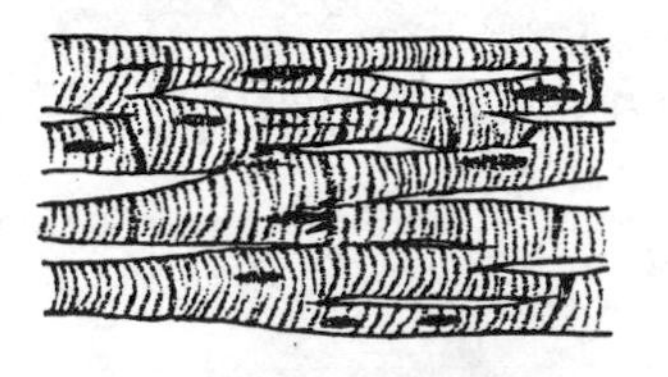

Cardiac Muscle

Visceral Muscle

Skeletal Muscle

Tendons. Tendons are composed of connective tissue. They are tough, inelastic, fibrous cords which attach muscles to bones.

Ligaments. Ligaments are composed of tough elastic connective tissue which is able to stretch only slightly during joint movement. Ligaments connect the ends of bones at movable joints such as the elbow, fingers, knee, and vertebral column.

Extended Area: ***Human Physiology***

B. Some Malfunctions Associated with Locomotion

Arthritis is an inflammation of the joints causing swelling and severe pain. Although arthritis does occur at young ages, it is a condition usually associated with the elderly.

Tendonitis is an inflammation of the tendon usually at the bone attachment caused by physical stress and irritation. This condition is common to athletes.

Questions

For each phrase in questions 1 through 6, select the human structure, chosen from the list below, that is best described by that phrase.

1) Bones
2) Cartilage Tissues
3) Ligaments
4) Smooth Muscles
5) Tendons
6) Voluntary Muscles

1 Cause peristalsis in the alimentary canal
2 Serve as extensors and flexors
3 Serve as levers for body movement
4 Bind the ends of bones together
5 Attach the muscles to bones
6 Manufacture red blood cells
7 In the human elbow joint, the bone of the upper arm is connected to the bones of the lower arm by flexible connective tissue known as
1 tendons 2 ligaments 3 muscles 4 neurons
8 In humans, the failure of a movable joint to function properly may be the result of damage to
1 cilia 2 ligaments 3 nephridia 4 spiracles
9 Which is *not* a major function of cartilage tissues in the human adult?
1 giving pliable support to body structures
2 cushioning joint areas
3 adding flexibility to joints
4 providing skeletal levers
10 Which statement most accurately describes human skeletal muscle tissue?
1 It is involuntary and striated.
2 It is involuntary and lacks striations.
3 It is voluntary and striated.
4 It is voluntary and lacks striations.

SELF–HELP: Unit III "*Core*" Questions

For each phrase in questions 1 through 4, select the organ, chosen from the list below, which is most closely related to that phrase.

1) esophagus
2) stomach
3) large intestine
4) small intestine

1 Where feces are formed.
2 Where protein digestion begins.
3 Where lipid digestion is completed.
4 Contains gastric glands.
5 In which human organ would the chemical digestion of a piece of meat normally begin?
   1 oral cavity
   2 small intestine
   3 stomach
   4 liver
6 Which process breaks down the large fat globules into smaller fat droplets?
   1 excretion 2 absorption 3 peristalsis 4 emulsification
7 A person who consumes large amounts of saturated fats may increase his or her chances of developing
   1 meningitis
   2 viral pneumonia
   3 hemophilia
   4 cardiovascular disease
8 In which organ does peristalsis occur?
   1 liver 2 pancreas 3 oral cavity 4 esophagus
9 A transport structure with nodes containing phagocytic cells that filter bacteria and dead cells is known as
   1 a capillary
   2 a lymph vessel
   3 an artery
   4 an aorta
10 Compared to a drop of blood taken from a vein in the arm, a drop of blood taken from an artery in the arm would normally contain
   1 more carbon dioxide and more oxygen
   2 more carbon dioxide and less oxygen
   3 less carbon dioxide and more oxygen
   4 less carbon dioxide and less oxygen
11 All living cells of the human body are surrounded by intercellular fluid. This fluid is most similar in chemical composition to
   1 gastric juice
   2 lactic acid
   3 adrenalin
   4 lymph
12 A type of "heart attack" in which a narrowing of the coronary artery causes an inadequate supply of oxygen to reach the heart muscle is known as
   1 anemia
   2 angina pectoris
   3 leukemia
   4 cerebral palsy
13 Which type of human cell is a phagocyte?
   1 white blood cell
   2 red blood cell
   3 muscle cell
   4 bone cell
14 Which blood vessel contains blood with the lowest concentration of oxygen?
   1 artery 2 arteriole 3 capillary 4 vein
15 In humans, circulation to and from the lungs is known as
   1 systemic circulation
   2 pulmonary circulation
   3 coronary circulation
   4 lymphatic circulation

16 Filtering dust particles out of the human respiratory tract is a function of the
1 cilia 2 bronchioles 3 epiglottis 4 alveoli

17 In humans, the exchange of respiratory gases occurs within the
1 nose 3 bronchial tubes
2 trachea 4 alveoli

18 Most carbon dioxide is carried in the plasma in the form of
1 hydrogen ions 3 lactic acid
2 bicarbonate ions 4 oxyhemoglobin

19 The principle muscle involved with breathing is the
1 biceps 3 abdominal
2 diaphragm 4 cardiac

20 Exchange of oxygen and carbon dioxide between the external environment and the blood occurs in the
1 pharynx 2 alveoli 3 trachea 4 bronchi

21 Which process reduces the concentration of urea in the blood of humans?
1 excretion 2 egestion 3 digestion 4 synthesis

22 In humans, the filtrate of the nephrons becomes urine and is stored in the
1 glomerulus 3 gallbladder
2 alveolus 4 urinary bladder

For each of questions 23 through 25, select the excretory structure, chosen from the list below, that best answers the question.

1) Alveolus 2) Nephron 3) Liver 4) Sweat Gland

23 Which structure forms urine from water, urea, and salts?

24 Which structure removes carbon dioxide and water from the blood?

25 Which structure is involved in the breakdown of red blood cells?

26 The functional unit of the human kidney is known as a
1 nephridium 3 nephron
2 Malpighian tubule 4 urinary bladder

27 In humans, sound waves received by receptors are converted to sensory impulses that are interpreted in the
1 medulla 3 cerebellum
2 spinal cord 4 cerebrum

28 The somatic nervous system contains nerves that run from the central nervous system to the
1 muscles of the skeleton 3 heart
2 smooth muscles 4 endocrine glands

29 In humans, which structure is primarily responsible for maintaining balance and coordinating motor activities?
1 cerebrum 3 medulla
2 spinal cord 4 cerebellum

30 Which is an example of an effector?
1 a taste bud of the tongue 3 the auditory nerve of the ear
2 the retina of the eye 4 a muscle of the arm

31 Which term refers to the chemical substance that aids in the transmission of the impulse at a synapse?
1 neurotransmitter 3 neuron
2 synapse hormone 4 nerve

32 A similarity of the human nervous and endocrine systems is that both normally
1 secrete chemical messengers
2 have the same rate of response
3 have the same duration of response
4 secrete hormones that travel by way of the neurons

33 Which part of the human central nervous system is involved primarily with sensory interpretation and thinking?
1 medulla
2 spinal cord
3 cerebrum
4 cerebellum

34 Which system of the body includes the thyroid gland, adrenal gland, pituitary gland, and the gonads?
1 excretory
2 nervous
3 respiratory
4 endocrine

35 Which gland produces the hormones insulin and glucagon?
1 pituitary
2 thyroid
3 pancreas
4 liver

36 In humans, which gland regulates the level of calcium in the bloodstream?
1 ovary
2 parathyroid
3 pancreas
4 salivary

37 A student's ability to think about a questions and answer it correctly is directly controlled by the
1 spinal cord
2 cerebellum
3 medulla
4 cerebrum

38 Ligaments of the knee are composed of
1 elastic cartilage tissue
2 smooth muscle fibers
3 striated muscle fibers
4 elastic connective tissue

39 The skeleton of the human embryo is made up primarily of
1 bone
2 muscle
3 cartilage
4 epithelium

40 Which statement most accurately describes human skeletal muscle tissue?
1 It is involuntary and striated.
2 It is involuntary and lacks striations.
3 It is voluntary and striated.
4 It is voluntary and lacks striations.

SELF–HELP: Unit III *"Extended"* Questions

1 Enlargement of the thyroid gland, resulting in the conditions known as goiter, is often associated with a
1 reduction in the capacity of the lungs
2 deposition of uric acid in the neck area
3 deficiency of iodine in the diet
4 deficiency of insulin in the body

2 Inflammation of the membranes surrounding the brain and spinal cord is known as
1 meningitis
2 leukemia
3 stroke
4 diabetes

For each statement in questions 3 through 6 select the organ indicated on the diagram at the right that is most closely associated with that statement.

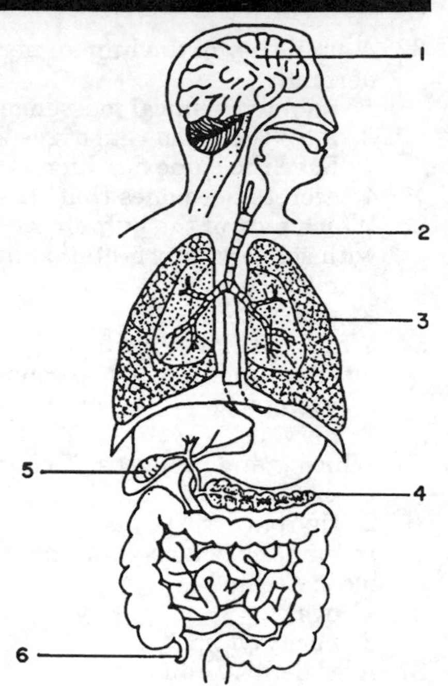

3 An inflammation of the membranes that line the tubes within this organ is called bronchitis.

4 Accumulations of hardened cholesterol deposits form gallstones in this organ.

5 An inflammation of this organ is called appendicitis.

6 A hemorrhage or blood clot in this organ may result in a stroke.

Base your answers to questions 7 through 11 on the diagram at the lower right and on your knowledge of biology. The diagram represents a human heart and some major blood vessels.

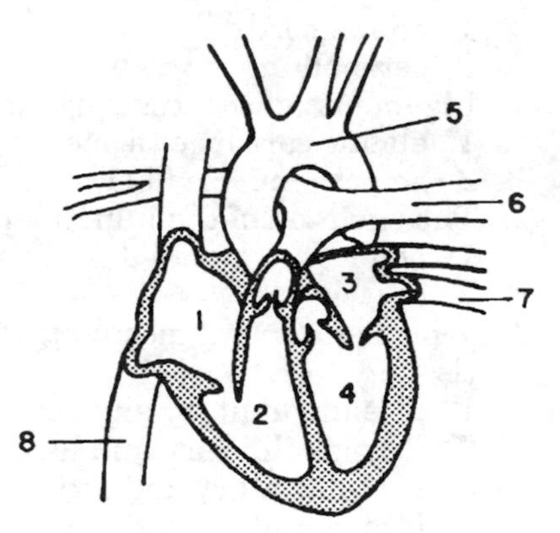

7 The blood vessel indicated by number 5 is known as the
 1 aorta
 2 pulmonary artery
 3 superior vena cava
 4 coronary artery

8 Deoxygenated blood from the lower part of the body returns to the heart by way of structure
 1 5
 2 6
 3 7
 4 8

9 Which numbers indicate the atria?
 1 1 and 3
 2 2 and 4
 3 3 and 4
 4 1 and 2

10 When blood is pumped out of chamber 2, it will circulate directly to the
 1 liver
 2 brain
 3 legs
 4 lungs

11 The heart chamber indicated by number 4 is the
 1 left atrium
 2 right atrium
 3 left ventricle
 4 right ventricle

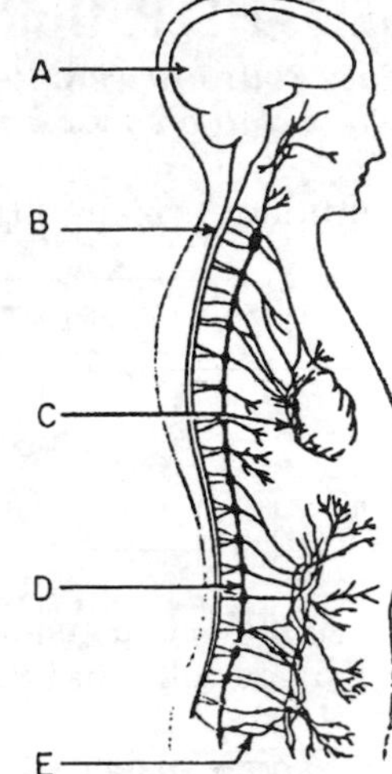

Base your answers to questions 12 through 14 on the diagram at the right of a portion of the human nervous system and on your knowledge of biology.

12 The structure labeled *B* would be most directly involved with the regulation of
1 conscious thought
2 reflex action
3 sense interpretation
4 reasoning ability

13 The human disorder known as meningitis would cause inflammation of which portions of the nervous system?
1 *A* and *B*
2 *B* and *C*
3 *C* and *D*
4 *D* and *E*

14 A stroke occurs in the area indicated by
1 *A*
2 *B*
3 *C*
4 *D*

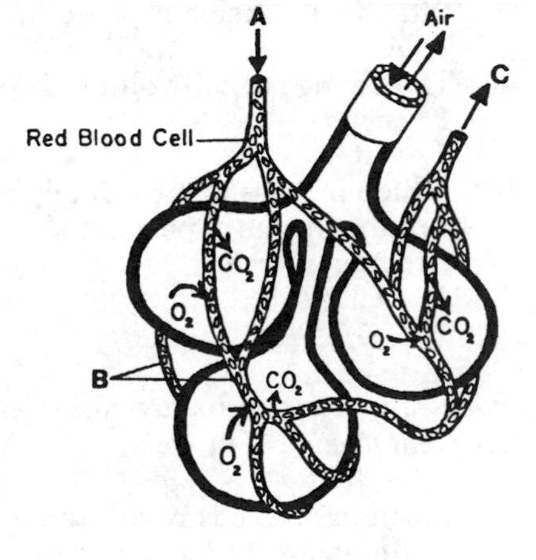

Base your answers to questions 15 through 18 on the diagram at the right which represents part of the human respiratory system.

15 The blood vessels (*B*) surrounding these air sacs are known as
1 arteries
2 capillaries
3 veins
4 lymphatic ducts

16 These air sacs are known as
1 alveoli
2 bronchi
3 bronchioles
4 tracheae

17 The heart chamber which most directly pumps blood to the vessel network at *A* is the
1 right atrium
2 left atrium
3 right ventricle
4 left ventricle

18 The process most directly involved with the exchange of gases between these air sacs and blood vessels is
1 active transport
2 pinocytosis
3 hydrolysis
4 diffusion

19 Improper metabolism of the body's calcium is most likely due to a malfunctioning of the
1 testes
2 parathyroids
3 liver
4 pancreas

20 The organ system in humans most directly involved with the production of antibodies for an immune reaction is the
1 digestive
2 nervous
3 excretory
4 circulatory

SELF–HELP: Unit III *"Skill"* Questions

Base your answers to the questions 1 through 5 on the laboratory investigation below and on your knowledge of biology.

A student prepared three different red blood cell suspensions as follows:

Suspension	Contents
A	red blood cells + normal blood serum
B	red blood cells + 10% salt solution
C	red blood cells + distilled water

1 Which suspension, when viewed under the microscope, would contain red blood cells that appear wrinkled and reduced in volume?
1 *A* 2 *B* 3 *C*

2 Which suspension, when viewed under the microscope, would contain red blood cells that had swollen and burst apart?
1 *A* 2 *B* 3 *C*

3 Which suspension represents the control in this investigation?
1 *A* 2 *B* 3 *C*

4 The change of cell volume is principally due to the loss or gain of
1 serum
2 water
3 hemoglobin
4 salt

5 Which process is most likely involved in the change in red blood cell volume in this experiment?
1 phagocytosis
2 cyclosis
3 osmosis
4 hydrolysis

Base your answers to questions 6 through 10 on the passage below and on your knowledge of biology.

Blood pressure is regulated by the interaction of the hormones renin, angiotensin, and aldosterone. Renin from the kidneys initiates production of angiotensin, which stimulates constriction of arterial walls and secretion of aldosterone by the adrenal glands. Aldosterone causes the body to store sodium, which results in increased water content in blood plasma. If too much renin is produced, too much water is retained and blood pressure increases.

Scientists have recently discovered that the upper chambers of the heart secrete a hormone that also appears to function in regulating blood pressure. This hormone, known as ANF, has been found to control excretion of sodium, a substance believed to be linked to hypertension. Studies underway at this time indicate that ANF reduces renin production, which affects angiotensin. This in turn slows the production of aldosterone. The result is an increase in the excretion of salt and water. This increased excretion has led to reduced blood pressure in volunteer patients.

6 The ANF hormone is produced by
1 adrenal glands
2 ventricles
3 kidneys
4 atria

7 High blood pressure in humans may be caused by
1 storing excess sodium in the body
2 excreting excess water
3 receiving injections of ANF
4 following a low-sodium diet

8 The interaction between the hormones and organs that regulate the salt and water content of the blood is an example of
1 maintenance of homeostasis
2 hypertension
3 excretion
4 cardiovascular disease

9 According to the reading passage, what effect does the presents of ANF have on the body?
1 It increases renin production.
2 It increases excretion of salt and water
3 It stimulates aldosterone production.
4 It stimulates production of angiotensis.

10 According to the reading passage, the heart secretes a hormone that appears to affect the
1 control of digestive secretions
2 transmission of nerve impulses to endocrine glands
3 control of all endocrine disorders
4 excretion of water by the kidneys

Base your answers to questions 11 through 15 on the information below:

An experiment was conducted to determine how long it takes for the pulse to return to a normal resting rate after exercise. An athlete with a resting pulse rate of 68 beats per minute exercised vigorously for a short period of time. His pulse rate was taken immediately after the exercise period and then at 60-second intervals. The following data were collected.

Using the information and the data table at the right, construct a graph, following the directions on the next page.

Data Table
Exercise Recovery Data

Time (sec)	Pulse (beats/15 sec)
0	38
60	31
120	22
180	19
240	17

11 Mark an appropriate scale on the axis labeled "Pulse (beats/15 sec)."

12 Mark and appropriate scale on the axis labeled "Time (sec)."

13 Plot the data and then connect the points.

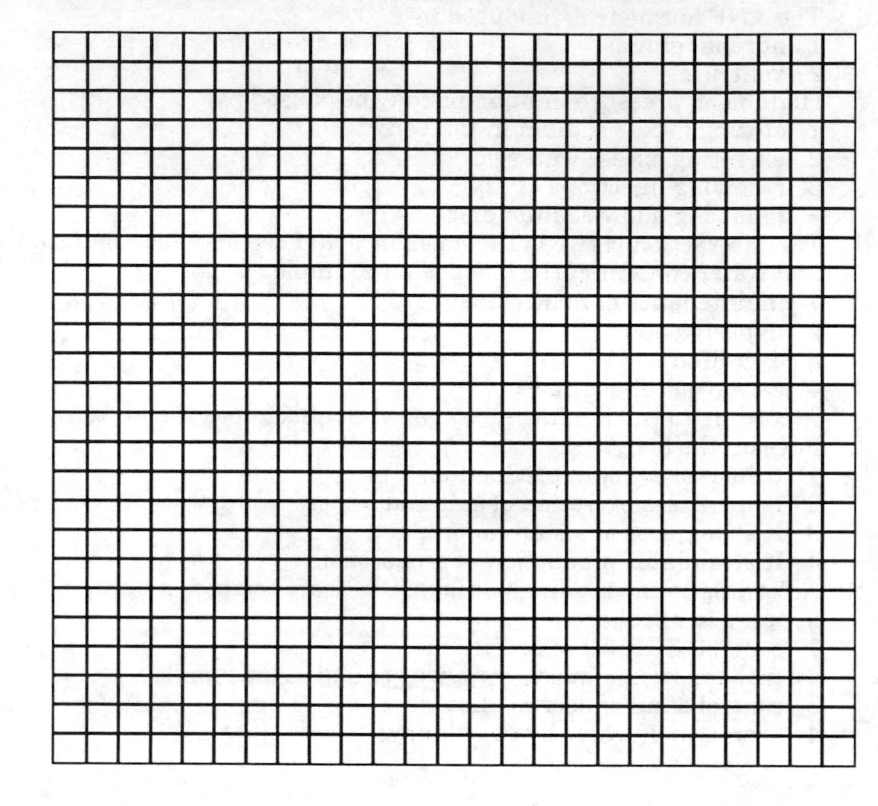

14 During which time interval of the recovery period did the greatest change in pulse rate occur?
1 0 to 60 sec
2 60 to 120 sec
3 120 to 180 sec
4 180 to 240 sec

15 How long did it take the athlete to reach his resting pulse rate?
1 1 min
2 2 min
3 3 min
4 4 min

Unit 4 **Reproduction and Development**

Objectives

The student should be able to:

- Describe the processes of mitosis, meiosis, and fertilization.
- Recognize the role of mitosis, meiosis, and fertilization in reproductive cycles.
- Compare the process of asexual and sexual reproduction in terms of methods and results.
- Compare the adaptations for sexual reproduction and development in both plants and animals.
- Explain the relationships among numbers of eggs, methods of fertilization, and sites of embryonic development, as they relate to species survival.
- Describe the development of plant and animal embryos.
- Describe hormonal interactions in the human male and female.

Asexual Reproduction – Terms & Concepts

Mitosis
  Replication of Chromosomes
Cytoplasmic Division`
Animal - Centrioles
  Membrane "pinching in"
Plant - Cell Plate "formation"
Cancer - Abnormal Cell Division

Asexual Reproduction
  Binary Fission
  Budding
  Sporulation
  Regeneration
  Vegetative Propagation

I. Asexual Reproduction

Asexual reproduction is the production of new organisms *without* the fusion of nuclei. The new organisms develop from a cell or cells of a *single parent* and have characteristics *identical* to the parent.

A. Mitotic Cell Division

According to the cell theory, all cells arise from other preexisting cells. This is done by cell division and involves both nuclear duplication and cytoplasmic division.

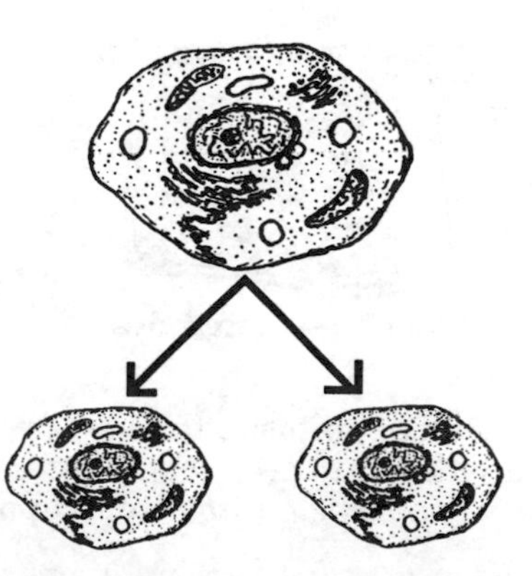

- **Mitosis** is an orderly series of complex changes in the nucleus, normally involving an **exact duplication** of the complete set of chromosomes (nuclei) and the separation of these chromosomes (nuclei) into two identical sets of chromosomes (nuclei).

- **Cytoplasmic Division** is the separation of cytoplasm which occurs either during or at the end of mitosis resulting in the formation of two daughter cells each containing an identical set of chromosomes.

1. Processes of Cell (Mitotic) Division

Mitosis

The process of mitosis involves:

- **Replication** (exact duplication) of each **single-stranded chromosome** during the non-dividing period (interphase), resulting in **double-stranded chromosomes**. Individual strands of a double-stranded chromosome are known as **chromatids** and are joined at a **centromere**.
- Disintegration (breakdown) of the nuclear membrane and nucleolus during the early stages of division.
- Synthesis of a **spindle apparatus** (a network of fibers).
- Attachment of double-stranded chromosomes to the spindle apparatus at the centromere region of the chromosomes.
- Replication of each centromere which results in the formation of two single-stranded chromosomes.
- Migration (movement) of single-stranded chromosomes toward opposite ends of the cell. As a result, there is one complete set of chromosomes at each of the two poles of the cell.
- Nuclear membrane formation around each set of chromosomes, forming two identical nuclei.

chromatids (2 strands)

centromere (kinetochore - connection region for chromatid strands)

Double Stranded Chromosome

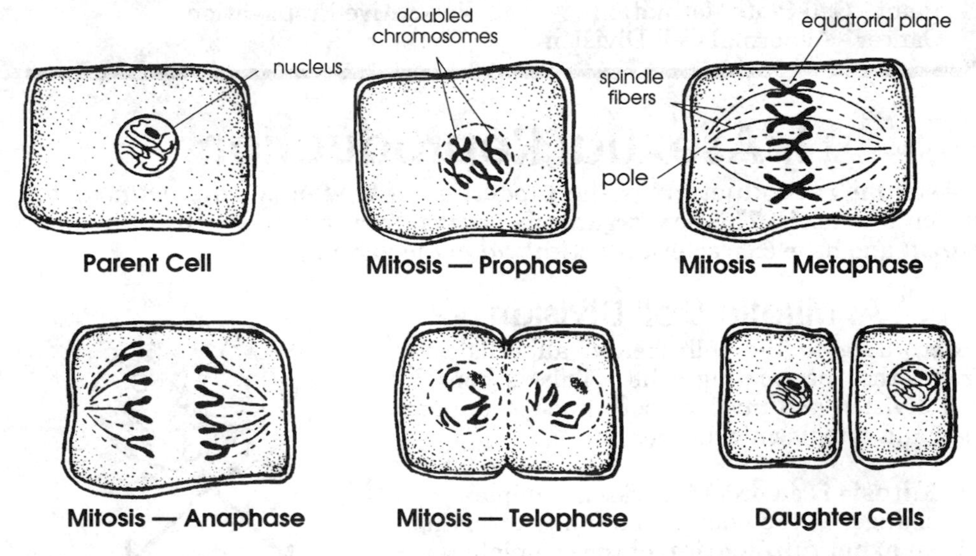

Parent Cell **Mitosis — Prophase** **Mitosis — Metaphase**

Mitosis — Anaphase **Mitosis — Telophase** **Daughter Cells**

During mitosis, the *hereditary material of the parent cell duplicates*, then is divided into two sets. This results in two daughter nuclei, each *identical in chromosome number* to the parent cell.

Cytoplasmic Division

Division of the cytoplasm usually, but not always, accompanies mitosis. Usually, cells divide into two relatively equal cytoplasmic parts, each one having a nucleus and called **daughter cells**. The methods of cytoplasmic division differ slightly in plant and animal cells.

2. Comparison Between Plant and Animal

Mitosis is similar in plant and animal cells. However, in animal cells, **centrioles** aid in the formation and orientation of the spindle apparatus. Cytoplasmic division is accomplished in animal cells by a "pinching" or "furrowing" of the outer surface of the cell membrane, thus separating the two nuclei. In plant cells, a **cell plate** is synthesized and grows from the center outward, separating the cytoplasm.

Self-regulated, normal mitotic cell division produces new cells used for growth and repair. However, uncontrolled, abnormal, and rapid mitotic cell division is known as **cancer**. Abnormally reproduced cells (cancer cells) do not function normally and may cause damage or death to the organism.

Questions

1 Which diagram most correctly represents the process of mitosis?

2 Which statement best describes a difference between cell division in plant and animal cells?
   1 In animal cells, cytoplasmic division is accomplished by a "pinching in" of the cell membrane, while in plant cells a cell plate is synthesized.
   2 In plant cells, cytoplasmic division is accomplished by a "pinching in" of the cell membrane, while in animal cells a cell plate is synthesized.
   3 In plant cells, centrioles have a distinct role in spindle formation, while in animal cells centrioles do not function during cell division.
   4 In animal cells, replication of chromosomes occurs during the non-dividing phase, while in plant cells replication occurs when the nuclear membrane disintegrates.

Base your answers to questions 3 and 4 on the diagram at the right which represents a microscopic structure observed during the process of cell division and on your knowledge of biology.

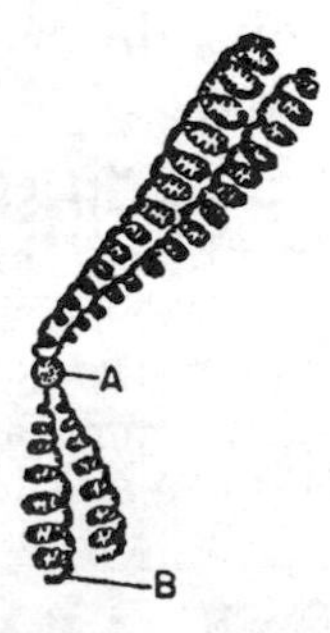

3 Letter A indicates a
   1 nucleolus
   2 ribosome
   3 centriole
   4 centromere

4 Letter *B* indicates a
   1 centrosome
   2 spindle fiber
   3 chromatid
   4 cell plate

5 Normally, a complete set of chromosomes (*2n*) is passed on to each daughter cell as a result of
1 reduction division
2 mitotic cell division
3 meiotic cell division
4 nondisjunction

6 Each strand of a double-stranded chromosome is known as a
1 tetrad
2 centromere
3 homolog
4 chromatid

7 Which event occurs in the cytoplasmic division of plant cells but not in the cytoplasmic division of animal cells?
1 cell plate formation
2 chromosome replication
3 centromere replication
4 centriole formation

8 When a cell with 24 chromosomes divides by mitotic cell division, the daughter cells will each have a maximum chromosome number of
1 6 2 12 3 24 4 48

9 Structures which hold chromatids together in double-stranded chromosomes are known as
1 centrioles 2 polar bodies 3 centromeres 4 spindle fibers

10 The following list describes some of the events associated with normal cell division.
A - Nuclear membrane formation around each set of newly formed chromosomes.
B - Separation of centromeres.
C - Replication of each chromosome.
D - Movement of single-stranded chromosomes to opposite ends of the cell.
What is the normal sequence in which these events occur?
1 A, B, C, D
2 C, B, D, A
3 C, D, B, A
4 D, C, A, B

B. Types of Asexual Reproduction

All types of asexual reproduction produce identical offspring from a single parent. The following are examples of different types of asexual reproduction.

1. Binary Fission

Binary fission involves the equal division of the nuclear materials and cytoplasm of an organism resulting in two new organisms. Binary fission is common in unicellular organisms, including the paramecium, ameba, and bacteria.

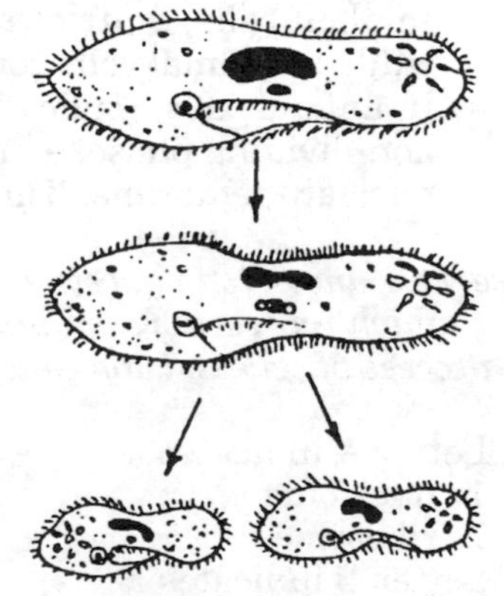

Binary Fission in Paramecium

2. Budding

In unicellular organisms such as yeast, **budding** is similar to fission except that the cytoplasmic division is unequal. The new cells may detach from each other or may remain together and form a **colony**.

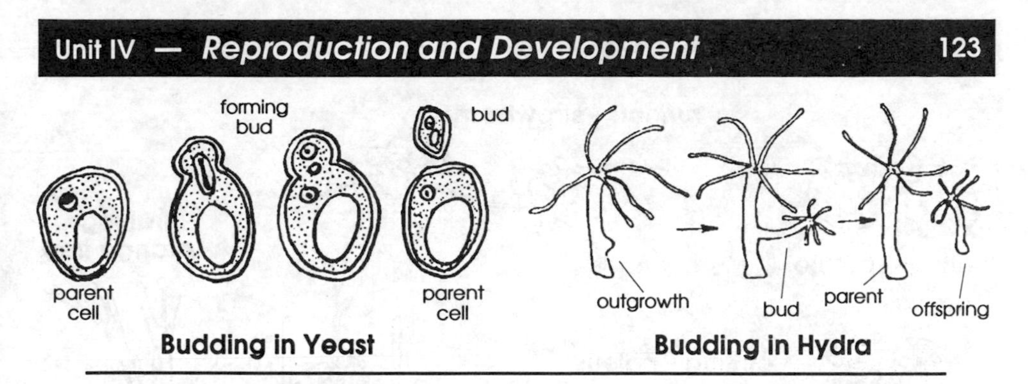

Budding in Yeast **Budding in Hydra**

In multicellular organisms such as the hydra, budding refers to the production of a multicellular outgrowth from the parent organism. The bud and parent may detach from each other or may remain together and form a colony.

3. Sporulation

Spores are mitotically produced and are single, specialized cells, released from the parent. Spores can develop into new individuals under favorable conditions of temperature and moisture. (Examples include bread mold and mushrooms)

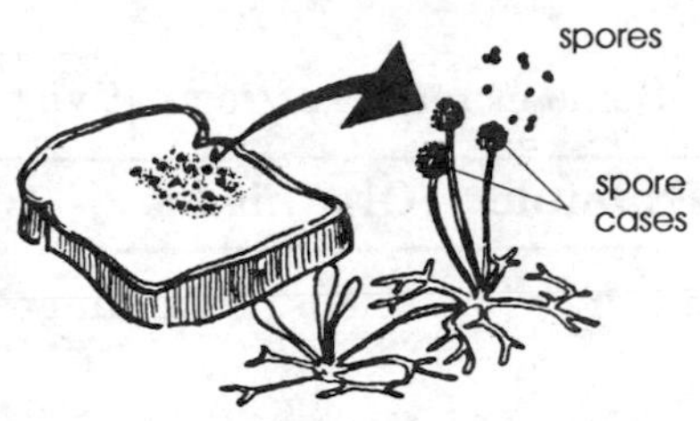

Sporulation - Bread Mold

4. Regeneration

Regeneration is the development of an entire new organism from a part of the original organism. An example is the starfish which may develop from a single ray and part of the central disc.

Regeneration also refers to the replacement of lost structures. A lobster may regenerate a lost claw. Generally, invertebrate animals possess more undifferentiated cells than do vertebrate animals. As a result, invertebrates exhibit a higher degree of regenerative ability than most vertebrates.

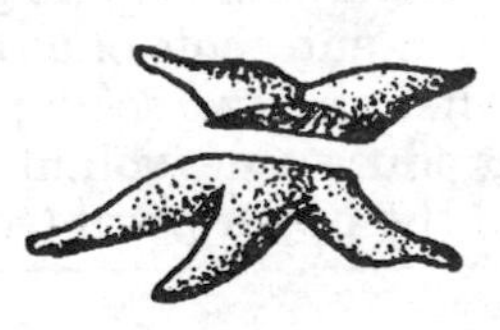

parent starfish
cut into two halves

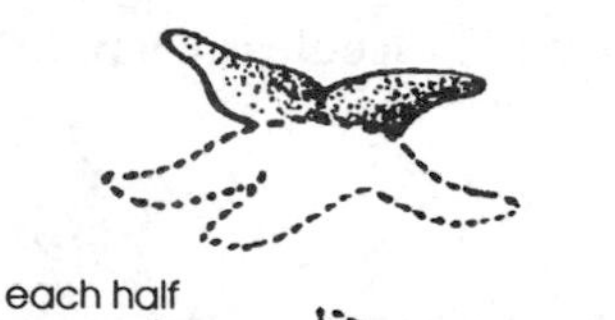

each half
regenerates
a new starfish

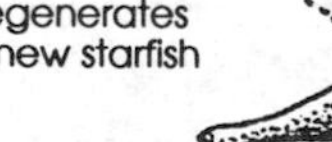

Regeneration - Starfish

5. Vegetative Propagation

Some multicellular plants reproduce asexually by vegetative propagation. In this process new plants develop from roots, stems, or leaves of the parent plant.

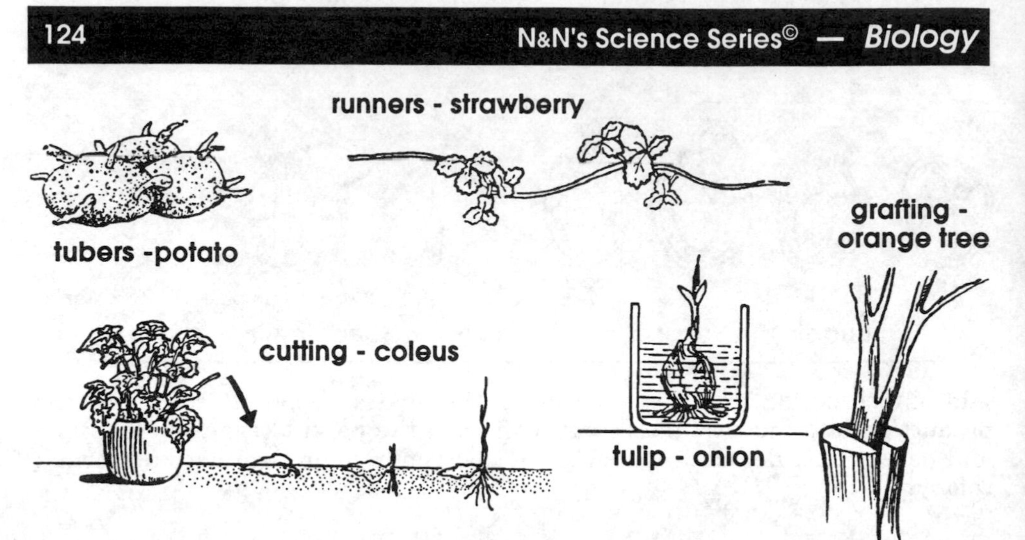

Some examples of vegetative propagation include:

Example	Organisms	Process
cuttings	coleus, geranium	• a part of a root, stem, or leaf; placed in a favorable environment develops into a new plant identical to the parent plant
bulbs	tulip	• underground stem with thick leaves; identical to and produced "like a bud" by the parent bulb
tubers	potato	• thick underground storage stem; placed in a favorable environment develops into a new plant identical to the parent plant
runners	strawberry	• stem-like structures that grow from the parent plant's stem on top of the ground; produce plants identical to the parent plant
grafting	seedless orange	• upper stem of one plant (scion) and lower stem and roots of a second plant (stock); when the two sections grow together, they produce a new plant with the characteristics of the upper plant (scion)

Questions

1 Organisms that can reproduce by budding are
  1 yeast and hydra
  2 yeast and earthworm
  3 bread mold and grasshopper
  4 grasshopper and goldfish

2 Budding, spore formation, and vegetative propagation are methods of reproduction that involve the process of
  1 mitosis
  2 meiosis
  3 fusion of gametes
  4 crossing-over

3 All types of asexual reproduction involve the process known as
1 mitosis
2 fertilization
3 artificial pollination
4 reduction division

4 Which is a type of asexual reproduction that commonly occurs in many species of unicellular protists?
1 external fertilization
2 tissue regeneration
3 binary fission
4 vegetative propagation

5 A florist discovers a mutant geranium that produces flowers of exceptional beauty. Geraniums that produce flowers most like those of the mutant plant may be produced by
1 binary fission
2 gamete fusion
3 sexual reproduction
4 vegetative propagation

6 The diagram at the right represents the nucleus of a leaf cell in a mature plant. A cutting was made from this plant. Which diagram best represents the nucleus of a cell from the plant grown from this cutting?

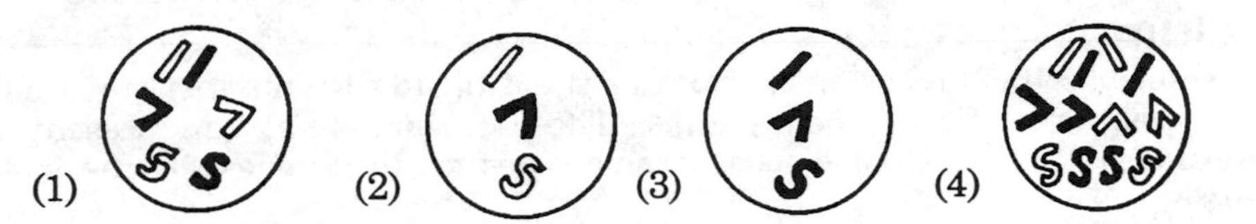

7 Which statement describes the process of budding in a unicellular organism?
1 Fertilization is preceded by meiosis.
2 There is an equal division of the cytoplasm.
3 The cytoplasm divides unequally.
4 Fertilization is preceded by mitosis.

8 Which type of reproduction is illustrated in the diagram at the right?
1 zygote formation
2 vegetative propagation
3 binary fission
4 spore formation

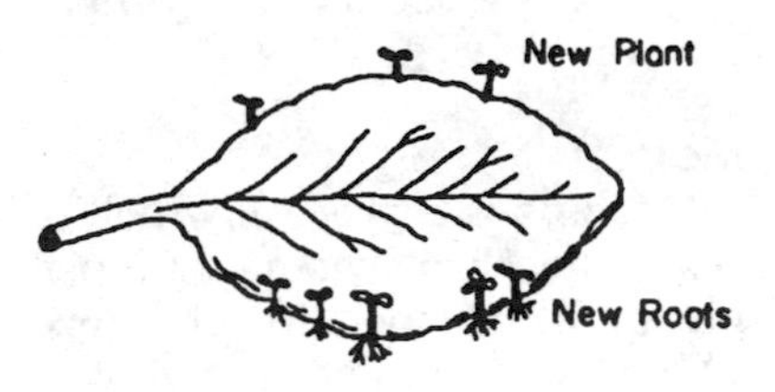

9 A certain fruit tree is found to have desirable characteristics. These characteristics could be propagated most easily and quickly by
1 planting seeds produced by the tree
2 grafting a cutting taken from the tree
3 cross-pollinating with another tree
4 culturing highly differentiated cells of the tree

10 One difference between budding and binary fission in unicellular organisms is that in budding the
1 genetic material is unequally divided
2 genetic material is equally divided
3 cytoplasm is unequally divided
4 cytoplasm is equally divided

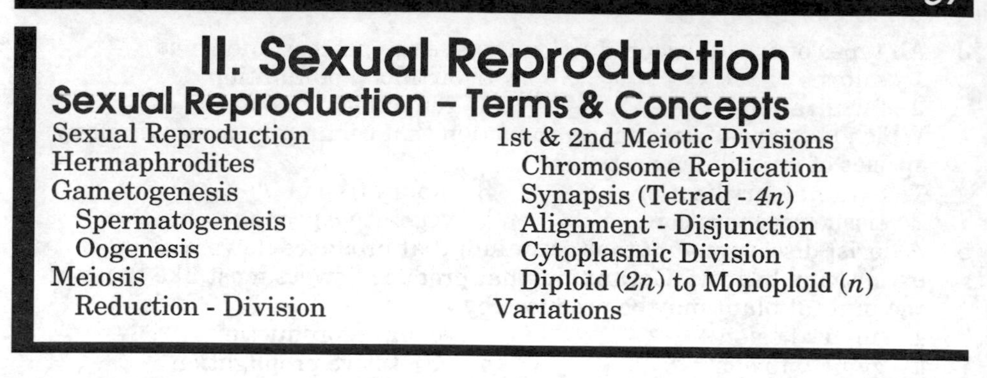

II. Sexual Reproduction

Sexual Reproduction – Terms & Concepts

Sexual Reproduction
Hermaphrodites
Gametogenesis
  Spermatogenesis
  Oogenesis
Meiosis
  Reduction - Division

1st & 2nd Meiotic Divisions
  Chromosome Replication
  Synapsis (Tetrad - $4n$)
  Alignment - Disjunction
  Cytoplasmic Division
  Diploid ($2n$) to Monoploid (n)
Variations

Sexual reproduction involves the production of specialized cells (**gametes**) and the fusion of their nuclei (**fertilization**) producing a fertilized egg cell (**zygote**). The zygote undergoes mitotic cell division during its growth and maturity.

A. Reproduction and Development in Animals

1. Gametogenesis

Each body cell of an organism contains the **diploid** (**$2n$**) number of chromosomes characteristic of that species. These chromosomes are present in **homologous** pairs. Homologous chromosomes contain genes for the same traits.

Gametogenesis is the process in which gametes are produced. It involves meiotic cell division and cell maturation. This process occurs in specialized organs called **gonads**. Some organisms have only male or female gonads while others have both, such as the earthworm, and are called **hermaphrodites**. The result of gametogenesis is the production of sperm (in the male) or egg cells (in the female) which have half (n) the number of chromosomes of a normal body cell ($2n$).

Meiosis

Meiosis is a process which involves **reduction division**. During this process, the chromosome number is reduced by one-half and **monoploid** (n) nuclei which contain one chromosome of each homologous pair are formed. The process of meiosis involves two separate and distinct divisions.

First Meiotic Division. The first meiotic division is the reduction division and involves:

- **Replication** of each single-stranded chromosome during the non-dividing period, results in the production of a pair of identical sister chromatids.
- **Synapsis** is the intimate pairing of sister chromatids fastened at their centromeres, each group of four chromatids is called a *tetrad*.
- **Alignment** occurs as the centromeres of the tetrads line up on the equator of the cell.
- **Disjunction** (separation) of the homologous chromosomes and their subsequent movement (migration) to opposite poles. Compared to the parent cell, there are half as many chromosomes at each pole, but each chromosome is double stranded.

- **Cytoplasmic division** forms two daughter cells each with half the number of the parent's cells chromosomes, already in replicated form.

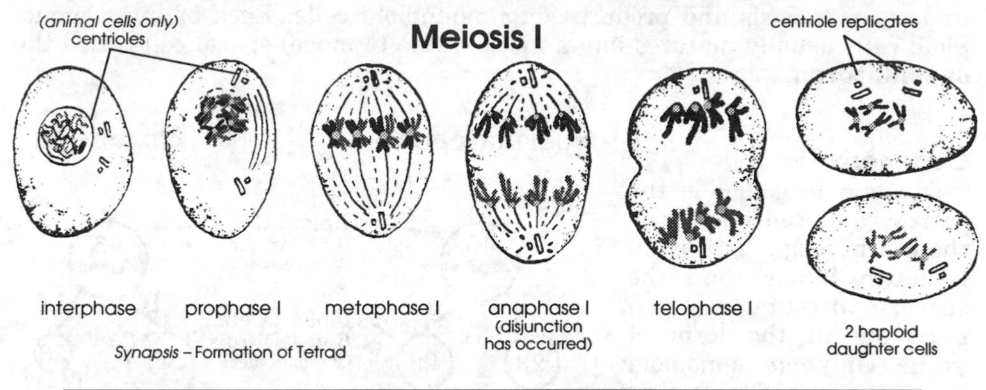

Second Meiotic Division. In each of the cells produced during the first meiotic division, the second meiotic division occurs:

- **Alignment** of pair of sister chromatids (double-stranded chromosomes) at their centromeres on the equatorial plane of the cell.
- **Replication** of the centromeres of the sister chromatids, followed by the separation of the two chromatids becoming single-stranded chromosomes.
- **Migration** of single-stranded chromosomes along the spindle apparatus toward opposite ends of the cell.
- **Cytoplasmic division** in which the two distinct cells divide and produce two additional distinct cells (total of 4 cells) each with a haploid set of single chromosomes (*n* or monoploid).

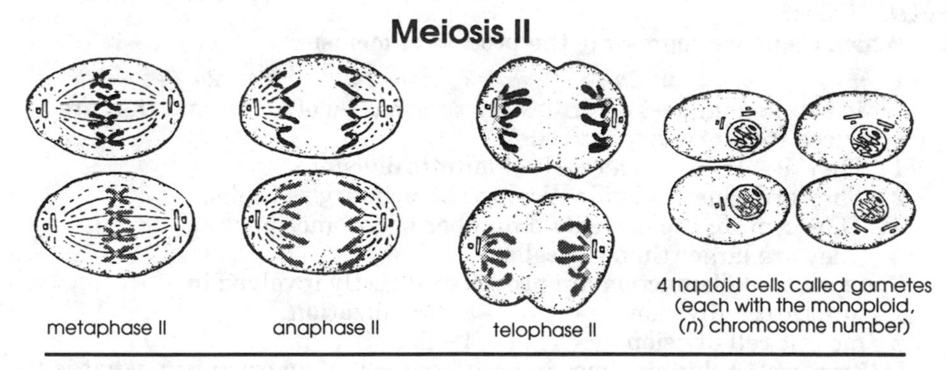

Meiosis Summary: As a result of meiosis, diploid (*2n*) **primary sex cells** divide and form monoploid (*n*) cells which mature into specialized reproductive cells (**gametes**). The distribution of the homologous chromosomes between the resultant nuclei is *random*, resulting in gene **variation** in the gametes.

Comparison With Mitotic Cell Division

Mitosis is associated with growth and asexual reproduction. *Meiosis*, is associated with sexual reproduction. As a result of *mitotic* cell division, the daughter cells are *identical* to the original cell. As a result of *meiotic* cell division, the resulting cells have *one-half* (*n*) the number of chromosomes of the original parent cell.

Spermatogenesis

Male gonads, **testes**, produce male gametes, **sperm**. The primary sex cell undergoes meiosis and produces *four* monoploid cells. Each of these monoploid cells usually matures into a motile (able to move) sperm cell called the **spermatozoa**.

Oogenesis

The female gonad is the **ovary**, and the gamete is the **ovum** (egg). The ovum is much larger than the sperm and contains stored nutrients in the form of **yolk**. Only *one* monoploid egg cell is usually formed from each primary sex cell that undergoes meiosis and maturation. The other cells produced are called **polar bodies**. The polar bodies result from unequal cytoplasmic divisions, are not fertilized, and degenerate.

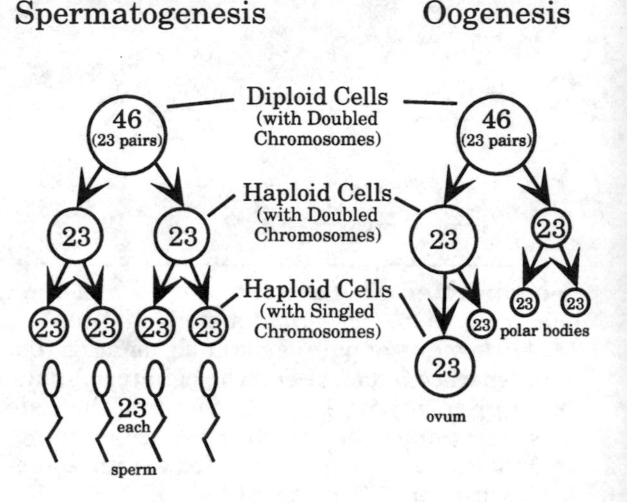

Human Gametogenesis

Questions

1 Which sequence represents the process of meiosis?
  1 $n \rightarrow n$ 2 $2n \rightarrow n$ 3 $n \rightarrow 2n$ 4 $2n \rightarrow 2n$

2 Which statement best describes the sperm cells of an animal that are produced from primary sex cells?
  1 They are diploid as a result of mitotic division.
  2 They are genetically identical to the primary sex cells.
  3 They contain the monoploid number of chromosomes.
  4 They are larger than egg cells.

3 Synapsis and disjunction are processes directly involved in
  1 mitotic cell division 3 fertilization
  2 meiotic cell division 4 fission

4 If there are 40 chromosomes in each body cell of an organism, what is the total number of chromosomes normally present in a gamete produced by that organism?
  1 10 2 20 3 40 4 80

5 Mitotic and meiotic cell division are similar in that both processes
  1 produce diploid gametes from monoploid cells
  2 produce monoploid gametes from diploid cells
  3 involve synapsis of homologous chromosomes
  4 involve replication of chromosomes

6 Which process reduces the chromosome number from diploid ($2n$) to monoploid (n)?
  1 meiosis 2 mitosis 3 polyploidy 4 fertilization

7 Only one member of each pair of homologous chromosomes is normally found in a
1 zygote
2 multicellular embryo
3 gamete
4 cheek cell

8 Ovaries produce reproductive structures known as
1 pollen grains
2 egg cells
3 sperm cells
4 zygotes

9 During the normal meiotic division of a diploid cell, the change in chromosome number that occurs is represented as
1 $4n$ to n
2 $2n$ to $4n$
3 $2n$ to n
4 n to $2n$

10 In animals, polar bodies are formed as a result of
1 mitotic cell division in males
2 meiotic cell division in males
3 mitotic cell division in females
4 meiotic cell division in females

2. Fertilization

Fertilization/Development – Terms & Concepts

Fertilization
- External - Internal

Embryonic Development
- Cleavage
- Gastrulation
- Differentiation
  - Endo-, Meso-, Ectoderm
- Growth

External Development
- In Water and On Land
- Yolk - Amnion
- Allantois - Chorion

Internal Development
- Placental Mammals
- Placenta - Umbilical Cord
- Marsupials - Pouch

Fertilization is the union (fusion) of a monoploid sperm nucleus (***n***) with a monoploid egg nucleus (***n***). During meiosis the chromosome number (***2n***) was reduced. In fertilization, the resulting diploid zygote has its species number of homologous chromosomes (***2n***) restored. Fertilization must occur in a moist environment.

Male gamete + Female gamete → Zygote
n + n → 2n

External Fertilization

Reproduction in many aquatic vertebrate animals such as fish and amphibians is characterized by **external fertilization**. Fertilization occurs *outside the body* of the female in the water. Usually, because of a lack of protection and a harsh environment, large numbers of eggs are required to insure species survival.

Internal Fertilization

Reproduction in most terrestrial vertebrate animals, such as birds, reptiles, and mammals, is characterized by internal fertilization. In order to provide a moist environment, the gametes fuse in the moist reproductive tract of the female.

3. Embryonic Development

Process. After internal or external fertilization, the zygote (fertilized egg) undergoes a series of rapid mitotic cell divisions called **cleavage**. Although there is very little growth in size, cleavage causes an increase in the number of cells with a decrease in the size of each cell. Once cleavage begins, the zygote becomes a developing embryo.

Extended Area: ***Reproduction & Development***

Stages of Development

Cleavage. Cleavage is a series of rapid mitotic cell divisions that leads to the formation of a mass of cells, the **morula**. Additional mitotic cell divisions leads to the formation of the **blastula**, a single layer of cells which is a hollow ball-type structure. During cleavage, there is little or no increase in individual cell size.

Gastrulation. In certain animals, one side of the blastula becomes indented (**gastrulation**) forming the **gastrula**, which has an inner layer, the **endoderm**, and an outer layer, the **ectoderm**. A third layer, the **mesoderm**, forms between the ectoderm and the endoderm.

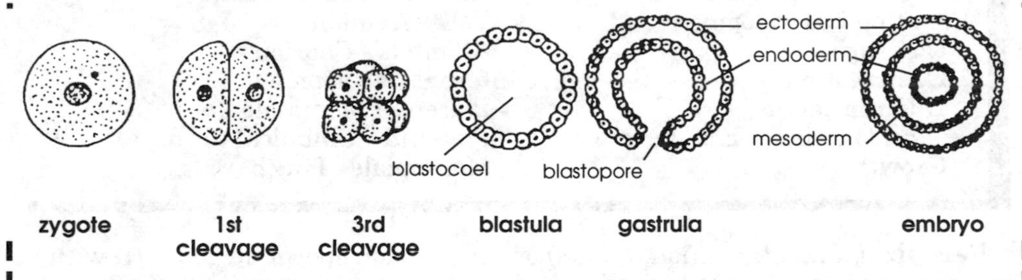

Differentiation. These three embryonic layers differentiate and give rise to the various tissues, organs, and systems of the multicellular animal.

- The nervous system and the skin originate from the **ectoderm** layer.
- The muscles, circulatory system, skeleton, excretory system, and gonads originate from the **mesoderm** layer.
- The lining of the digestive and respiratory tracts, and portions of the liver and pancreas, originate from the **endoderm**.

Growth. Growth is an increase in cell number as well as in cell size. Early development consists chiefly of the differentiation, growth and development of specialized cells, tissues, and organs.

4. Site of Development

Internal development occurs inside of the female's (mother's) body.

External Development

External development occurs *outside* of the female's body in both terrestrial and aquatic environments.

Extended Area: ***Reproduction & Development***

• **In Water** (aquatic environment). The eggs of many fish and amphibians are fertilized externally and develop externally in an aquatic environment. The survival rate is generally low, which accounts for the large number of fertilized eggs produced. The developing embryo's source of food is the yolk stored in the egg.

• **On Land** (terrestrial environment). Eggs of birds, many reptiles, and a few mammals develop externally on a land environment after internal fertilization. Since there is a better survival rate, a somewhat fewer number of fertilized eggs are produced. The developing embryo's source of food is the yolk.

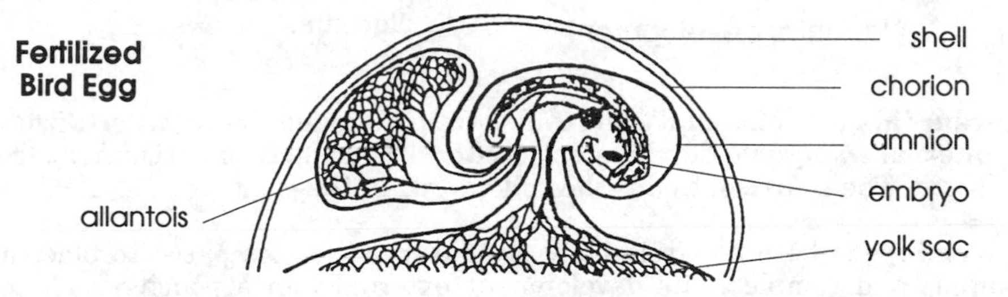

Some adaptations for animals which develop externally on land are a **shell** which provides protection and **membranes** which help provide a favorable environment for embryonic development. These embryonic membranes include:

- The **amnion** contains the amniotic fluid. This fluid provides a watery environment, protects the embryo from shock, and prevents adhesion of the embryonic tissues to the shell.
- The **yolk sac** surrounds the yolk. Blood vessels which penetrate the yolk sac transport food to the developing embryo.
- The **allantois** functions as a respiratory membrane and a storage site for the nitrogenous waste, uric acid.
- The **chorion** is an outer membrane surrounding the other embryonic membranes and separates them from the environment.

Internal Development

Internal development involves the growth of the embryo within the body of the parent which provides nutrition and protection. Generally, a relatively high survival rate allows the production of fewer fertilized eggs.

Placental Mammals

Mammals have internal fertilization. The embryo(s) develops internally within a structure called the **uterus**. The eggs of mammals have relatively little yolk and therefore are very small. Within the uterus, a specialized organ, the **placenta**, is formed from embryonic and maternal tissues. It is through this structure that the exchange of nutrients, wastes, and respiratory gases between the mother and the embryo occurs.

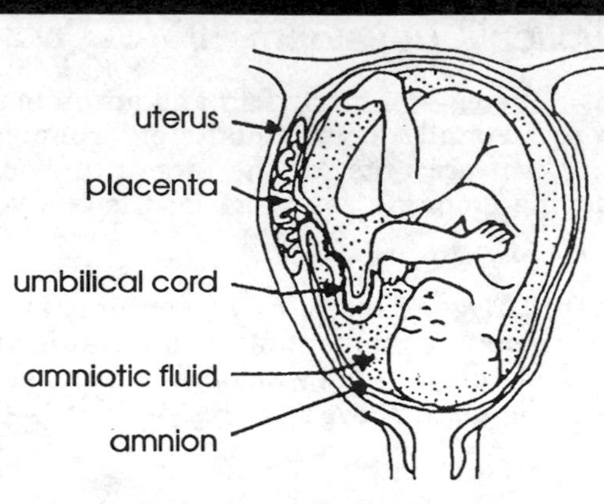

Placental Development

There is no direct connection between maternal and embryonic bloodstreams. Transport is accomplished by diffusion and active transport. An **umbilical cord**, containing blood vessels, attaches the embryo to the placenta.

Humans are placental mammals. In a few mammals, such as the marsupials, there is no placenta.

Marsupials (nonplacental mammals). In marsupials, internal fertilization and internal *embryonic* development occur without direct nourishment from the parent. The source of food is the yolk stored in the egg.

The embryo is born at a relatively premature stage compared to placental mammals and completes its development externally in a pouch which contains mammary (milk producing) glands.

Questions

1 The gamete produced in the ovary of an animal is the
  1 egg cell 3 spore
  2 sperm cell 4 zygote

2 Which organ of a cow is used to feed a newborn calf?
  1 a mammary glands 3 a placenta
  2 the uterus 4 the gonad

3 The fusion of monoploid gametes restores the diploid condition in organisms that reproduce
  1 asexually 2 sexually 3 vegetatively 4 mitotically

4 The diploid chromosome number in a certain species of fish is 20. How many chromosomes would normally be found in a bone cell of this fish?
  1 10 2 20 3 23 4 40

5 In animals, monoploid gametes are usually formed as a direct result of
  1 fertilization 3 meiotic cell division
  2 fusion 4 mitotic cell division

6 In most multicellular animals, meiotic cell division occurs in specialized organs known as
  1 gonads 3 gametes
  2 kidneys 4 cytoplasmic organelles

7 A structure common to both placental animals and birds is the
  1 uterus 3 umbilical cord
  2 placenta 4 amnion

8 A male fish produces gametes called
  1 testes 3 zygotes
  2 sperm cells 4 egg cells

9 Invertebrates exhibit a greater degree of regeneration than vertebrates because, compared to vertebrates, invertebrates have more
 1 undifferentiated cells
 2 complex structures
 3 chloroplasts and mitochondria
 4 cells in their nervous systems

10 In animals, the process which results in monoploid gametes is known as
 1 meiosis
 2 mitosis
 3 fertilization
 4 fusion

Extended Area: ***Reproduction & Development***

4. Reproduction and Development in Humans

Human Reproduction – Terms & Concepts

Gametogenesis
Male - Testis in Scrotum
 Urethra in Penis
 Semen - Sperm and Fluids
 Testosterone
Female - Ovaries within Body
 Ova Produced in Follicles
 Oviduct (Fallopian Tube)
 Uterus
 Cervix - Vagina
 Estrogen
 Progesterone
In Vitro Fertilization
Menstrual Cycle
 Puberty to Menopause
 Follicle Stage
 Ovulation
 Corpus Luteum Stage
 Menstruation
Hormone Interaction
 Hypothalamus, Pituitary, Ovaries
Fertilization in Oviduct
Cleavage - Implantation
Twins - Identical and Fraternal
Development
 Prenatal - Birth - Postnatal

Male Reproductive System

The male reproductive system performs two major functions:

- the *production* of sperm cells
- the *deposition* of these cells within the female reproductive tract

Sperm production occurs in the **testes**. The testes are located in the **scrotum** where the temperature is 1-2 degrees Celsius cooler than normal body temperature. This provides the optimum temperature for sperm production and storage. From the testes, sperm move through the sperm duct and the urethra. The **urethra** is a tube contained within the penis. The **penis** is a structural adaptation for internal fertilization.

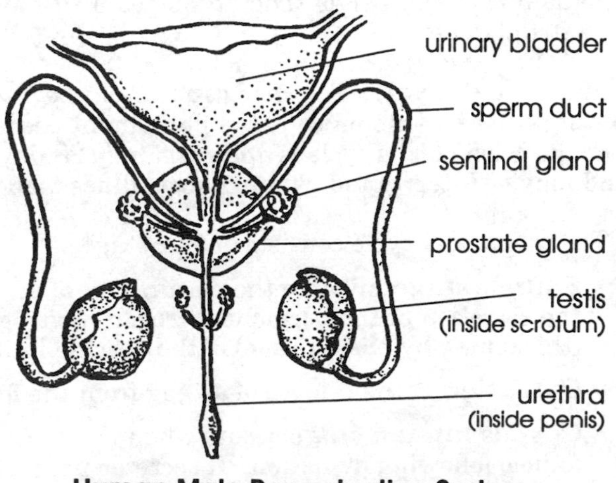

Human Male Reproductive System

Glands secrete a liquid into these tubes. The liquid serves as a transport medium for the sperm which is an adaptation for life on land. This liquid and the sperm cells constitute **semen**.

In addition to producing sperm, the testes also produce the male sex hormone, **testosterone**. This regulates the *maturation of sperm* and the *development of secondary sex characteristics*, such as beard development and the lower voice pitch.

Female Reproductive System

Ovaries are paired structures located within the lower portion of the body cavity. Ovaries produce **eggs** (ova) in tiny cavities called **follicles**. Following **ovulation** (release of the egg from the ovary), the egg cell is transported through an **oviduct (Fallopian tube)** to the **uterus**.

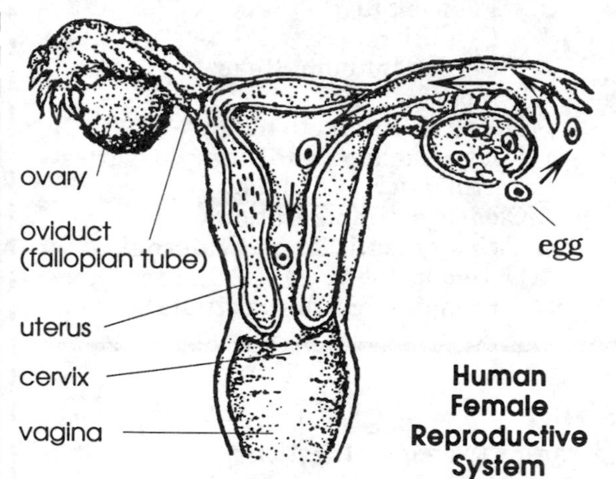

Human Female Reproductive System

At the lower end of the uterus, known as the cervix, is a muscular tube, the **vagina**. The vagina receives the sperm from the male and is also the birth canal.

At the birth of a female, all of the potential eggs that she will ovulate are present in *immature* form. Usually only *one* egg is released at the time of each ovulation.

In addition to eggs, the ovaries also produce the female sex hormones **estrogen** and **progesterone**. These regulate the *development of secondary sex characteristics* such as the development of the mammary glands and the broadening of the pelvis. The hormones also have a *coordinating role* in the menstrual cycle.

Menstrual Cycle. The menstrual cycle *begins* at **puberty** and *ceases* at **menopause**. Menopause is the permanent cessation of the menstrual cycle. The duration of this cycle is *approximately* 28 days but may vary considerably and may be interrupted by pregnancy, illness, and other factors.

The menstrual cycle consists of four stages:

- **Follicle stage** involves the maturation of an egg within the follicle and the secretion of the hormone estrogen. Estrogen initiates vascularization (thickening by blood tissue) of the uterine lining.
- **Ovulation** is the release of an egg from the follicle.
- **Corpus luteum stage** occurs when the corpus luteum forms from the follicle following ovulation. It secretes progesterone which enhances the vascularization of the uterine lining.

- **Menstruation** is the periodic shedding of the thickened uterine lining. Generally, menstruation lasts for a few days and occurs when fertilization does not take place.

Role of Hormones in the Menstrual Cycle. The reproductive cycle of the human female involves the interaction of the hormones from the hypothalamus, pituitary gland, and ovaries.

During the menstrual cycle, under the influence of the hormones from the hypothalamus, the pituitary gland releases hormones (**FSH** and **LH**) which influence the functioning of the ovaries. FSH stimulates follicle growth and the ovary to produce estrogen. LH stimulates the growth of the corpus luteum (in the ovary) to secrete progesterone which inhibits estrogen production. The ovaries, in turn, secrete hormones (**estrogen** and **progesterone**) which produce changes in the uterus. Estrogen causes the release of FSH to be stopped. An increase in progesterone causes a reduction in the production of FSH and LH. In addition, the hormones from the ovaries regulate the secretion of hormones by the pituitary gland and the hypothalamus. This is an example of a **negative feedback mechanism.**

Fertilization

Fertilization usually occurs in the upper one third of the **oviduct**. If the egg is not fertilized within approximately 24 hours after ovulation, it deteriorates. If fertilization occurs, cleavage of the fertilized egg (zygote) begins in the oviduct, and six to ten days later, the resulting embryo may become implanted in the uterine lining. At this stage of development, the yolk of the egg has been depleted. But with implantation, the embryo can obtain nutrients from the uterus.

If more than one egg is released and fertilized, multiple births may result. **Identical twins** develop from *one* zygote which separates during cleavage. **Fraternal twins** develop from *two* eggs, *each* fertilized by *separate* sperm cells.

The technique of *in vitro* fertilization and subsequent implantation that has been perfected in animals has now been applied to humans with some degree of success. *In vitro* fertilization is commonly called the "test tube baby" process.

Development

Prenatal Development. Prenatal development includes the following processes:

- **Cleavage** occurs in the oviduct.
- **Gastrulation** usually occurs after the embryo is implanted in the uterus.
- **Differentiation** and **growth** result in the formation of specialized tissues and organs from the embryonic layers of the gastrula.

Prenatal development is dependent upon several structures that are similar to the structures found inside of the externally developing egg.

- The **placenta** is the "connection" between the mother and the developing embryo which allows nutrients, respiratory gases, and wastes to be exchanged. Note that there is *no direct connection*, therefore, there is no mixing of blood between the embryo and the mother. Substances move between the two separate circulatory systems by active and passive transport.

- The **amnion** is the membrane that surrounds the embryo and contains the amniotic fluid which protects the developing embryo from shock and temperature changes.

- The **umbilical cord** ("life line") holds the blood vessels that carry materials between the embryo and the placenta.

The process involved in prenatal development is dependent upon the supplying of a proper balance of nutrients to the developing embryo. Good nutrition during pregnancy for both the mother and the developing fetus is essential for proper and safe development.

Birth. Birth usually occurs after a gestation period of approximately nine months. During birth, strong contractions of the uterine muscles (labor) forces the baby through the cervix and vagina (birth canal) and to the outside of the mother's body. The expulsion of the placenta occurs after the delivery of the baby.

Postnatal Development. Development continues with various parts of the body growing at different rates. Although the development of the organism is often assumed to conclude with the mature adult, it actually continues throughout the life of the organism and terminates with death.

The term **aging** is applied to the complex developmental changes that occur naturally with the passage of time. The cause or causes of the aging process are still not fully understood. It appears that the aging process results from the interplay of hereditary and environmental factors. One recent definition of *death is the irreversible cessation of all brain functions*.

Questions

1 In humans, the fusion of the nuclei of two functional gametes results in the formation of a
   1 monoploid cell
   2 zygote
   3 cotyledon
   4 polyploid cell

2 If the first stage of an uninterrupted human menstrual cycle is the follicle stage, then the last stage would include the
   1 formation of the sperm cells in the testis
   2 release of a mature egg
   3 buildup of the uterine lining
   4 shedding of the uterine lining

3 Which structure of the female reproductive system "catches" the released egg from the ovary?
1 oviduct 2 uterus 3 vagina 4 cervix

4 Within which structure of the human female system would sperm cells normally be deposited?
1 vagina 2 cervix 3 uterus 4 oviduct

5 When an ovum is released from the ovary, it is referred to as
1 cleavage 3 ovulation
2 fertilization 4 pregnancy

6 In humans, gestation normally occurs while a developing organism is in the
1 uterus 3 follicle
2 vagina 4 ovary

7 Which of the following structures is not directly associated with the interaction of hormones in the human female?
1 ovaries 3 pituitary gland
2 hypothalamus 4 vagina

8 The normal gestation period for the human is
1 2 weeks 2 2 months 3 5 months 4 9 months

9 Which stage of human development is most directly the result of the interplay of hereditary and environmental factors?
1 embryo development 3 fetus development
2 birth 4 aging process

10 The human male semen is best referred to as
1 sperm, only 3 liquid, only
2 body cells, only 4 sperm cells and liquid

11 Which order represents the correct sequence of stages in the human menstrual cycle?
1 ovulation - follicle stage - menstruation - corpus luteum stage
2 follicle stage - ovulation - corpus luteum stage - menstruation
3 menstruation - corpus luteum stage - ovulation - follicle stage
4 corpus luteum stage - ovulation - follicle stage - menstruation

12 A function of the umbilical cord is to
1 absorb nutrients from the yolk sac for the mother
2 carry wastes from the urinary bladder to the kidney
3 mix the blood of the embryo with the blood of the mother
4 transport nutrients to the embryo

13 The human male testes are located in an outpocketing of the body wall known as the scrotum. An advantage of this adaptation is that
1 the testes are better protected in the scrotum than in the body cavity
2 sperm production requires contact with atmospheric air
3 a temperature lower than the body temperature is best for the sperm
4 the sperm cells can enter the urethra directly from the testes

14 Which human female structure is best prepared for implantation of a fertilized egg as a result of the action of reproductive hormones?
1 follicle 2 ovary 3 oviduct 4 uterus

15 During which stage of the menstrual cycle does the uterine tissue break down?
1 menstruation 3 ovulation
2 follicle stage 4 corpus luteum stage

B. Flowering Plants: Reproduction and Development

Plant Reproduction – Terms & Concepts

Flower - Sexual Reproduction
  Female - Pistil
    Stigma - Style - Ovary
  Male - Stamen
    Anther - Filament
Pollination - Self and Cross
  Pollen Grain and Tube
  Monoploid (sperm) Nucleus
Fertilization and Development
  Zygote
  Seed from Ovule
  Seed Coat and Embryo
Fruit from Ripened Ovary
Plant Embryo
  Hypocotyl - Root, Lower Stem
  Epicotyl - Leaves, Upper Stem
  Cotyledon - Stored Nutrients
Germination - Seed Dispersal
  Sufficient Moisture
  Proper Temperature - Enzymes
  Sufficient Oxygen
Growth - Meristems
  Apical - Root and Stem Tips
  Lateral - Cambium

The processes of meiosis and fertilization occur in the flower, a plant structure specialized for sexual reproduction.

1. Flower Structure

The flower may contain both the male reproductive organ, the **stamen**, and female reproductive organ, the **pistil**. Flowers with both reproductive organs are called "perfect" or "complete" flowers. In some species, certain flowers contain only stamens, while others contain only pistils. Flowers having only one reproductive organ are called "imperfect" or "incomplete" flowers. Accessory structures, petals and sepals may also be present in flowers.

- The **stamen** is composed of an **anther** and **filament**. As a result of meiosis, the diploid cells of the anther produce pollen grains which contain monoploid nuclei (*n* chromosome number).
- The **pistil** is composed of the **stigma**, **style**, and the **ovary**. As a result of meiosis, ovule(s), developing within the ovary, contain the monoploid egg nucleus (*n* chromosome number).

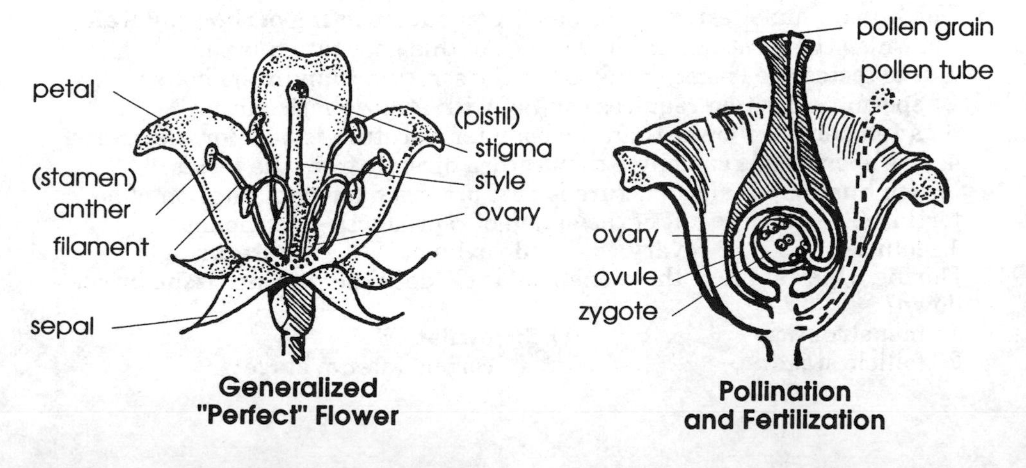

Generalized "Perfect" Flower

Pollination and Fertilization

2. Pollination

Pollination is the transfer of pollen grains from the anther to the stigma. Pollination may be accomplished by wind, insects, and birds. In some instances, the colored petals act as a visual attractant for insects while nectar acts as a chemical attractant.

- **Self-pollination** is the transfer of pollen from the anther to the stigma of the *same* flower or to the stigma of another flower on the *same* plant.
- **Cross-pollination** is the transfer of pollen from an anther on one flower to a stigma of a flower on a *different* plant. Cross-pollination is an adaptation which enhances variations (increases variety).

In flowering plants, the problem of reproduction in a dry, external environment is partially solved by the presence of the thick wall of the pollen grain. This prevents the dehydration of its contents during its transfer to the female reproductive organ.

Following pollination, the pollen grain **germinates** (is activated) on the stigma and forms a **pollen tube** which extends into the ovule. Sperm nuclei are formed at this time in the pollen tube from the monoploid nucleus in the pollen grain. The pollen tube is an adaptation for internal fertilization.

3. Fertilization and Embryo Development

The union (fusion) of the male and female nuclei in the ovule results in a **zygote**. The zygote undergoes development resulting in the formation of the **embryo**. The **ripened ovule**, containing the embryo, develops into the seed.

A **seed** consists of a seed coat, which develops from the outer coverings of the ovule, and an embryo. The **ripened ovary** develops into the **fruit**. The plant embryo consists of three parts: hypocotyl, epicotyl, and cotyledon(s).

- The **hypocotyl** develops into the root and, in some species, the lower portion of the stem.
- The **epicotyl** develops into the leaves and upper portions of the stem.
- The **cotyledons** contain stored food which provide the nutrients for the germinating plant.

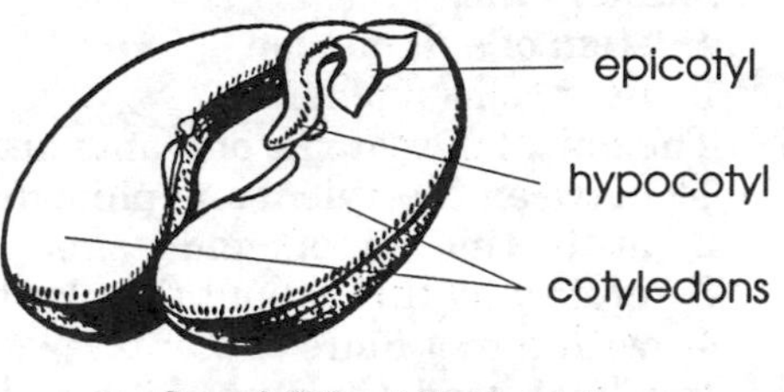

Seed (Plant Embryo)

4. Germination

In flowering plants, seeds develop inside a fruit. Fruits are specialized structures which aid in seed dispersal. Under suitable environmental conditions, the dispersed seeds **germinate**. For most seeds, these conditions include:

- *sufficient moisture*
- *proper temperature*
- *sufficient oxygen*

The development of a seed into a mature plant, which is capable of reproduction, involves cell division, differentiation, and growth.

5. Growth

Growth in higher plants is restricted largely to specific regions known as **meristems**.

- **Apical meristems** are found in the tips of roots and stems and are responsible for the growth in the length of the plant.
- Some plants also contain an active **lateral meristem** region, the **cambium**, located between the xylem and phloem. This is responsible for the growth in diameter of roots and stems.

The meristematic (growth) regions contain undifferentiated cells which undergo mitotic cell division and elongation. As a result of differentiation, the various tissues and organs are developed.

Questions

1 In flowering plants, pollen formation occurs in the
1 stigma 2 anther 3 pistil 4 ovary

2 In which part of a flower do both meiosis and fertilization occur?
1 ovule 2 stigma 3 anther 4 petal

3 When seeds from flowering plants germinate, they obtain food most directly from
1 soil 2 cotyledons 3 water 4 air

4 In tulips, stamens produce
1 ovules 2 seeds 3 egg cells 4 pollen grains

5 In most flowering plants, which structures aid in the transport of sperm nuclei to egg nuclei?
1 sepals 2 pollen tubes 3 anthers 4 phloem tubes

6 Cells with the monoploid number of chromosomes would normally be found in the
1 stem of a dandelion
2 liver cells of a horse
3 skin cells of a human
4 anthers of a rose

7 The apical meristems of plants are located
1 between the xylem and phloem tissue of woody stems
2 at the tips of roots and stems
3 uniformly throughout the plant
4 within root hairs of plants

8 In a bean seed, the part of the embryo which develops into the leaves and upper portion of the stem is known as the
1 seed coat 2 epicotyl 3 hypocotyl 4 cotyledon

9 Which represents the proper sequence in the development of a flowering plant?
1 germination - pollination - fruit formation - fertilization
2 fertilization - zygote - gastrula - embryo
3 fertilization - gastrula - blastula - embryo - fetus
4 pollination - fertilization - seed formation - seed germination

10 The growing point at the tip of a root consists of an actively dividing tissue known as the
1 epidermis 2 xylem 3 meristem 4 phloem

SELF–HELP: Unit IV "Core" Questions

1 The diagrams represent various processes associated with reproduction.

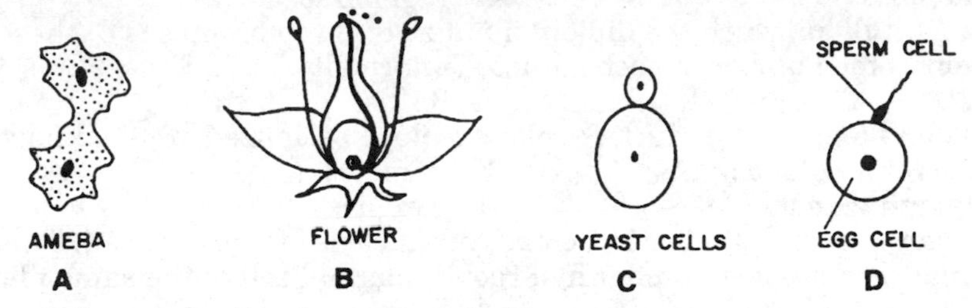

Asexual reproduction is represented by
1 *A*, only 2 *B*, only 3 *A* and *C* 4 *B* and *D*

2 One difference between mitotic cell division in animals and in plants is that in plants
1 chromosomes are duplicated, whereas in animals chromosomes are not duplicated
2 chromosomes are separated, whereas in animals chromosomes are not separated
3 spindle fibers are formed, whereas in animals spindle fibers are not formed
4 cell plates are formed, whereas in animal cells plates are not formed

3 Which process is represented by the diagram?
1 germination
2 fertilization
3 mitotic cell division
4 meiotic cell division

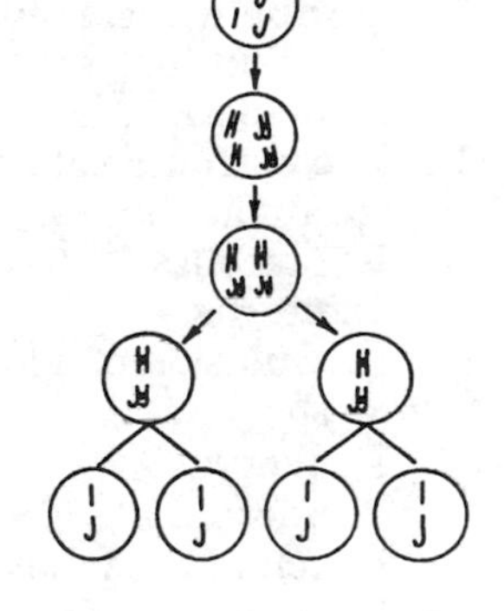

4 Which statement is true of budding?
1 It produces zygotes by fertilization.
2 It involves an equal division of cytoplasm.
3 It produces gametes by meiosis.
4 It is an asexual form of reproduction.

5 The reproduction of a new plant from a runner is an example of
1 cleavage 3 zygote formation
2 fission 4 vegetative propagation

6 Each cell in the leaf bud of a plant contains 26 chromosomes. How many chromosomes would be found in each cell of the mature leaf formed from the bud by mitotic cell division?
1 13 2 26 3 28 4 52

7 In Hydra, asexual reproduction differs from sexual reproduction in that asexual reproduction does not involve
1 mitotic cell division 3 cytoplasmic division
2 fusion of nuclei 4 division of the nucleus

8 Mitotic cell division in a geranium plant differs from mitotic cell division in a mouse. One difference is that in a mouse
1 spindle fibers are produced 3 centrioles are replicated
2 cell plates are synthesized 4 chromosomes are paired

9 The ameba demonstrates a type of asexual reproduction known as
1 budding
2 meiosis
3 binary fission
4 regeneration

10 In a fruit fly in which the diploid number of chromosomes is 8, the chromosome number in each gamete is normally
1 16
2 2
3 8
4 4

11 Male and female reproductive cells develop specialized organs known as
1 excretory glands
2 lymph glands
3 gametes
4 gonads

12 The chromosome number in an egg cell nucleus of a plant is 14. The normal chromosome number in a root epidermal cell of the same plant is
1 7
2 14
3 21
4 28

13 In which organ does meiosis take place?
1 kidney
2 cervix
3 testis
4 urethra

14 Uncontrolled cell division is characteristic of
1 cancer
2 meiosis
3 budding
4 sporulation

15 In humans, a single primary sex cell may produce four gametes. These gametes are known as
1 diploid egg cells
2 monoploid egg cells
3 polar bodies
4 sperm cells

16 In human females, the maximum number of functional egg cells that is normally produced from each primary sex cell is
1 one
2 two
3 three
4 four

17 During which menstrual cycle stage is the lining of the uterus shed?
1 ovulation
2 menstruation
3 follicle stage
4 corpus luteum stage

18 The exchange of oxygen, food, and wastes between mother and fetus occurs at the
1 placenta
2 uterus
3 umbilical cord
4 amniotic sac

19 Stimulation of a follicle in the human female involves a hormone secreted by the
1 ovary
2 adrenal gland
3 pituitary gland
4 uterus

20 Which of the following least affects the human female menstrual cycle?
1 ovary
2 corpus luteum
3 pancreas
4 pituitary gland

21 Which portion of a bean seed contains the greatest percentage of starch?
1 seed coat
2 epicotyl
3 cotyledon
4 hypocotyl

22 Two flowering holly plants grew in the same yard. Each year fruit developed on one plant, but never on the other. A correct explanation for this observation would be that the plant that never produced fruit
1 had flowers with stamens, only
2 had flowers with pistils, only
3 produced cones
4 produced spores

23 In apple blossoms, the function of the stigma is to
1 form sperm nuclei
2 pollinate the ovule
3 produce the pollen
4 receive the pollen

24 Which reproductive structures are produced within the ovaries of plants?
1 pollen grain
2 sperm nuclei
3 egg nuclei
4 pollen tubes

25 Which of the following structures are collectively known as the pistil?
1 stigma and ovary
2 anther and filament
3 stigma, style, and ovary
4 stigma, style, and filament

SELF–HELP: Unit IV *"Extended"* Questions

Base your answers to questions 1 through 3 on the diagram of the reproductive system of a pregnant female shown at the right and on your knowledge of biology.

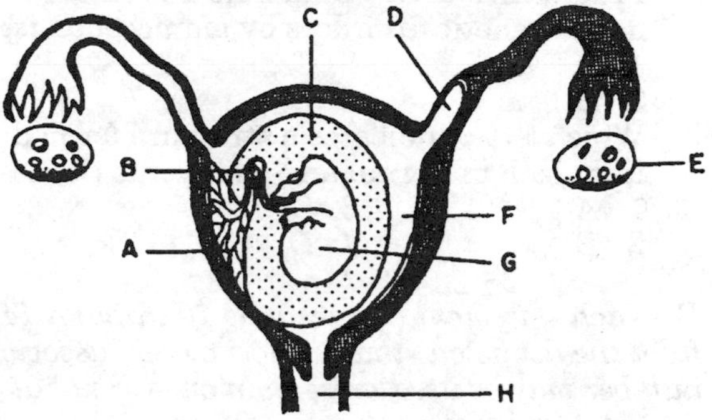

1 Female gametes are produced within the structure labeled
1 *A*
2 *E*
3 *C*
4 *H*

2 When sperm cells are deposited inside the female, the pathway they follow to reach the egg is from
1 *A* to *B* to *E*
2 *H* to *F* to *D*
3 *E* to *D* to *F*
4 *H* to *B* to *D*

3 Nutrients and oxygen needed by the developing fetus are obtained by diffusion between blood vessels of the mother and the fetus. This exchange occurs through the structure labeled
1 *A* 2 *G* 3 *C* 4 *D*

4 The cyclical nature of the menstrual cycle in the human female is maintained by a hormonal feedback system between the ovaries and the
1 pituitary
2 placenta
3 vagina
4 oviducts

Base your answers to questions 5 through 7 on the diagram at the right which represents some events in the reproduction of a typical vertebrate and on your knowledge of biology.

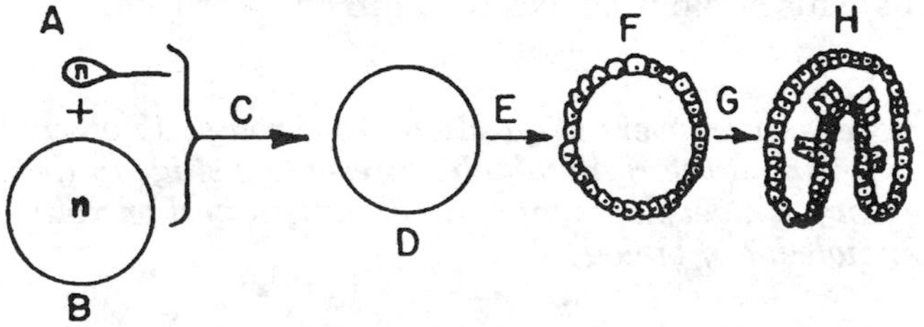

5 The process represented by arrow *E* is known as
1 feedback
2 cleavage
3 meiosis
4 differentiation

6 The process represented by arrow *C* is known as
1 fertilization
2 spermatogenesis
3 oogenesis
4 mitosis

7 Which structure directly results from the process of gastrulation?
1 *F* 2 *B* 3 *H* 4 *D*

Base your answers to questions 8 and 9 on the diagram at the right which represents a fetus developing within a human female and on your knowledge of biology.

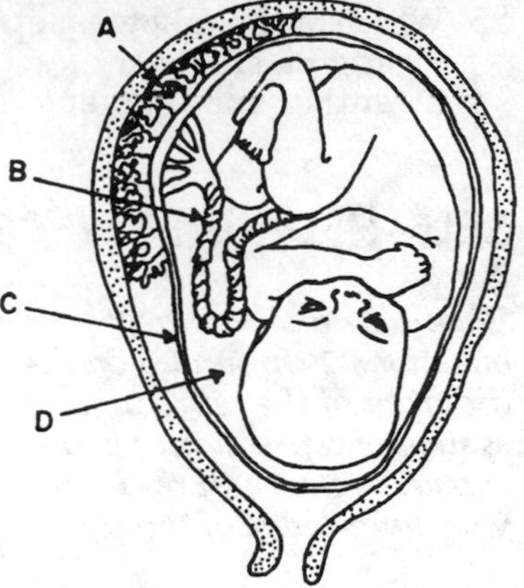

8 From which area would fluid be removed to detect genetic disorders by amniocentesis?
 1 *A* 3 *C*
 2 *B* 4 *D*

9 Which letter indicates a structure formed from both maternal and embryonic tissues?
 1 *A* 3 *C*
 2 *B* 4 *D*

For each statement in questions 10 through 14 select the structure, chosen from the list below, that is most closely associated with that statement. [A number may be used more than once or not at all.]

Structures

1) Uterus 3) Scrotum 5) Oviduct
2) Ovary 4) Urethra

10 This organ produces monoploid cells.

11 This structure provides an optimum temperature for sperm production and storage

12 In males, this structure serves both an excretory and reproductive function.

13 This structure serves as the normal site of fertilization.

14 This organ produces the hormone estrogen.

Base your answers to questions 15 through 17 on the diagram at the right which represents a stage in the embryonic development of an organism and on your knowledge of biology.

A
B
C
D

15 In the development of this organism, which event follows the stage represented in the diagram?
 1 blastula formation 3 differentiation
 2 cleavage 4 zygote formation

16 The cells represented by region *C* are known as
 1 ectodermal cells 3 mesodermal cells
 2 endodermal cells 4 epidermal cells

17 The nervous system develops from cells represented by which region?
 1 *A* 3 *C*
 2 *B* 4 *D*

Base your answers to questions 18 through 20 on the diagrams below of four different animal species and on your knowledge of biology.

18 Which of these species produce the largest number of mature egg cells during a single reproductive cycle?
1 *A* and *C* 2 *B* and *C* 3 *C* and *D* 4 *A* and *B*

19 In which species is complete embryonic development dependent upon placental tissue?
1 *A* 2 *B* 3 *C* 4 *D*

20 In which species does embryonic development occur externally on land?
1 *A* 2 *B* 3 *C* 4 *D*

For each phrase in questions 21 through 24, select the extra-embryonic membrane, chosen from the list below, which is best described by that phrase.

Extra-embryonic Membranes

1) Chorion 2) Amnion 3) Allantois 4) Yolk Sac

21 An outer membrane which surrounds the other extra-embryonic membranes.

22 A membrane that contains fluid which protects the embryo from shock.

23 A membrane which surrounds the major source of food for the developing embryo.

24 A membrane that serves as a respiratory membrane.

25 In humans, immediately after ovulation the egg normally enters the
1 follicle sac 2 cervix 3 oviduct 4 uterus

SELF–HELP: Unit IV *"Skill"* Questions

Base your answers to questions 1 through 4 on the experiment described below and on your knowledge of biology.

Two groups of 100 carrot seeds each were used in an investigation to test for the influence of temperature on germination of seeds. One group was kept at a temperature of 20°C, and the other at 10°C. All other conditions were the same. Observations made during the investigation were used to construct the data table.

Data Table

Day of Observation	Total Number of Seeds That Germinated	
	10° C	20° C
7	0	5
10	20	35
15	40	70
20	45	80
25	45	80

Directions (1-4): Using the information in the data table, construct a line graph, following the directions on the next page.

1 **Mark an appropriate scale on each of the labeled axes.**

2 **Plot the data for germination at 10°C on the grid. Surround each point with a small triangle and connect the points.**

3 **Plot the data for germination at 20°C on the grid. Surround each point with a small circle and connect the points.**

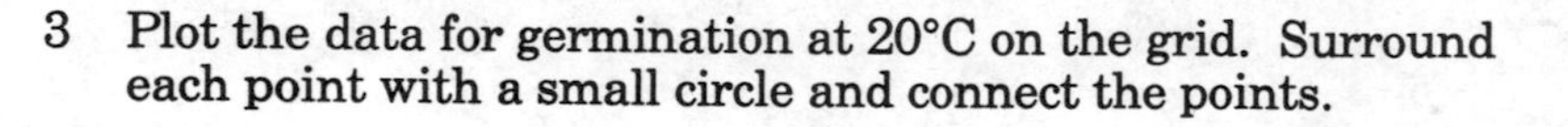

4 **According to this investigation, what is the difference in the number of seeds germinated at the two temperatures on day 10?**
1 15
2 20
3 30
4 35

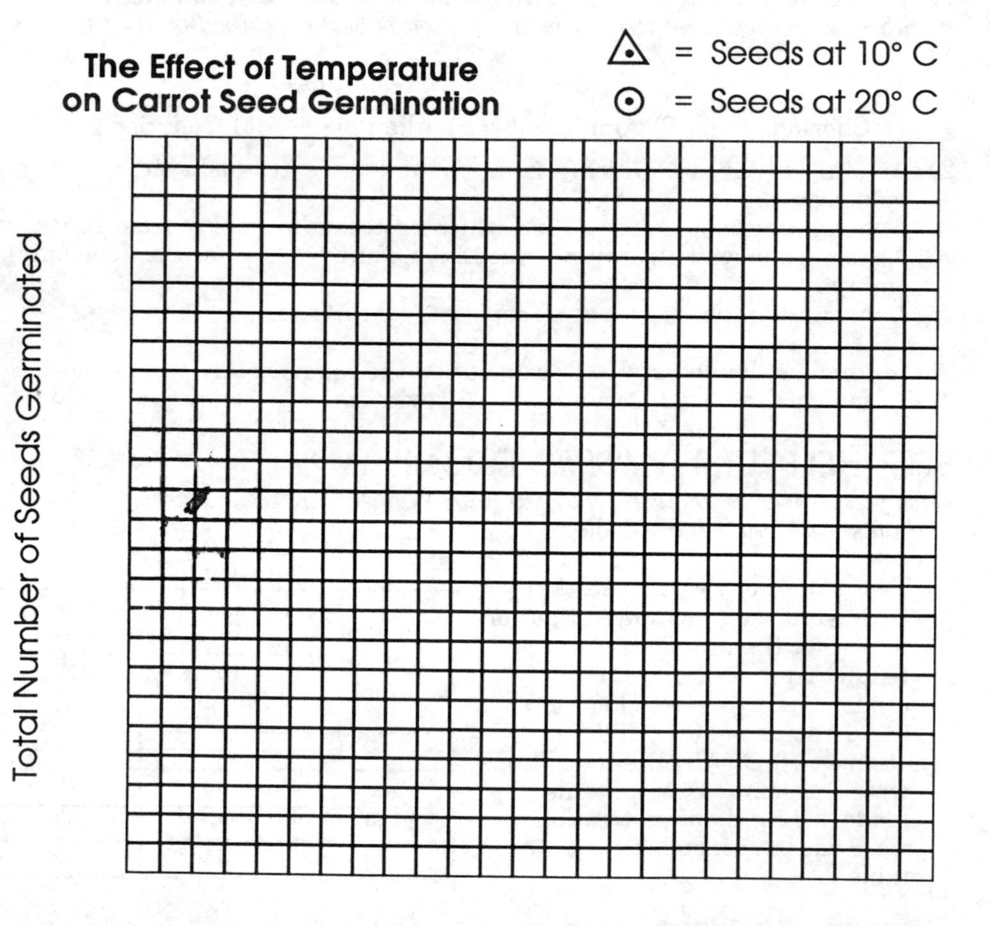

Unit 5

Transmission of Traits: Generation to Generation

Objectives

The student should be able to:

- Explain transmission of traits with the gene-chromosome theory.
- Predict the probable results of genetic crosses.
- Identify patterns of inheritance by interpreting pedigree charts.
- List various mutations and describe their consequence.
- List several practical applications of the principles of genetics.
- Recognize the role of heredity and environment in gene expression.
- Describe some techniques used in genetic research.
- Describe some genetically-related disorders in humans.
- Describe the basic structure of DNA and its assumed role in heredity.
- Explain changes in a population in the Hardy-Weinberg Principle.

Historical Genetics – Terms & Concepts

Mendel - Principles of Heredity
Gene - Chromosome Theory
  Morgan and Drosophila
  Meiosis - Genes - Alleles
Dominance
  Dominant / Recessive Alleles
Genotype - Phenotype
  Homozygous - pure
  Heterozygous - hybrid

Segregation - Recombination
  Monoploid (n) and Diploid ($2n$)
Intermediate Inheritance
  Co-dominance / Incomplete
Independent Assortment
Gene Linkage - Crossing Over
Multiple Alleles
Chromosomes - Autosomes
  Sex Linkage - Chromosomes

I. Foundations of Genetics

A. Mendelian Principles

In the 19th century, an Austrian monk and teacher, **Gregor Mendel**, developed some basic principles of heredity without any knowledge of genes or chromosomes. His principles of **dominance**, **segregation**, and **independent assortment** were established through the mathematical analysis of large numbers of offspring. He experimented in his laboratory (a garden) with pea plants, cross-pollinating plants with specific, contrasting characteristics.

Gregor Mendel

As a result of analyzing specific mathematical ratios associated with certain characteristics in the offspring, Mendel proposed that *characteristics were inherited as a result of the transmission of hereditary factors.*

B. Gene-Chromosome Theory

The significance of Mendel's work was not immediately recognized. After his work was rediscovered, evidence from the microscopic study of dividing cells and breeding experiments (by Thomas H. Morgan at Columbia University) with *Drosophila* (fruit fly) enabled scientists to link the presence of chromosomes and their migration during meiosis with the hereditary factors which Mendel proposed in his principles.

allelic pairs
centromere
chromatids
Chromosome (Double Stranded)

Mendel's hereditary factors, called **genes**, exist at definite *loci* (permanent locations) in a linear fashion on chromosomes. Two genes associated with a specific characteristic are known as **alleles** and are located on homologous chromosomes. The **gene-chromosome theory** provides the mechanism to account for the hereditary patterns which Mendel observed.

II. Major Genetic Concepts

A. Dominance

In some patterns of heredity, if only one of the genes in an allelic pair is expressed, it is called a **dominant** allele. In this case, the dominant gene characteristic "masks" the other gene characteristic. The gene which is present but not expressed is called the **recessive** allele. In order for a recessive gene characteristic to be observed, there must be two recessive genes present for that characteristic.

By convention, a *capital letter* symbolizes a **dominant allele**. The *lower case* form of the *same letter* symbolizes the **recessive allele**. For example, in certain pea plants, the allele for tallness (T) is dominant, and the allele for shortness (**t**) is recessive.

If two genes of an allelic pair are the same, the genetic makeup (gene combination), **genotype**, is said to be **homozygous** or **pure** (**TT** - *tall* or **tt** - *short*).

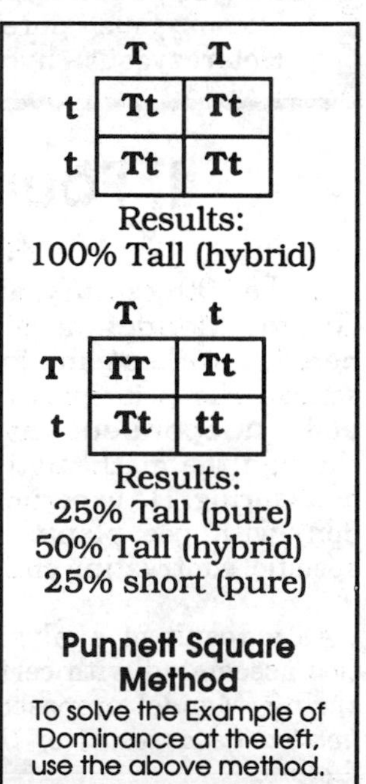

T T
t Tt Tt
t Tt Tt
Results:
100% Tall (hybrid)
T t
T TT Tt
t Tt tt
Results:
25% Tall (pure)
50% Tall (hybrid)
25% short (pure)

Punnett Square Method
To solve the Example of Dominance at the left, use the above method.

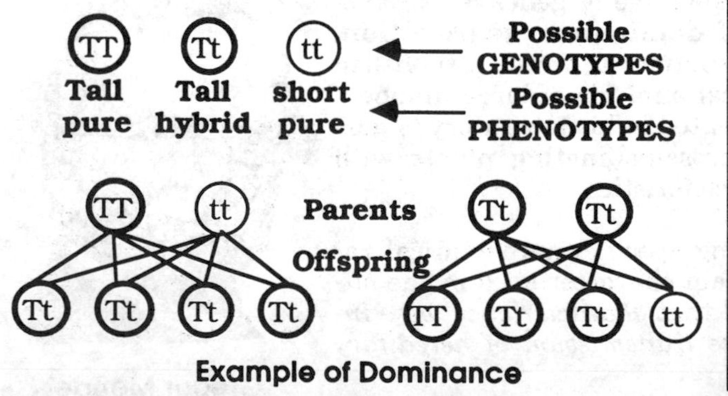

Example of Dominance

If two genes of an allelic pair are different, the genetic makeup is said to be **heterozygous** or **hybrid** (**Tt** - *tall*).

When an individual homozygous for the dominant trait is crossed with an individual homozygous for the recessive trait, the physical appearance (**phenotype**) of this offspring, known as the F_1 (first filial) generation, is like that of the dominant parent. However, the genotype of these offspring is heterozygous. Offspring resulting from the cross between members of the F_1 (first filial) generation comprise the F_2 (second filial) generation.

B. Segregation and Recombination

When gametes are formed during meiosis there is a **random segregation** (separation and movement) of homologous chromosomes.

As a result of fertilization, alleles **recombine**. Although the results can be predicted, the gametes come together in a *random mating*. As a consequence of this random mating, new allelic gene combinations are likely to be produced.

Segregation and recombination is illustrated by the diagram below right which represents a cross between two guinea pigs (1st generation) heterozygous for color (black and white).

Assuming large numbers of such crosses, the phenotypic ratio of black offspring to white offspring is always 3:1, and the genotypic ratio of homozygous black offspring to heterozygous black offspring to homozygous recessive offspring is always 1:2:1.

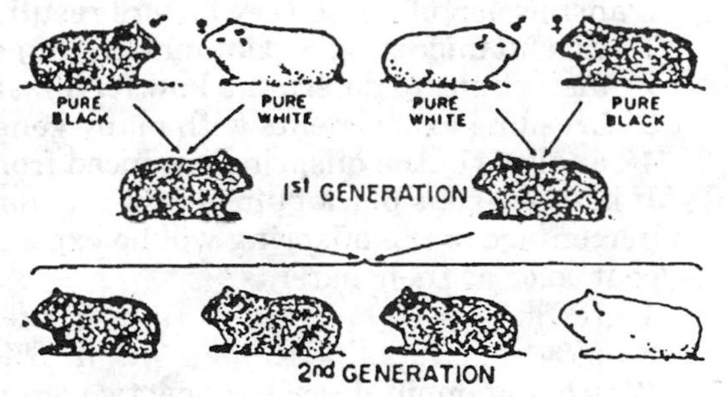

Heterozygous Guinea Pig Cross

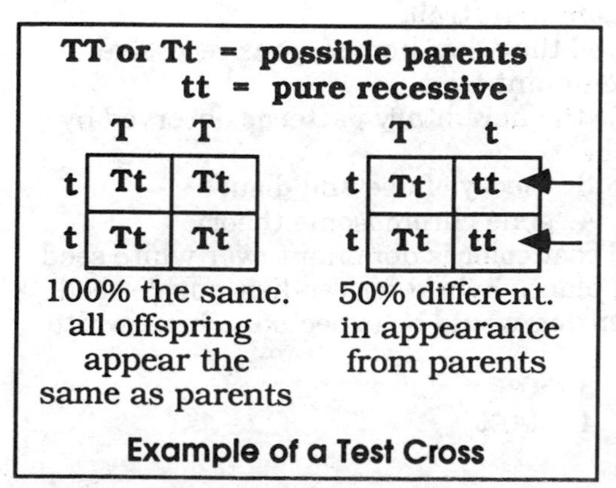

TT or Tt = possible parents
tt = pure recessive
T T
t Tt Tt
t Tt Tt
T t
t Tt tt
t Tt tt
100% the same. all offspring appear the same as parents
50% different. in appearance from parents

Example of a Test Cross

It is sometimes possible to determine the genotypes of the parents (F_1 generation) when the genotypes of the offspring (F_2 generation) are known. To determine the genotype of an individual showing the dominant phenotype, it is crossed with a homozygous-recessive individual. Recessive phenotypes among the offspring indicate a heterozygous parent. This procedure is known as a **test cross**.

Questions

1 Which genetic concept was proposed by Mendel?
1 chromosome nondisjunction
2 independent assortment
3 multiple alleles
4 sex linkage

2 In pea plants, the allele for tallness is dominant over the allele for shortness. Tall pea plants are crossed with short pea plants, and the cross results in 271 tall plants and 268 short plants. Based on this cross, the genotype of the parental tall pea plants can be correctly described as
1 homozygous
2 heterozygous
3 pure recessive
4 pure dominant

3 Two mice with black fur were crossed and produced offspring with brown fur and offspring with black fur. If (*B*) represents the dominant allele for black fur and (*b*) represents the allele for brown fur, which would represent the most probable genotypes of the parental mice?
1 *BB* x *Bb*
2 *BB* x *BB*
3 *Bb* x *Bb*
4 *BB* x *bb*

4 Allelic pairs of genes are located only
1 on homologous pairs of chromosomes
2 in a DNA nucleotide
3 in organisms that are homozygous
4 on non-homologous chromosomes

5 Mendel's discovery that characteristics are inherited due to the transmission of hereditary factors resulted from his
1 careful microscopic examinations of genes and chromosomes
2 dissections to determine how fertilization occurs in pea plants
3 breeding experiments with many generations of fruit flies
4 analysis of the offspring produced from many pea plant crosses

6 If heterozygous black guinea pigs are mated with each other, what percentage of the offspring will be expected to have the same genotype of coat color as their parents?
1 100%
2 75%
3 50%
4 0%

7 Which statement describes how two organisms may show the same trait yet have different genotypes for that phenotype?
1 One is homozygous dominant and the other heterozygous.
2 Both are heterozygous for the dominant trait.
3 One is homozygous dominant and the other homozygous recessive.
4 Both are homozygous for the dominant trait.

8 The mechanism which accounts for the hereditary patterns observed by Mendel is presently described in the
1 heterotroph hypothesis
2 cell theory
3 theory of use and disuse
4 gene-chromosome theory

9 In pea plants, reddish-purple seed coat color is dominant over white seed coat color. In a cross between pea plants hybrid for reddish-purple seed coat color, what percent of the offspring would be expected to have white seed coats?
1 0%
2 25%
3 50%
4 100%

10 In cats, the gene for short hair (*A*) is dominant over the gene for long hair (*a*). A short-haired male cat is mated with a long-haired female, and four kittens are produced, two short-haired and two long-haired. The genotypes of the parent cats are most probably
1 *Aa* x *aa*
2 *AA* x *Aa*
3 *Aa* x *Aa*
4 *AA* x *aa*

11 In pea plants, round seed is dominant over wrinkled seed and yellow seed is dominant over green seed. When hybrid round, yellow plants are crossed, the offspring may include some round, green-seeded plants and some wrinkled, green-seeded plants. This phenomenon illustrates the genetic principle of
1 independent assortment
2 incomplete dominance
3 polyploidy
4 gene mutation

12 In horses, black color is dominant over chestnut color. Two black horses produce both a black and a chestnut-colored offspring. If coat color is controlled by a single pair of genes, it can best be assumed that
1 in horses, genes for hair color frequently mutate
2 one of the parent horses is homozygous dominant and the other is heterozygous for hair color
3 both parent horses are homozygous for hair color
4 both parent horses are heterozygous for hair color

13 When a strain of fruit flies homozygous for light body color is crossed with a strain of fruit flies homozygous for dark body color, all of the offspring have light body color. This illustrates the principle of
1 segregation
2 incomplete dominance
3 dominance
4 independent assortment

14 In pea plants, the trait for smooth seeds is dominant over the trait for wrinkled seeds. When two hybrids are crossed, which results are most probable?
1 100% smooth seeds
2 100% wrinkled seeds
3 75% smooth, 25% wrinkled
4 50% smooth, 50% wrinkled

15 When a white guinea pig is crossed with a black guinea pig, all four of the resulting offspring are black. Assuming that the black coat color condition is the result of a dominant gene, what would represent the genotype of the offspring?
1 *BB*, only
2 *Bb*, only
3 *bb*, only
4 *BB* or *Bb*, only

16 A student tossed 2 pennies at the same time and recorded the following results: both tails, 23; one head and one tail, 53; both heads, 21 Which genotypes represent a cross resulting in approximately the same ratio?
1 *AA* x *aa*
2 *Aa* x *aa*
3 *Aa* x *AA*
4 *Aa* x *Aa*

17 For a given trait, the two genes of an allelic pair are not alike. An individual possessing this gene combination is said to be
1 homozygous for that trait
2 recessive for that trait
3 heterozygous for that trait
4 pure for that trait

18 In cattle, black color is dominant over red color. Which statement describes the offspring produced when a homozygous black bull is mated with several red cows?
1 100% of the offspring will be red.
2 100% of the offspring will be black.
3 75% of the offspring will be black and 25% will be red.
4 50% of the offspring will be black and 50% will be red.

19 Curly hair in humans, white fur in guinea pigs, and needle-like spines in cacti all partly describe each organism's
1 alleles
2 autosomes
3 chromosomes
4 phenotype

20 A student crossed wrinkled-seeded (*rr*) pea plants with round-seeded (*RR*) pea plants. Only round seeds were produced in the resulting plants. This illustrates the principle of
1 independent assortment
2 dominance
3 segregation
4 incomplete dominance

C. Intermediate Inheritance

As molecular genetics and classical genetics begin to merge, long held concepts of classical genetics become less definitive. The actual interactions of gene products shed new light on our concepts of dominance and recessiveness. *Dominance and recessiveness may, in fact, be relative terms.*

Sometimes traits are not clearly dominant or recessive due to the complex nature of gene action. In **intermediate inheritance**, the heterozygous offspring are phenotypically different than their homozygous parents. In all instances of intermediate inheritance, the F_2 offspring of heterozygous parents exhibit a 1:2:1 phenotypic ratio.

There are different degrees of intermediate inheritance.

• **Incomplete dominance**, as exhibited by pink snapdragons and four o'clocks, is a type of intermediate inheritance in which the heterozygous individuals exhibit a phenotype *intermediate* between either homozygous parent.

When a pure white parent (*WW*) is crossed with a pure red parent (*RR*), the resulting offspring are all pink (*RW*).

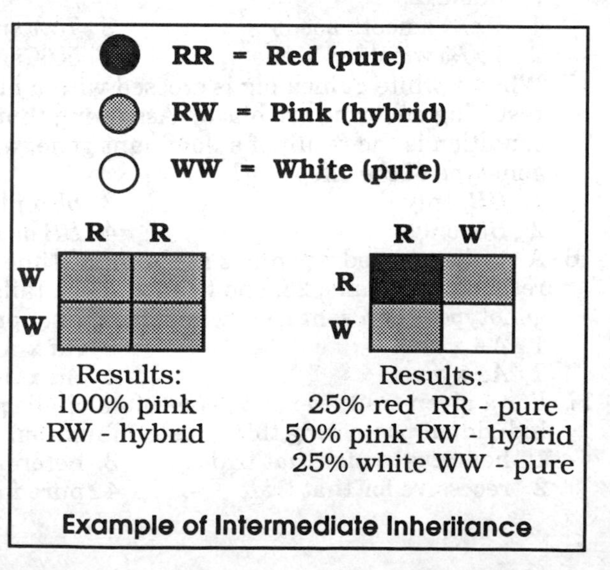

Example of Intermediate Inheritance

- **Co-dominance** involves the expression of two dominant alleles. This results in the simultaneous expression of both alleles in the phenotype of the heterozygous individual. For example, in the inheritance of coat color in roan cattle, the following symbols can be used: C^RC^R represents the genotype of the homozygous red coat. C^WC^W represents the genotype of the homozygous white coat. C^RC^W represents the genotype of the heterozygous roan coat. The coat of the roan animal is composed of a mixture of red and white hairs.

In humans, examples of co-dominance include sickle-cell anemia and blood groups.

D. Independent Assortment

If the genes for two *different* traits are located on *different* chromosome pairs (nonhomologous chromosomes), they segregate *randomly* during meiosis. Therefore, they may be *inherited independently* of each other producing much of the genetic variation observed in living organisms.

E. Gene Linkage

All genes on the same chromosome make up a linkage group. If the genes for two different traits (**non-allelic genes**, i.e. red hair and freckles) are located on the same chromosome, they are said to be linked. These genes are most often *inherited together*, resulting in an offspring with red hair and freckles However, not all individuals with red hair, have freckles. Exceptions can be explained by crossing over.

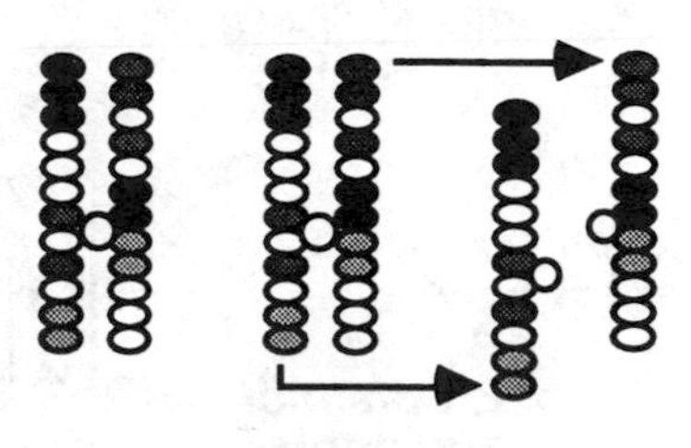

Gene Linkage

F. Crossing Over

There are exceptions to linked gene inheritance. For example, not all persons having red hair have freckles. This "variation from the expected or normal" may be due to **crossing-over**. During synapsis in the first meiotic division, the chromatids in a homologous pair of chromosomes often twist around each other, break, exchange segments, and then rejoin. This results in a rearrangement of linked genes and increases the variability of offspring.

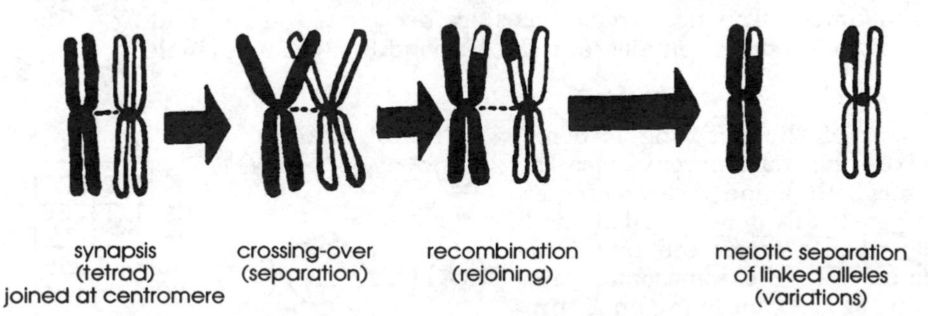

Crossing-over During Meiosis

G. Multiple Alleles

In some instances, an observed pattern of heredity cannot be explained on the basis of a single pair of alleles. Experimental evidence indicates that in such patterns **multiple alleles** are involved. In this pattern no more than two of these alleles for the given trait may be present within each cell.

ABO blood types is an example of the expression of multiple alleles. In a human population, the inheritance of the ABO blood group may be explained using a model that employs three alleles (I^A, I^B and i). In this model, alleles I^A (type "A") and I^B (type "B") are co-dominant with each other and i (type "O") is recessive to both I^A and I^B.

Genotypes associated with each blood type in the ABO group are:

Genotype	Phenotype	Antigens Made
$I^A I^A$	A	A *only*
$I^A i$	A	A *only*
$I^B I^B$	B	B *only*
$I^B i$	B	B *only*
$I^A I^B$	AB	*Both* A *and* B
ii	0	*Neither* A *nor* B

Example of Blood Type Crosses

	B	B
A	AB	AB
A	AB	AB

	B	O
A	AB	AO
O	BO	OO

	A	B
A	AA	AB
B	AB	BB

	A	O
B	AB	BO
B	AB	BO

	O	O
A	AO	AO
A	AO	AO

	O	O
B	BO	BO
O	OO	OO

H. Sex Determination

Diploid cells of many organisms contain two types of chromosomes: **autosomes** ("body" traits) and **sex chromosomes** ("male" and "female" traits). In each human cell, there are 22 pairs of **autosomes** and one pair of **sex chromosomes**. The sex chromosomes are designated as "*X*" and "*Y*." The ***XX*** condition produces females, and the ***XY*** condition produces males.

The sex of the offspring is determined by the male parent since the male has both *X* and *Y* chromosomes. Sex is genetically determined at fertilization when a sperm cell containing either an *X* or *Y* chromosome unites with an egg cell containing an *X* chromosome.

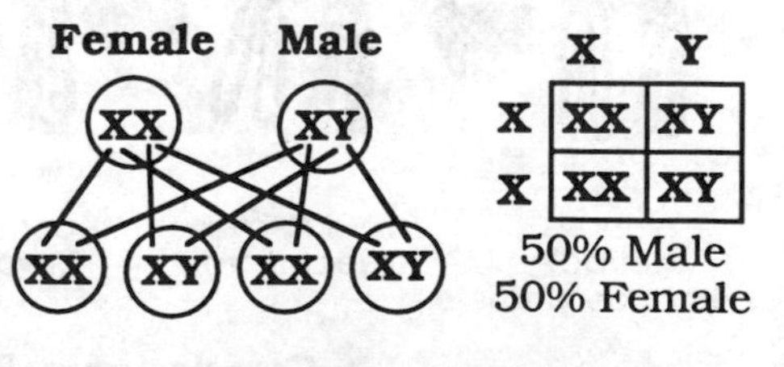

Sex Determination

I. Sex Linkage

Thomas Hunt Morgan's work with *Drosophila* demonstrated that genes for certain traits are located on the *X* (sex) chromosomes. These genes are said to be **sex-linked** and do not appear to have corresponding alleles on the *Y* (sex) chromosome.

Many sex-linked genes are recessive. Therefore, they are expressed more frequently in males than in females. Since the male has only one *X* chromosome, if that *X* chromosome is carrying the recessive trait then he is "pure" for the trait and expresses it. Since the female has two *X* chromosomes, if only one of her *X* chromosomes has the recessive trait, she appears normal and does not express the trait.

However, she is a **carrier** for that trait, meaning that she is likely to produce half of her eggs with the recessive trait. In order for a female to express the sex-linked trait, she must carry the recessive gene on both of her *X* chromosomes. Hemophilia and color-blindness are examples of sex-linked traits that are found in humans.

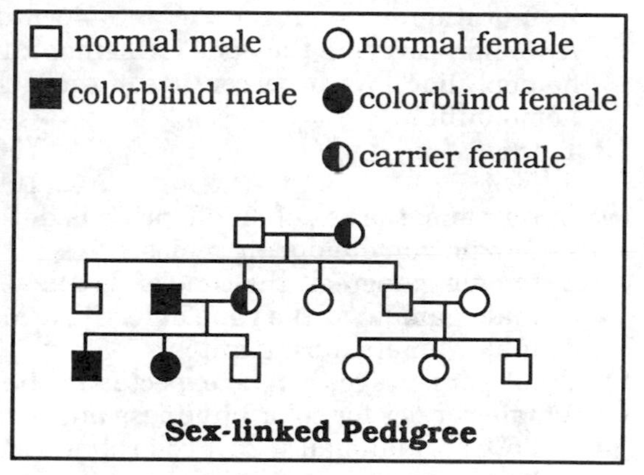

Sex-linked Pedigree

Questions

1 If two roan cattle are crossed, what percent of the offspring are expected to show the parental phenotype for coat color?
1 25% 2 50% 3 75% 4 100%

2 A man who has blood type AB marries a woman who has blood type B. This couple would not normally have a child with which genotype?
1 I^Ai
2 I^BI^B
3 I^AI^B
4 ii

3 When a colorblind woman marries a male with normal vision, all their daughters have normal vision and all their sons are colorblind. This is an example of which type of inheritance?
1 multiple alleles
2 co-dominance
3 sex linkage
4 autosomal dominance

4 What percent of the male offspring produced by a father with normal vision and a colorblind mother are expected to be colorblind?
1 0%
2 33%
3 75%
4 100%

5 White mice with fluffy (tufted) tails are mated with brown mice with hairless tails. In the F_2 generation, some of the white offspring have hairless tails, while some of the brown offspring have tufted tails. These results best demonstrate
1 independent assortment
2 sex linkage
3 gene mutation
4 intermediate inheritance

6 The letters in the following crosses represent parental blood types. Which cross could produce offspring that represent all four blood types of the ABO blood group?

1 I^AI^A x I^AI^B
3 I^AI^B x I^AI^B
2 ii x I^Ai
4 I^Ai x I^Bi

7 The chances of a YY chromosome combination occurring in humans as a result of normal meiotic division and normal gametic fusion is

1 0%
3 50%
2 24%
4 100%

8 A cross of a red cow with a white bull produces all roan offspring. This type of inheritance is known as

1 co-dominance
3 sex linkage
2 mutation
4 multiple alleles

9 A woman carrying the gene for hemophilia marries a man who is a hemophiliac. What percentage of their children can be expected to have hemophilia?

1 0%
3 75%
2 50%
4 100%

10 In humans, the sex of an offspring is determined by the

1 female gamete during meiosis
2 female gamete at the time of fertilization
3 male gamete at the time of fertilization
4 male gamete during mitosis

11 What percentages can be expected in the offspring of a cross between a female carrier for color blindness and a male with normal color vision?

1 25% normal males, 25% colorblind males, 25% normal females, 25% carrier females
2 25% normal males, 25% colorblind males, 25% carrier females, 25% colorblind females
3 75% normal males, 25% carrier females
4 50% colorblind males, 50% colorblind females

12 Geneticists have observed that fruit flies that commonly inherit vestigial wings also inherit lobed eyes. Observations such as this have helped to develop the genetic concept known as

1 segregation
3 gene linkage
2 dominance
4 crossing-over

13 When red coat cattle (C^RC^R) are crossed with white coat cattle (C^WC^W), all the offspring are roan coat. How many different genotypes can be produced when these roan coat cattle are crossed with white coat cattle?

1 1
3 3
2 2
4 4

14 Two genes are linked if they are

1 located on separate sex chromosomes
2 members of an allelic pair
3 located on the same chromosome
4 able to segregate at random

15 In humans, sex is normally determined at fertilization by

1 1 pair of sex chromosomes
3 2 pairs of sex chromosomes
2 11 pairs of autosomes
4 22 pairs of autosomes

III. Mutations

Mutations – Terms & Concepts

Mutations
Chromosomal Alterations
Gene Mutations (DNA)
Mutagenic Agents
Radiation - Chemicals
Inbreeding and Hybridization
Nondisjunction
Polyploid Condition
Translocation
Artificial Selection

Changes in the genetic material are called **mutations**. If these mutations occur in sex cells they may be transmitted (passed on) to the next generation. Mutations occurring *only* in the body cells may be perpetuated in the individual but will *not* be passed on to the offspring by sexual reproduction.

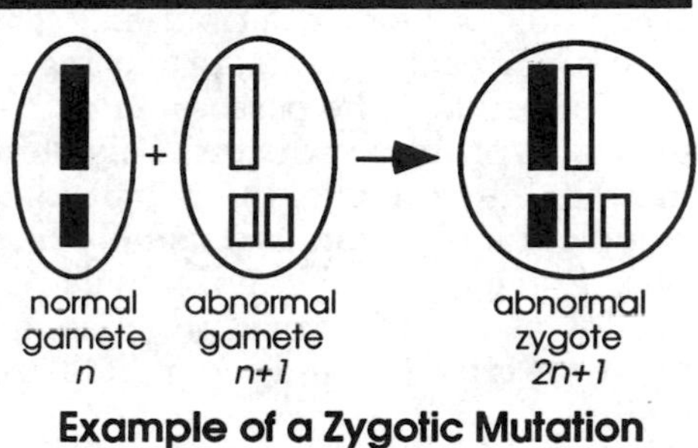

Example of a Zygotic Mutation

A. Types of Mutations

Mutations may be classified as **chromosomal alterations** or **gene mutations**.

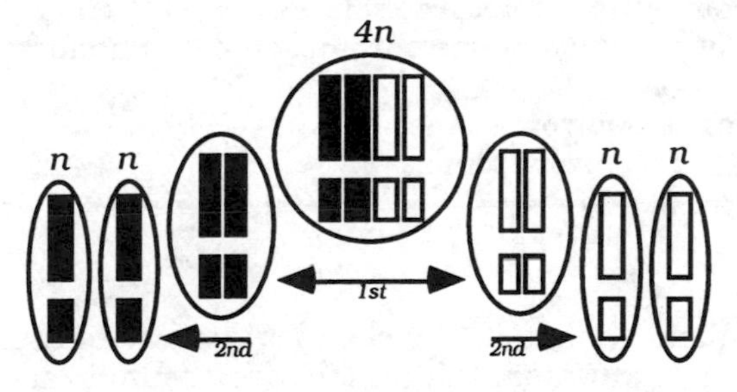

Normal Separation During Meiosis

1. Chromosomal Alterations

A **chromosomal alteration** is a change in the *number* of chromosomes or in the *structure* of the chromosomes. The effects of chromosomal alterations are often visible in the phenotype of an organism because *many* genes are usually involved.

Change in Chromosome Number

During meiosis, pairs of sister chromatids separate during Meiosis I. This separation is known as **disjunction**. However, sometimes a pair of sister chromatids *fails to separate* from its other sister pair (**nondisjunction**). This results in gametes with more (or less) than their normal (*n*) chromosome number. If these gametes are involved in fertilization, the resulting zygote may have more (or less) than the normal (*2n*) chromosome number.

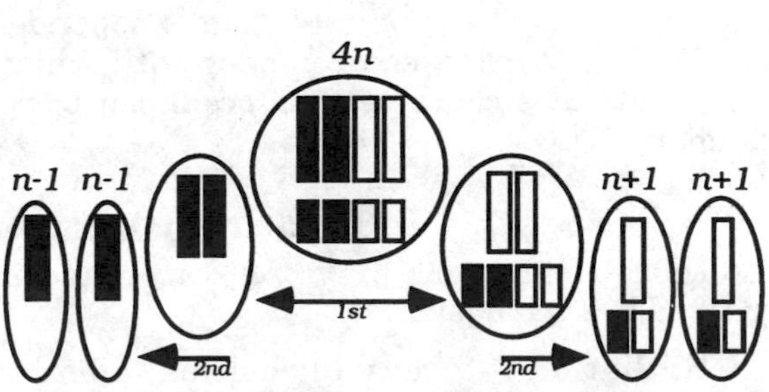

Nondisjunction During Meiosis

Often this abnormal condition results in the death of the affected zygote. However, when the zygote is viable, the resulting offspring develops with unusual characteristics.

For example, in humans, **Down's syndrome** usually results from the possession of an extra chromosome (*2n +1*). This is usually due to the nondisjunction of chromosome number 21 in one of the parents.

Occasionally, the disjunction of a complete set of chromosomes (*2n*) fails to occur during gamete formation. The resulting *2n* gamete sometimes fuses with a normal *n* gamete, producing a *3n* zygote. If two *2n* gametes fuse, a *4n* zygote results. The possession of extra whole sets of chromosomes is known as a **polyploid condition**. Polyploid is rather common in plants but rare in animals. In animals, the polyploid condition most often is lethal. In plants, polyploid individuals are usually larger or more vigorous than diploid varieties. Some polyploid individuals are sterile (seedless). Examples include strains of cotton, wheat, potatoes, alfalfa, apples, tobacco and zinnias. Polyploid watermelons and cucumbers are nearly seedless.

Change in Chromosome Structure

Structure alterations in chromosome composition may result from random breakage and recombination of chromosome parts.

Extended Area: ***Transmission of Traits***

Some examples of changes in chromosome structure include:
- **Translocation** is the transfer of a section of one chromosome to a nonhomologous chromosome.
- **Addition** is the *gain* of a portion of a chromosome.
- **Deletion** is the *loss* of a portion of a chromosome.

2. Gene Mutations

A **gene mutation** involves a random change in the chemical nature of the genetic material (DNA). When DNA is mutated, the control over cellular activities is changed. This is likely to cause alterations in the phenotype of the organism. While the effects of some gene mutations, such as **albinism**, are obvious, the effects of other gene mutations may not be readily noticed.

Although most gene mutations cause changes that are detrimental for the organism or produce no obvious effect, some changes may be beneficial. The *adaptive value* of a gene mutation is dependent upon the nature of the mutation and the type of environment with which the organism interacts. These gene mutations could help an organism to better survive in a changed environment.

B. Mutagenic Agents

Although mutations may occur spontaneously, their incidence may be increased by such agents as:

- **Radiation** — x-rays, ultraviolet, radioactive substances, and cosmic rays
- **Chemicals** — formaldehyde, benzene, PCB's, and asbestos fibers

IV. Genetic Applications: Animal And Plant Breeding

Artificial selection, including **inbreeding** and **hybridization**, and the maintenance of desirable mutations by **vegetative propagation** (means) are methods used by scientists to improve, produce, and maintain new varieties of animals and plants.

In animals, artificial selection through inbreeding and cross-breeding (hybridization) is used to produce and maintain certain desired traits. Breeders of dogs, cattle, and horses often use artificial selection. Many plants, such as seedless oranges, hybrid roses, and apple varieties, are produced and maintained through artificial selection.

V. Interaction Of Heredity And Environment

The environment interacts with genes in the development and the expression of inherited traits. The relationship between gene action and environmental influence has been studied in many organisms. Examples include:

- **Effect of light** on chlorophyll production. Without light, most plants only produce a yellow pigment. However, in the presence of light, the plant will develop chlorophyll, turn green, and be able to carry on photosynthesis.

- **The effect of temperature** on hair color in the Himalayan rabbit. In their normally Arctic environment, the hair of Himalayan rabbits is white with black only on their extremities. In a warm climate, all of their hair develops as white. In experiments where ice was applied to the normally white hair areas of the rabbit, the hair grew black, indicating that temperature plays a role in the production of hair color in the Himalayan rabbit.

- **Identical twin studies**. Since identical twins develop from a single zygote that separates into two cell masses during cleavage, their genetic information is the same, and they should develop in an identical manner. However, studies indicate that identical twins often show changes in their phenotype as they mature. These changes have been shown to be dependent upon the environment.

Questions

1 Occasionally during meiosis, a single homologous chromosome pair may fail to separate. A human gamete produced by such a nondisjunction would have a chromosome number of
1 23 2 24 3 25 4 26

2 Which may occur in meiotic division I of a primary sex cell?
 1 fertilization 3 crossing-over
 2 polyploidy 4 differentiation

3 Cosmic rays, X-rays, ultraviolet rays, and radiation from radioactive substances may function as
 1 pollinating agents 3 plant auxins
 2 mutagenic agents 4 animal pigments

4 In a particular variety of corn, the kernels turn red when exposed to sunlight. In the absence of sunlight, the kernels remain yellow. Based on this information, it can be concluded that the color of these corn kernels is due to the
 1 effect of sunlight on transpiration
 2 law of incomplete dominance
 3 principle of sex linkage
 4 effect of environment on gene expression

5 The Himalayan rabbit has white fur over most of its body, but it has black fur on its tail, ears, and the tips of its legs and nose. Two rabbits that are homozygous for this hair pattern are mated. When their offspring are exposed to normal temperatures, they exhibit normal Himalayan hair pattern. However, when their offspring are exposed to low temperatures (10°C), they have black hair covering their entire bodies. This illustrates
 1 that mutations can be caused by heat radiation
 2 that the traits of an organism are determined by its genes
 3 the importance of the environment in gene expression
 4 the law of incomplete dominance

6 Substances that cause a chemical change in the DNA of a cell are known as
 1 glycogens 3 chromatids
 2 mutagens 4 chromosomes

7 Race horses show many variations from the wild horse ancestors from which they were derived. It is most likely that these variations between race horses and their ancestors are due to
 1 use and disuse of organs 3 gene cloning
 2 artificial selection 4 chromosomal nondisjunction

8 The process by which homologous chromosomes exchange segments of DNA is
 1 segregation
 2 crossing-over
 3 fertilization
 4 independent assortment

9 During egg cell production in a human female, the 21st pair of chromosomes may fail to separate. This failure to separate is known as
 1 crossing-over 3 polyploidy
 2 gene mutation 4 nondisjunction

10 Which term best describes most mutations?
 1 dominant and disadvantageous to the organism
 2 recessive and disadvantageous to the organism
 3 recessive and advantageous to the organism
 4 dominant and advantageous to the organism

11 Some studies of identical human twins show that their IQ, height, and talents may be different. The best explanation for these differences is that
  1 the environment interacts with genes in the development and expression of inherited traits
  2 heredity and environment have no influence on the expression of phenotypes
  3 the genotype of twins depends on the interaction of diet and hormone control
  4 people are considered identical if at least half their genes are the same

12 Which process most likely accounts for the sudden appearance of a white-flowered plant in a population of pure self-pollinating red-flowered plants?
  1 mutation
  2 nondisjunction
  3 segregation
  4 independent assortment

13 Animal breeders often cross members of the same litter in order to maintain desirable traits. This procedure is known as
  1 hybridization
  2 natural selection
  3 inbreeding
  4 vegetative propagation

14 Corn seeds were germinated in a dark room. Not one of the seedlings was green. When placed in the light, 75 percent of these seedlings turned green. Which conclusion about chlorophyll production in corn plants can most reasonably be drawn from this information?
  1 Light is the only factor that controls the production of chlorophyll.
  2 Darkness is the only factor that prevents the production of chlorophyll.
  3 Light and vitamins are necessary for chlorophyll production.
  4 Light and some other factor are necessary for chlorophyll production.

15 The alteration of genetic material resulting in the appearance of a new trait which continues to appear in a species is known as
  1 disjunction
  2 segregation
  3 linkage
  4 mutation

VI. Human Heredity

Human Genetics – Terms & Concepts

Human Genetics	Genetic Disorders
Counseling - Screening	Down's Syndrome - Phenylketonuria
Karyotyping - Amniocentesis	Sickle Cell Anemia - Tay Sachs

The principle of genetics appears to hold for all organisms including humans. The acquisition of knowledge of human genetics has been limited because humans are not suitable subjects for experimentation.

An increased knowledge of human genetics has resulted from the expansion of the field of **genetic counseling**. Genetic counseling involves discussions often between a physician and the family and include probabilities of inheritance patterns, predictions of genetic disorders, and family planning.

Extended Area: ***Transmission of Traits***

A. Techniques for Detection

The presence of many genetic disorders can be detected either before or after birth. In some instances, carriers of genetically-related disorders may also be identified. These genetic disorders are often detected through genetic counseling and the use of the following techniques.

- **Screening** is the chemical analysis of body fluids such as blood and urine. Analysis may indicate the presence of chemicals associated with genetically-related disorders.
- **Karyotyping** is the preparation of an enlarged photograph showing paired homologous chromosomes from a cell. Karyotyping may show such chromosome abnormalities as an extra or missing chromosome.
- **Amniocentesis** involves the removal of amniotic fluid for chemical and/or cellular analysis. This technique can show the chromosome content of the cells of the embryo. Down's syndrome may be identified by this technique.

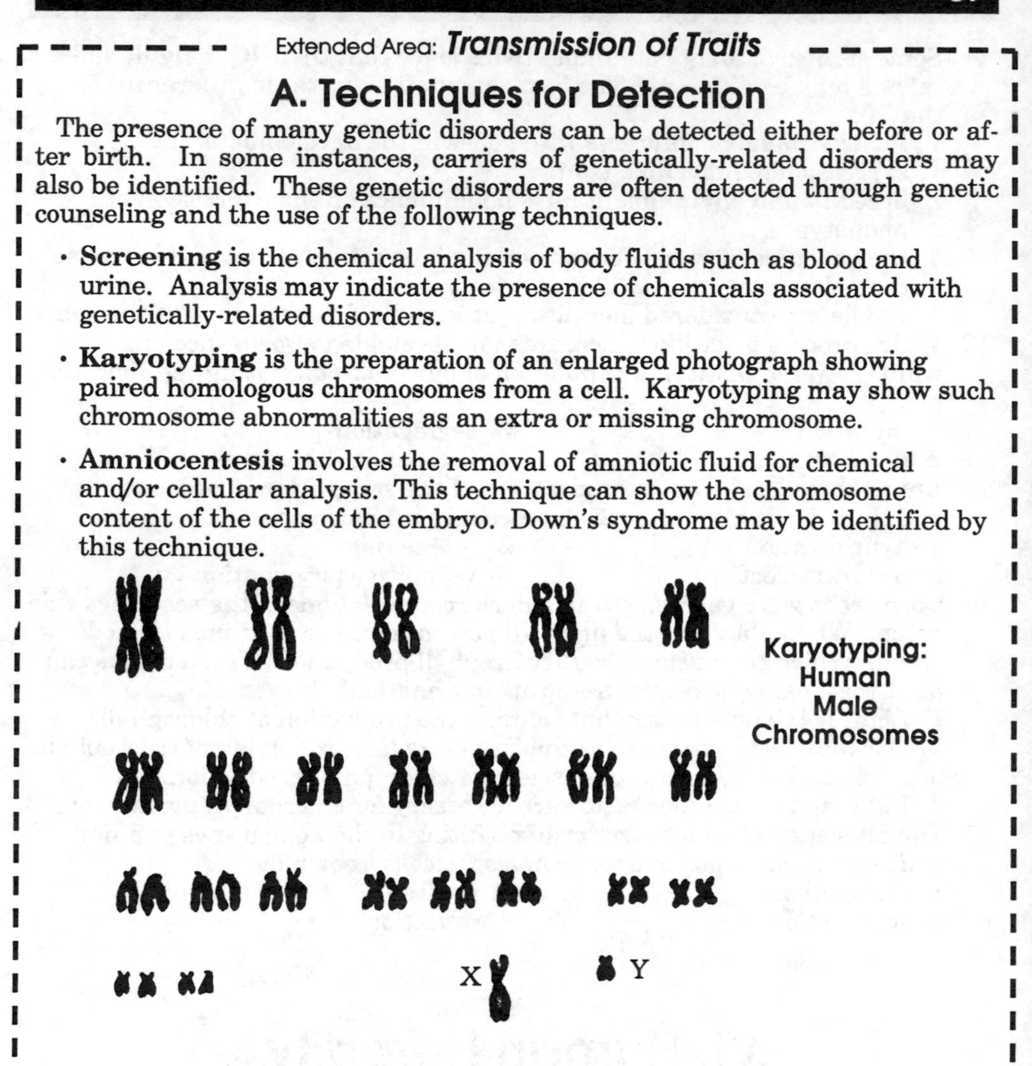

Karyotyping: Human Male Chromosomes

B. Genetically Related Disorders

Some genetically related disorders in humans include the conditions of Phenylketonuria, sickle-cell anemia, and Tay-Sachs.

Phenylketonuria (PKU). PKU is a condition characterized by the development of mental retardation and has been shown to occur when an individual is homozygous with respect to a recessive mutant gene. The symptoms of the disorder apparently result from the inability of a gene to synthesize a *single* enzyme necessary for the normal metabolism of *phenylalanine*. Urine analysis of newborns allows for proper dietary treatment to prevent the mental retardation associated with the disorder.

Sickle-Cell Anemia. Sickle-cell anemia is a homozygous condition resulting in the formation of abnormal hemoglobin and sickled red blood cells. It is characterized by severe pain due to obstructed blood vessels and anemia caused by the fragility of red blood cells. This disorder is more likely to occur among individuals of African descent. Heterozygous and homozygous individuals may be detected by blood screening. An afflicted fetus may be detected by amniocentesis.

Tay-Sachs. Tay-Sachs is a recessive genetic disorder characterized by the malfunctioning of the nervous system, caused by the deterioration of nervous tissue. This deterioration is due to the accumulation of fatty material as a result of the inability to synthesize a specific enzyme. This fatal disorder is more likely to occur among Jewish people of Central European descent. Carriers may be detected through blood screening. Chemical analysis of amniotic fluid can detect an afflicted fetus.

Questions

1 Down's syndrome is a genetic disorder caused by the presence of an extra chromosome number 21 in the body cells of humans. This extra chromosome most likely is a result of
 1 mitotic cell division in brain
 2 development of infertile egg
 3 multiple allelic pairs of genes
 4 nondisjunction during meiosis

2 All the children of a hemophiliac male and a normal female are normal with respect to blood clotting. However, some of their grandsons are hemophiliacs. This is an example of the pattern of heredity known as
 1 sex determination
 2 incomplete dominance
 3 sex-linkage
 4 multiple alleles

3 Which is a genetic disorder in which abnormal hemoglobin leads to fragile red blood cells and obstructed blood vessels?
 1 phenylketonuria
 2 sickle-cell anemia
 3 leukemia
 4 Down's syndrome

4 Human disorders such as PKU and sickle-cell anemia, which are defects in the synthesis of individual proteins, are most likely the result of
 1 gene mutations
 2 nondisjunction
 3 crossing-over
 4 polyploidy

5 Which is not a current technique for genetic disorder detection?
 1 screening
 2 amniocentesis
 3 karyotyping
 4 blood typing

6 Which technique can be used to examine the chromosomes of a fetus for possible genetic defects?
 1 pedigree analysis
 2 karyotyping
 3 analysis of fetal urine
 4 cleavage

7 Genetically related diseases are generally the result of
 1 a homozygous recessive gene state
 2 a heterozygous recessive gene state
 3 a homozygous dominant gene state
 4 a heterozygous dominant gene state

8 Genetic counseling can be best used to
 1 cure genetic defects
 2 produce genetically perfect offspring
 3 avoid unwanted genetic diseases in offspring
 4 correct a defective genotype

9 Abnormal hemoglobin is a characteristic of which genetic disorder?
 1 Tay-Sachs
 2 Phenylketonuria
 3 sickle-cell anemia
 4 Down's syndrome

10 The inability to synthesize a specific enzyme is generally due to which genetic "error"?
 1 crossing-over
 2 a gene mutation
 3 nondisjunction
 4 recombinant DNA

VII. Modern Genetics

Modern Genetics – Terms & Concepts

Modern Genetics
 Deoxyribonucleic Acid (DNA)
 Ribonucleic Acid (RNA)
Nucleotide
 Phosphate - 5 carbon sugar
 Nitrogenous Bases
  Adenine - Thymine
  Guanine - Cytosine
Watson - Crick Model
 Double Helix - DNA Replication
RNA - Uracil, Ribose
 mRNA, tRNA and rRNA
Genetic Code - Triplet (Codon)
 Amino Acid Sequence
Protein Synthesis
"One Gene - One Polypeptide"
Gene Mutations
 Deletion, Addition, Substitute
Genetic Research
 Cloning - Genetic Engineering
Population Genetics
 Population - Gene Pool
 Gene Frequency
 Hardy Weinberg Principle

A. DNA as the Hereditary Material

In the 1940's, biochemists Oswald Avery, Colin MacLeod, and Maclyn McCarty found that **deoxyribonucleic acid (DNA)** is the genetic material which replicates and is passed from generation to generation. Ten years of complicated chemical studies followed. Then in 1952, biologist Alfred Hershey and Martha Chase concluded that DNA controls cellular activity by influencing the production of enzymes.

1. DNA Structure

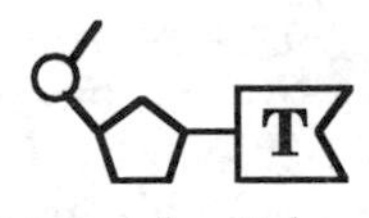

DNA, found in the nucleus of cells, is a *polymer*. It is a large molecule consisting of thousands of smaller, repeating units known as **nucleotides**.

DNA Nucleotide
A DNA nucleotide is composed of three parts:

- **Phosphate group** made up of oxygen, hydrogen, and phosphorus.
- **Deoxyribose** (5-carbon sugar) molecule made up of carbon, hydrogen, and oxygen.
- **Nitrogenous base** made up of carbon, oxygen, hydrogen, and nitrogen. There are four: **adenine**, **thymine**, **guanine**, or **cytosine**.

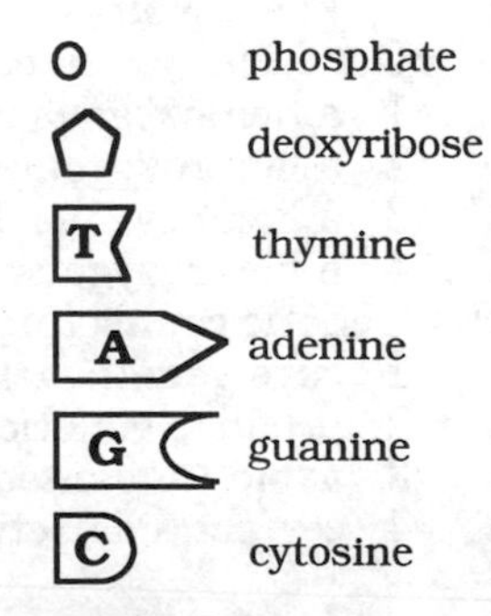

Watson-Crick Model

In 1953, James Watson, an American biologist, and Francis Crick, a British biophysicist, developed a model of the DNA molecule. This most important scientific breakthrough won them the Nobel Prize and opened the door to modern genetics. In this model the DNA molecule consists of two complementary chains of nucleotides.

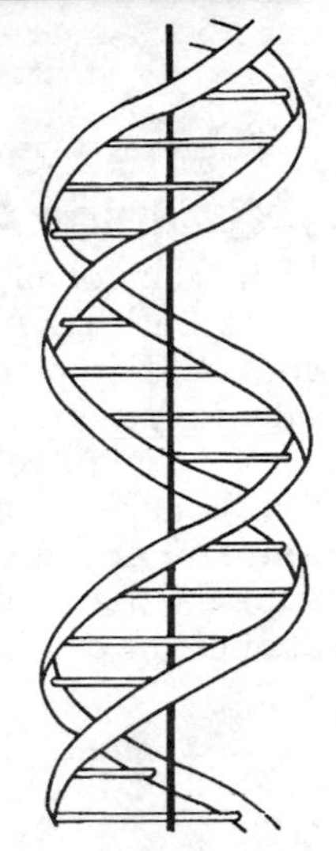

Double Helix Model of DNA

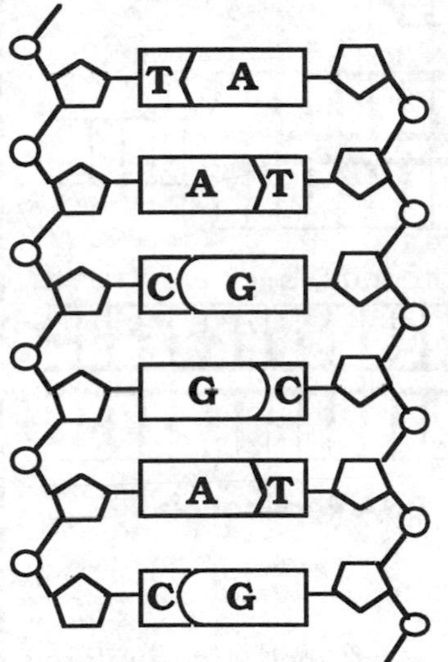

DNA Double Helix
("untwisted")

- The DNA molecule has a "ladder" type organization. The uprights of the "ladder" are composed of alternating phosphate and deoxyribose molecules.

  Each rung of the "ladder" is composed of bases held together by relatively weak hydrogen bonds.

  Base pair combinations are restricted to:

- adenine (A) - thymine (T), and guanine (G)-cytosine (C); thus the combinations: A—T and C—G.

- This "ladder" is thought to be twisted around a protein "framework" in the form of a **double helix**.

Extended Area: ***Transmission of Traits***

The Watson-Crick model can be used to explain the principal actions of genes:

- Maintaining characteristic (genotype) continuity from generation to generation by means of the replication of the genetic code.
- Controlling cellular activity by controlling the production of specific enzymes; therefore, the phenotype of the organism.

2. DNA Replication

The exact self-duplication of the genetic material is accomplished through DNA replication. DNA is believed to replicate in the following manner during the processes of mitosis and meiosis:

- Double-stranded DNA unwinds and "unzips" along weak hydrogen bonds between the base pairs.
- Free nucleotides in the nucleus are incorporated by each of the unwound strands of DNA, in sequence. This forms two new double strands of DNA which are identical to each other and to the original DNA molecule. Exact self-duplication is accomplished.

3. Gene Control of Cellular Activities

The control of cellular activities involves both DNA and RNA. DNA provides the information needed by the cells to produce the specific enzymes needed for all cellular activities. RNA functions to carry this information from the nucleus to the cytoplasm and helps in the protein synthesis mechanism of the cell.

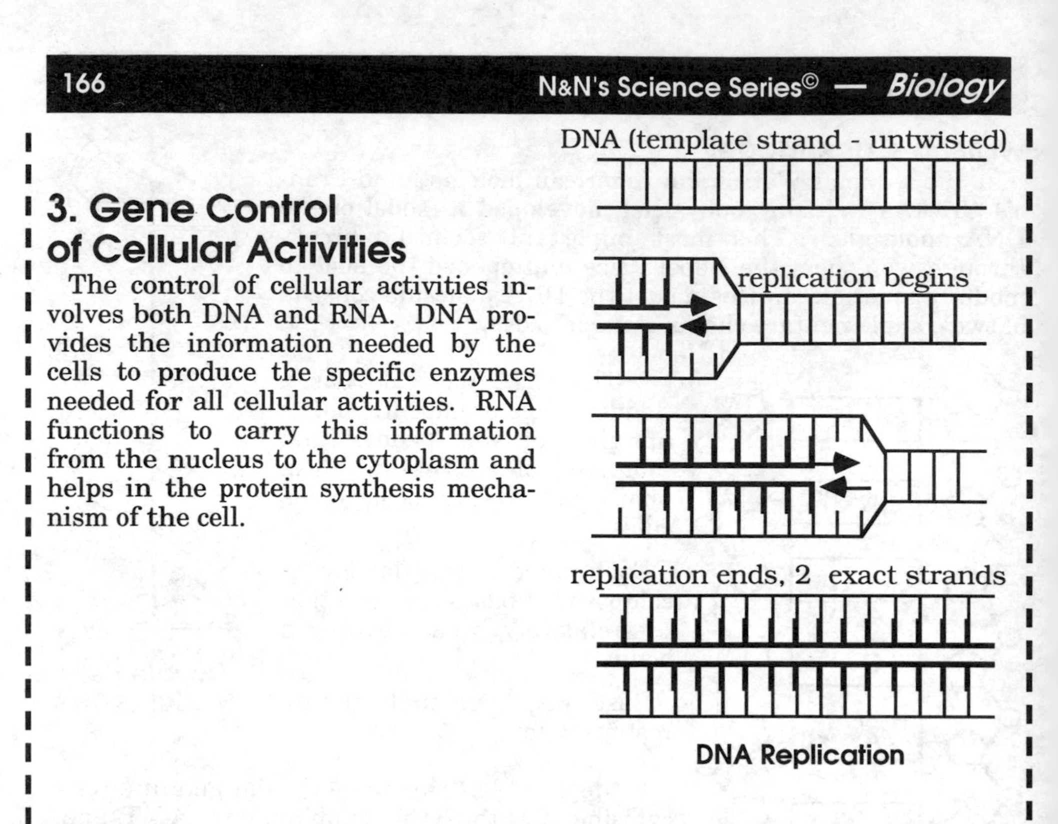

DNA Replication

RNA

Ribonucleic acid (**RNA**), like DNA, is composed of nucleotide building blocks. However, there are three major differences between the structure of DNA and RNA molecules. In RNA:

- **Ribose** is substituted for deoxyribose.
- **Uracil** is substituted for thymine.
- There is a **single chain** of nucleotides.

There are three types of RNA:

- Messenger RNA (**mRNA**) — "carries" the DNA message from the DNA in the nucleus to the sites of protein synthesis in the cytoplasm (ribosomes).
- Transfer RNA (**tRNA**) — "transports" amino acids within the cytoplasm to the ribosomes.
- Ribosomal RNA (**rRNA**) — makes up the "identification" code of each ribosome for specific protein manufacturing.

RNA Nucleotide Symbols and Structure

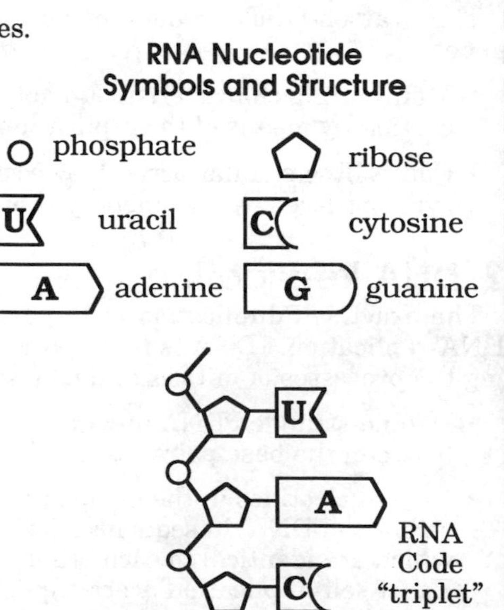

Genetic Code

Evidence strongly suggests that a **genetic code** exists. This code contains the information for the sequence of amino acids in a particular protein. The code is present in the messenger RNA molecules which are complementary to DNA molecules.

The RNA code is a **triplet code** (**codon**), having just three nitrogen bases, complimentary to and based on various sequences of three bases in the DNA molecules.

Protein Synthesis

DNA serves as a **template** for the synthesis of messenger RNA from free RNA nucleotides. The **messenger RNA** molecules, carrying a specific code determined by the base sequence of the original DNA molecules, move from the nucleus to the cytoplasm where they become associated with ribosomes composed in part of ribosomal RNA.

DNA
(template strand - untwisted)
transcripted
RNA
Messenger RNA
DNA
(template strand - untwisted)

RNA Transcription

During the process of **translation**, it is generally thought that **ribosomal RNA** temporarily binds messenger RNA to the ribosomes insuring that a specific protein will be manufactured at a specific ribosome.

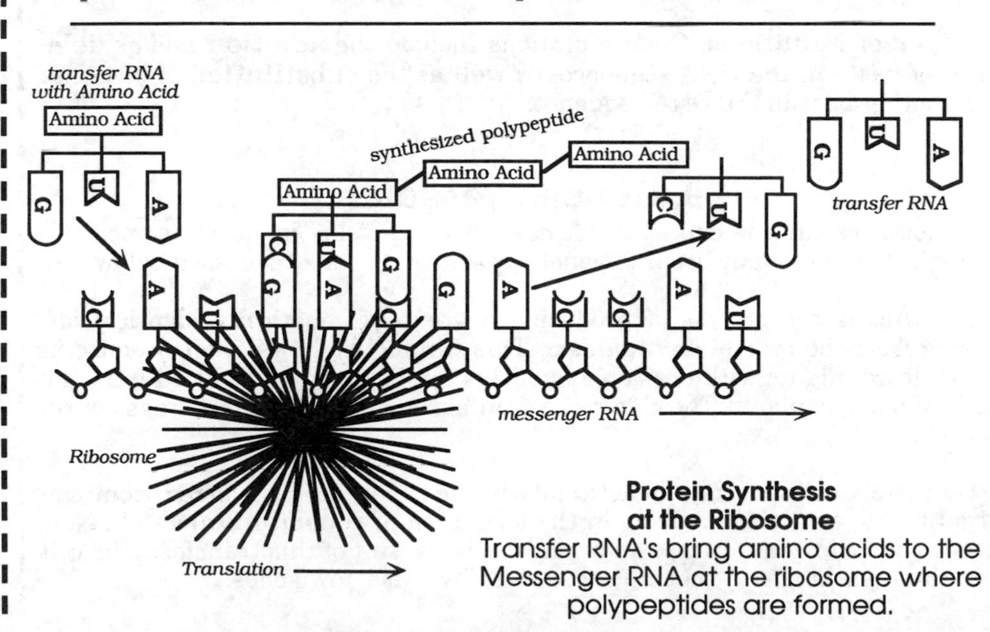

Protein Synthesis at the Ribosome
Transfer RNA's bring amino acids to the Messenger RNA at the ribosome where polypeptides are formed.

Specific **transfer RNA** molecules (arranged as **anticodons**, which are complimentary to mRNA codons) pick up and transfer to the ribosomes specific amino acid molecules found in the cytoplasm. At the ribosomes, each specific transfer RNA molecule bonds to a particular codon. Particular peptide chains are formed as the amino acids associated with the transfer RNA molecules are bonded in a sequence determined by the base sequence of the messenger RNA.

The "One Gene - One Polypeptide Hypothesis"

A hypothesis called the **"one gene-one enzyme hypothesis"** proposed that the synthesis of each enzyme in a cell was governed by the action of a single gene (a specific sequence of DNA nucleotides.) However, the name of this hypothesis has been changed to the **one gene -one polypeptide hypothesis** since it is now known that a single enzyme may be composed of several polypeptides and the synthesis of each polypeptide is governed by a different gene. A modern definition of the **gene** is the sequence of nucleotides in a DNA molecule necessary to synthesize a polypeptide.

Individuality of Organisms as Related to Their DNA

Since the sequence of nucleotides in DNA determines the sequence of nucleotides in messenger RNA, DNA ultimately determines the sequence of the amino acids in specific proteins. The specificity of enzymes is dependent on their protein makeup, and, since the individuality of a cell is largely a function of the enzymes it possesses, it is evident that DNA determines the individuality of an organism.

Gene Mutations

Description of Mutations. Gene mutations may be interpreted biochemically as any change in the base sequence of an organism's DNA.

Types of Mutations. Gene mutations include the **addition** and/or **deletion** of bases in the DNA sequence, as well as the **substitution** of one base for another base in the DNA sequence.

B. Genetic Research

Genetic research is one of the fastest growing fields in all of science. Although there are many areas of genetic research, two are discussed below.

Cloning is the process of producing a group of genetically identical offspring from the cells of an organism. This technique shows great promise in agriculture. Plants with desirable qualities can be rapidly produced from the cells of a single plant. The cloning of animals is still in the early stages of research.

Genetic Engineering. Genetic information may be transferred from one organism to another, resulting in the formation of **recombinant DNA**. New genes can be introduced into an organism as a result of this transfer. The cell can then synthesize the chemical coded for by these new genes.

Genetic engineers have been successful in taking human genetic information and transferring these genes into certain bacterial cells. The "genetically engineered" bacterial cells reproduce in great numbers, producing the human chemicals useful in treating human disorders such as hormonal deficiencies, diabetes, and auto-immune diseases. Examples of these chemicals include: human growth hormone (HGH), interferon, and insulin. In addition, some helpful dietary products have been produced. For example, aspartame is an artificial no-calorie sweetener manufactured by Nutrasweet.

In the future, genetic engineering has potential for the correction of genetic defects and the development of agriculturally desirable plants and animals.

C. Population Genetics

The study of factors which affect gene frequencies in populations of sexually reproducing organisms is known as **population genetics**.

- A **population** is the basic unit for population genetics and includes all members of a species inhabiting a given location. For example, a population would include all of the sunfish living in a pond or all of the goldenrod in a country field.

- The **gene pool** of a population consists of the sum total of all the heritable genes for the traits in a given population. For example, in a population of humans, all of their characteristics make up the gene pool, hair, skin, and eye color, etc.

- The **gene frequency** is the percentage of each allele for a particular trait in a population. Knowing how often a particular gene appears in a population, it is possible to predict how often that characteristic will be seen in the offspring of that population. It is also possible to predict what part of that population is homozygous or heterozygous for that particular trait.

The Hardy-Weinberg Principle

In the early 1900's, Godfrey Hardy, a British mathematician, and Wilhelm Weinberg, a German physician, studied populations of sexually reproducing individuals. Their studies resulted in the formulation of the **Hardy-Weinberg Principle**. This principle states that the gene pool of a population tends to remain stable (that is, gene frequencies remain constant) if certain conditions are met.

These conditions for gene pool stability include:

- **Large populations** with equal numbers of sexes
- **Random mating** between the sexes
- **No migration** or "mixing" with other populations
- **No mutation** of any chromosome or gene within the population

These conditions should be considered as "ideal" for keeping change out of a population. However, this "ideal" does not seem to exist in a natural environment. Therefore, since some or all of the conditions for gene pool stability may not be met, the stability of a population suggested by the Hardy-Weinberg Principle rarely if ever occurs. The unstable gene pool that results tends to produce change in the population.

Questions

1 With respect to normal base pairing, when a molecule of DNA replicates, thymine will most likely pair with
 1 adenine 2 cytosine 3 guanine 4 uracil

2 The diagram represents the building block of a large molecule known as a
 1 protein
 2 fatty acid
 3 carbohydrate
 4 nucleic acid

3 The genetic material in living organisms is composed of organic molecules known as
 1 starches 3 fatty acids
 2 lipids 4 nucleic acids

4 If one strand of a DNA molecule has the base sequence *A-G-C-T-A*, the complementary strand of DNA would have the base sequence
 1 *A-G-C-T-A* 3 *U-C-G-A-U*
 2 *U-C-G-A-T* 4 *T-C-G-A-T*

5 The presence of DNA is important for cellular metabolic activities because DNA
 1 directs the production of enzymes
 2 is a structural component of cell walls
 3 directly increases the solubility of nutrients
 4 is the major component of cytoplasm

6 Which is a structural component of a DNA nucleotide?
 1 glucose 3 deoxyribose
 2 an enzyme 4 an amino acid

7 In a DNA molecule, the number of guanine nucleotides will most likely equal the number of
 1 adenine nucleotides 3 cytosine nucleotides
 2 thymine nucleotides 4 ribose nucleotides

8 During the replication of a DNA molecule, separation or "unzipping" of the DNA molecule will normally occur when hydrogen bonds are broken between
 1 thymine and thymine 3 guanine and uracil
 2 adenine and cytosine 4 cytosine and guanine

9 Which set of conditions would most likely cause a change in gene frequency in a sexually reproducing population?
 1 mutations and small populations
 2 large populations and no migrations
 3 random matings and large populations
 4 no mutations and no migrations

10 The formation of recombinant DNA results from the
1 addition of messenger RNA molecules to an organism
2 transfer of genes from one organism to another
3 substitution of a ribose sugar for a deoxyribose sugar
4 production of a polyploid condition by a mutagenic agent

11 RNA carries out the original instructions found in molecules of
1 ATP
2 DNA
3 PTC
4 ADP

12 The gene pool in a population of *Rana pipiens* in a pond remained constant for many generations. The most probable reason for this stable gene pool is that
1 it was a small population with nonrandom mating and variations
2 random mating occurred in a small population with many mutations
3 no mutations occurred in a large, migrating population
4 no migration occurred in a large population with random mating

13 The "one-gene, one-enzyme" hypothesis deals most directly with the relationship of genes to the synthesis of
1 polypeptides
2 polysaccharides
3 lipids
4 carbohydrates

14 The messenger RNA genetic codes for 3 different amino acids are:

UUU = phenylalanine, GCU = alanine, and GGU = glycine

Using this information, the strip of messenger RNA (GCUUUUGGU) would result in an amino acid sequence consisting of

1 phenylalanine-alanine-glycine
2 alanine-glycine-phenylalanine
3 alanine-glycine-glycine
4 alanine-phenylalanine-glycine

15 Watson and Crick described the DNA molecule as a
1 straight chain
2 single strand
3 double helix
4 branching chain

16 A segment of DNA has nine (9) nucleotides. How many codons are represented in this segment?
1 1
2 2
3 3
4 4

17 What are the basic structural units of a DNA molecule?
1 glucose
2 lipids
3 amino acids
4 nucleotides

18 DNA and RNA molecules are similar in that they both contain
1 nucleotides
2 a double helix
3 deoxyribose sugars
4 thymine

19 The presence of which nitrogen base indicates that the molecule associated with the ribosome is RNA?
1 guanine
2 uracil
3 cytosine
4 adenine

20 The replication of a double-stranded DNA molecule begins when the strands separate at the
1 phosphate bonds
2 deoxyribose molecules
3 ribose molecules
4 hydrogen bonds

SELF–HELP: Unit V "*Core*" Questions

1 In guinea pigs, black coat color (*B*) is dominant over white coat color (*b*). When two black guinea pigs were mated, the ratio of black-coated offspring to white-coated offspring was 3:1. In this cross, the parental genotypes were most likely
   1 *Bb* x *Bb*
   2 *BB* x *bb*
   3 *Bb* x *bb*
   4 *BB* x *Bb*

2 In a certain species of mice, brown fur is dominant over white fur and long tails are dominant over short tails. Both of these traits are inherited independently of each other. With respect to only these traits, how many different phenotypes would be present in a large population of mice?
   1 1
   2 2
   3 3
   4 4

3 A person with type O blood marries a person with type AB blood. Possible blood genotypes of their children are
   1 $I^A i$ and $I^B I^B$
   2 $I^B I^B$ and $I^A I^A$
   3 $I^A i$ and $I^B i$
   4 $I^A I^B$ and ii

4 In humans there are 23 pairs of chromosomes. One pair of the 23 are sex chromosomes. The other 22 pairs are known as
   1 autosomes
   2 homozygotes
   3 heterozygotes
   4 centrioles

5 Fruit flies that have gray bodies usually have long wings. What is the most probable reason for the inheritance of this combination of characteristics?
   1 The genes for these traits are linked on the same chromosome.
   2 Both traits are determined by co-dominant genes.
   3 The traits are inherited independently of each other.
   4 Nondisjunction occurs in more than one pair of chromosomes.

6 Which statement correctly describes the normal number and type of chromosomes present in human body cells of a particular sex?
   1 Males have 22 pairs of autosomes and 1 pair of sex chromosomes known as *XX*.
   2 Females have 23 pairs of autosomes.
   3 Males have 22 pairs of autosomes and 1 pair of sex chromosomes known as *XY*.
   4 Males have 23 pairs of autosomes.

7 Based on the pattern of inheritance known as sex linkage, if a male is a hemophiliac, how many genes for this trait are present on the sex chromosomes in each of his diploid cells?
   1 1
   2 2
   3 3
   4 0

8 When pure white and pure red four-o'clocks are crossed, all of the offspring are pink. The phenotype of the offspring illustrates the pattern of inheritance known as
   1 dominance
   2 incomplete dominance
   3 segregation
   4 multiple alleles

9 A color-blind man marries a woman with normal vision. Her mother was color-blind. They have one child. What is the chance that this child is color-blind?
   1 0%
   2 25%
   3 50%
   4 100%

10 A boy has brown hair and blue eyes, and his brother has brown hair and brown eyes. The fact that they have different combinations of traits is best explained by the concept known as
1 multiple alleles
2 sex linkage
3 incomplete dominance
4 independent assortment

11 A cross between two tall garden pea plants produced 314 tall plants and 98 short plants. The genotypes of the tall parent plants were most likely
1 *TT* and *tt*
2 *TT* and *Tt*
3 *Tt* and *Tt*
4 *TT* and *TT*

12 Only red tulips result from a cross between homozygous red and homozygous white tulips. This illustrates the principle of
1 independent assortment
2 segregation
3 dominance
4 incomplete dominance

13 The condition known as Down's syndrome may result from
1 nondisjunction of chromosome pair number 21
2 disjunction of chromosome pair number 18
3 crossing-over between homologous chromosomes
4 alteration of a single base pair of DNA

14 The presence of only one *X*-chromosome in each body cell of a human female produces a condition known as Turner's syndrome. This condition most probably results from the process known as
1 polyploidy
2 crossing-over
3 nondisjunction
4 hybridization

15 Which process, occurring during synapsis, results in the chromosomal changes illustrated in the diagram?
1 replication
2 independent assortment
3 crossing-over
4 segregation

16 During synapsis in meiosis, portions of one chromosomes may be exchanged for corresponding portions of its homologous chromosome. This process is known as
1 nondisjunction
2 polyploidy
3 crossing-over
4 hybridization

17 In many humans, exposing the skin to sunlight over prolonged periods of time results in the production of more pigment by the skin cells (tanning). This change in skin color provides evidence that
1 ultraviolet light can cause mutations
2 gene action can be influenced by the environment
3 the inheritance of skin color is an acquired characteristic
4 albino is a recessive characteristic

18 Plant and animal breeders usually sell or destroy undesirable specimens and use only desirable ones for breeding. This practice is referred to as
1 vegetative propagation
2 natural breeding
3 artificial selection
4 random mating

19 Down's syndrome is a condition which occurs as a result of
1 crossing-over
2 gene mutation
3 polyploidy
4 nondisjunction

20 Amniocentesis involves the analysis of which body fluid?
 1 blood
 2 urine
 3 amniotic
 4 uterine

21 The modern day descendants of Central European Jewish people are most likely to be affected by which genetic disease?
 1 Phenylketonuria
 2 Down's syndrome
 3 Sickle-cell Anemia
 4 Tay-Sachs

22 Which nitrogenous bases tend to pair with each other in a double-stranded molecule of DNA?
 1 adenine-uracil
 2 thymine-adenine
 3 cytosine-thymine
 4 guanine-adenine

23 The function of transfer RNA molecules is to
 1 transport amino acids to messenger RNA
 2 transport amino acids to DNA in the nucleus
 3 synthesize more transfer RNA molecules
 4 provide a template for the synthesis of messenger RNA

24 When bonded together chemically, deoxyribose, phosphate, and an adenine molecule make up
 1 a DNA nucleotide
 2 an RNA nucleotide
 3 a DNA molecule
 4 an RNA molecule

25 The genetic material, responsible for the individuality of an organism, that is passed from parent to offspring
 1 DNA
 2 RNA
 3 both DNA and RNA
 4 neither DNA nor RNA

SELF–HELP: Unit V *"Extended"* Questions

For each phrase in questions 1 through 5, select the type of nucleic acid molecules, chosen from the list below, that is best described by that phrase.

Types of Nucleic Acid Molecules
1) DNA molecules, only
2) RNA molecules, only
3) Both DNA and RNA molecules
4) Neither DNA nor RNA molecules

1 May contain the four bases adenine, cytosine, guanine, and thymine

2 Present in the nucleus of a cell

3 Carry genetic information from the nucleus to the ribosomes

4 Contain the nitrogenous compound known as urea

5 Consist of chains of nucleotides

6 Plants with desirable qualities can be rapidly produced from cells of a single plant by a process known as
 1 cloning
 2 gamete fusion
 3 reduction division
 4 gametogenesis

Base your answers to questions 7 through 9 on your knowledge of biology and on the charts at the right which show human chromosomes arranged in pairs

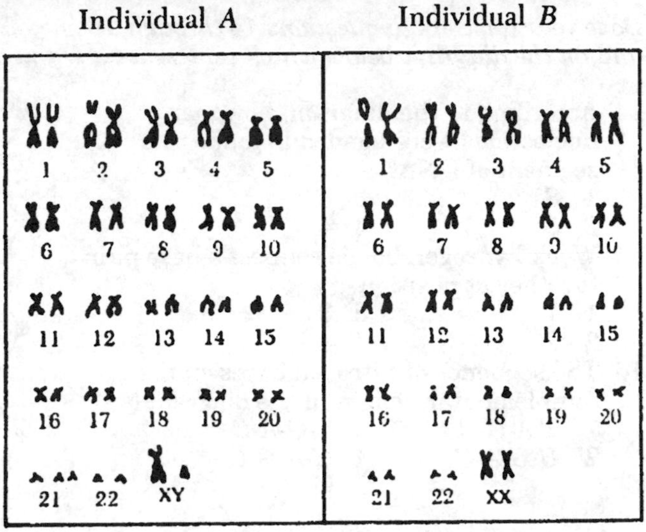

Individual *A* Individual *B*

1 2 3 4 5
6 7 8 9 10
11 12 13 14 15
16 17 18 19 20
21 22 XY

1 2 3 4 5
6 7 8 9 10
11 12 13 14 15
16 17 18 19 20
21 22 XX

7 The chromosomes numbered 1 through 22 are known as
1 ribosomes
2 lysosomes
3 centrosomes
4 autosomes

8 The preparation of these charts for individuals *A* and *B* is known as
1 microsurgery
2 karyotyping
3 blood typing
4 chemical screening

9 Which genetic disorder in individual A is indicated by the number of chromosomes labeled 21?
1 Phenylketonuria (PKU)
2 Tay-Sacks
3 sickle-cell anemia
4 Down's syndrome

Base your answers to questions 10 through 12 on the diagram at the right which represents portions of nucleic acid molecules and on your knowledge of biology.

A C E
B D F

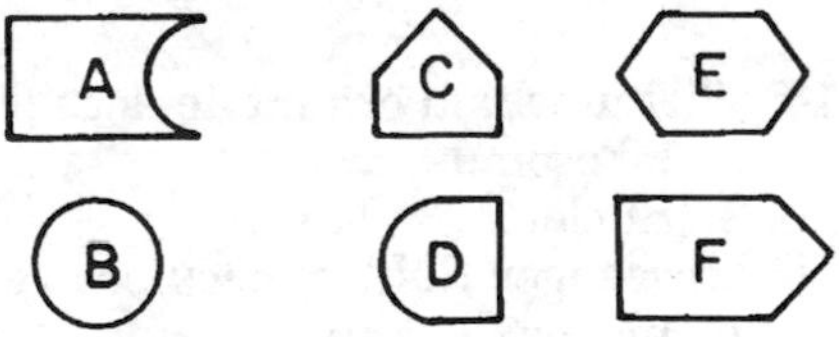

10 Deoxyribose and ribose could be represented by letter
1 *A*
2 *B*
3 *C*
4 *E*

11 Two nitrogenous bases that could combine by forming hydrogen bonds are represented by letters
1 *A* and *D*
2 *B* and *C*
3 *A* and *E*
4 *C* and *D*

12 An RNA nucleotide could be constructed by combining the molecules represented by letters
1 *A, E,* and *F*
2 *B, C,* and *F*
3 *B, C,* and *E*
4 *A, B,* and *D*

13 Certain genetic disorders can be detected by preparing and studying an enlarged photograph of paired chromosomes from a cell. The preparation of this photograph is known as
1 genetic screening
2 karyotyping
3 genetic counseling
4 amniocentesis

Base your answers to questions 14 through 16 on your knowledge of biology and on the diagram below which represents a segment of a DNA molecule.

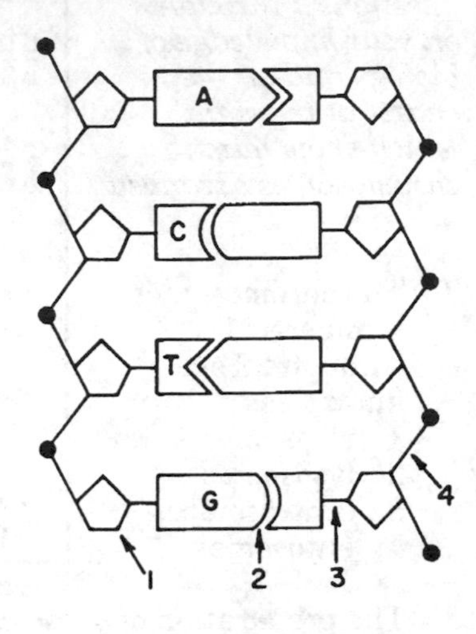

14 According to the diagram, how many nucleotides were bonded to make this segment of DNA?
1 8
2 6
3 3
4 4

15 Weak hydrogen bonds connect a base pair together at position
1 1
2 2
3 3
4 4

16 The sequence of nitrogen bases on the complementary chain in the diagram is
1 *T-A-G-C*
2 *U-G-A-C*
3 *T-G-A-C*
4 *U-A-G-C*

Base your answers to questions 17 through 19 on the diagram below and on your knowledge of biology.

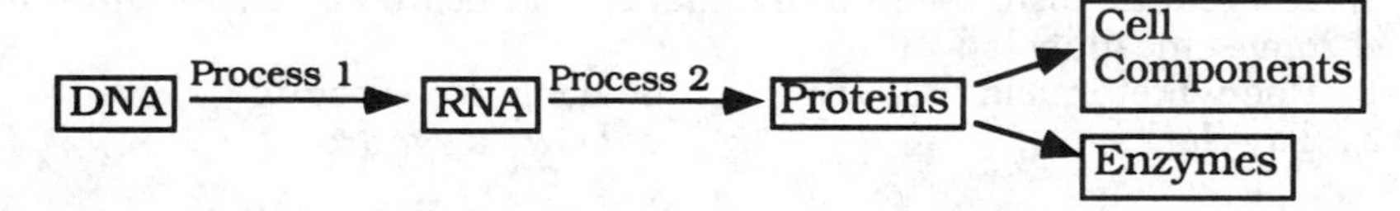

17 Within which organelle does process 1 occur?
1 ribosome
2 nucleus
3 centriole
4 lysosome

18 Messenger RNA molecules are formed as a result of
1 process 1, *only*
2 process 2, *only*
3 *both* process 1 *and* process 2
4 *neither* process 1 *nor* process 2

19 Within a living cell, which organelles are necessary for process 2 to occur?
1 centrioles
2 lysosomes
3 Golgi bodies
4 ribosomes

20 According to the Hardy-Weinberg principle, which factors tend to keep a population's gene frequencies constant?
1 high mutation rate and geographic isolation
2 large population size and random mating
3 nonrandom mating and frequent migrations
4 small population size and changing environmental conditions

SELF–HELP: Unit V "*Skill*" Questions

1 The data below are based on laboratory studies of male *Drosophila* showing the inherited bar-eye phenotype.

Culture Temperature (°C) During Development	15	20	25	30
Number of Compound Eye Sections	270	161	121	74

Which is the best conclusion to be made from an analysis of these data?
1 The optimum temperature for culturing *Drosophila* is 15°C.
2 *Drosophila* cultured at 45°C will show a proportionate increase in the number of compound eye sections.
3 Temperature determines eye shape in *Drosophila.*
4 As temperature increases from 15°C to 30°C, the number of compound eye sections in male *Drosophila* with bar-eyes decreases.

Base your answers to questions 2 through 6 on the passage below.

Gene Splicing

Recent advances in cell technology and gene transplanting have allowed scientists to perform some interesting experiments. Some of these experiments have included splicing a human gene into the genetic material of bacteria. The altered bacteria express the added genetic material.

Bacteria reproduce rapidly under certain conditions. This means that bacteria with the gene for human insulin could multiply rapidly, resulting in a large bacterial population which could produce large quantities of human insulin.

The traditional source of insulin has been the pancreases of slaughtered animals. Continued use of this insulin can trigger allergic reactions in some humans. The new bacteria-produced insulin does not appear to produce these side effects.

The bacteria used for these experiments are *E. coli,* bacteria common to the digestive systems of many humans. Some scientists question these experiments and are concerned that the altered *E. coli* may accidentally get into water supplies.

For each statement on the next page, write the number 1 if the statement is true according to the paragraph, the number 2 if the statement is false according to the paragraph, or the number 3 if not enough information is given in the paragraph.

____ 2 Transplanting genetic material into bacteria is a simple task.

____ 3 Under certain conditions bacteria reproduce at a rapid rate.

____ 4 Continued use of insulin from other animals may cause harmful side effects in some people.

____ 5 The bacteria used in these experiments are normally found only in the nerve tissue of humans.

____ 6 Bacteria other than *E. coli* are unable to produce insulin.

Base your answers to questions 7 through 10 on the diagram at the right which represents part of an organic molecule and on your knowledge of biology.

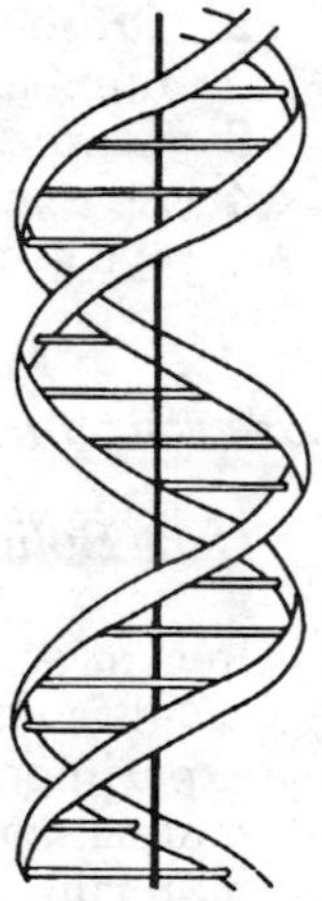

7 The diagram represents a molecule of
  1 DNA
  2 RNA
  3 ATP
  4 FSH

8 Which two scientists proposed the double helix arrangement of this molecule?
  1 Hardy and Weinberg
  2 Darwin and Lamarck
  3 Watson and Crick
  4 Mendel and De Vries

9 Which statement describes the substance represented in the diagram?
  1 It is a polymer found in chromosomes.
  2 It is a small molecule found in ribosomes.
  3 It is an energy-releasing molecule located in the cytoplasm.
  4 It is a double lipid layer molecule with connecting proteins.

10 The diagram at the right represents the base sequence on a single strand of this molecule. The complementary base sequence is represented in which diagram below?

A
G
T
C

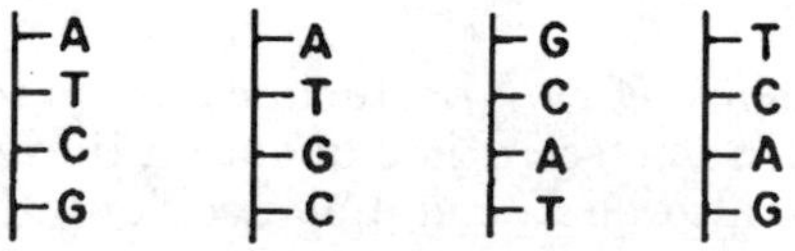

(1)	(2)	(3)	(4)
A	A	G	T
T	T	C	C
C	G	A	A
G	C	T	G

Unit 6

Evolution

Objectives

The student should be able to:

- Understand that evolution is a process of change.
- Recognize that evolutionary theory is supported by observations and inferences from many branches of science.
- Describe some of the supporting data for evolutionary theory.
- Discuss the historical development of evolutionary theory.
- Describe a hypothesis which attempts to explain how primitive environmental conditions may have contributed to the formation of the first forms of life.

Evolution – Terms & Concepts

Evolution
  Observations - Conclusions
  Theory - Supportive Evidence
Geologic Record
  Radioactive Dating - Fossils
Comparative Cytology
Comparative Biochemistry
Comparative Anatomy
  Homologous Structures
Comparative Embryology
Diversity - Adaptations
Lamarck - "Use and Disuse"
  Acquired Characteristic Trans.
Charles Darwin - Theory of Evolution
  Overproduction - Competition
  Survival of the Fittest
  Reproduction - Speciation
Modern Evolution
  Variations and Mutations
Evolution Time Frame
  Gradualism - Punctuated Equil.
Heterotroph Hypothesis
  "Hot, Thin Soup" - Primitive Earth
  Miller's Experiment - Aggregates
  Reproduction Adaptation
  Anaerobes - Aerobes

I. Evolution

Evolution is a process of change through time. All living things have changed to some extent over geologic time, and the study of the mechanisms believed to cause these changes is called **organic evolution**. These mechanisms have produced changes including variations within a particular species and the production of new species.

For centuries, philosophers, scientists, politicians, and religious authorities have designed theories as to how this change has occurred through time. Since most of these theories, including the supernatural accounts of the origins of life and the like, are outside of the domain of science, meaning that these accounts cannot be tested through scientific investigation, they have not been included in this discussion of evolution. It should also be noted that the scientific theories of evolution discussed here, as is the case with other scientific theories, are and continue to be subjected to verification by scientific processes.

II. Evolution Theory

Evolution theory is a unifying principle for the biological sciences. It provides an explanation for the differences in structure, function, and behavior among life forms. Evolution includes the change in characteristics of populations through generations. Thus, existing life forms have evolved from earlier life forms.

A. Supporting Observation

Observations supporting the theory of organic evolution can be made through the study of the geologic record and comparative studies in the fields of cytology, biochemistry (including modern genetic studies), anatomy, and embryology.

Fossil Evidence

1. Geologic Record

Geologists have dated the Earth to be between four and one-half and five billion years old. This age was determined by radioactive dating of rocks. It may be assumed that the Earth is at least as old as the oldest rocks and minerals composing its crust.

Fossils are direct or indirect remains of organisms preserved in media such as sedimentary rocks, amber, ice, or tar. Fossils have been found which indicate that organisms existed over three billion years ago. Fossils of prokaryotic life forms (Monerans) indicate an age of 3.4 billion years or greater.

Upper, undisturbed strata generally contain fossils of more complex organisms, whereas, the lower strata contain fossils of simpler life forms. When comparing fossils in undisturbed strata, fossils can be found in upper strata which, although different from fossils in lower strata, resemble those fossils. This suggests *links* between modern forms and older forms, as well as divergent pathways from **common ancestors**. Common ancestry, the divergence of modern forms of living things from preexisting life forms, is the central concept in the science of evolution.

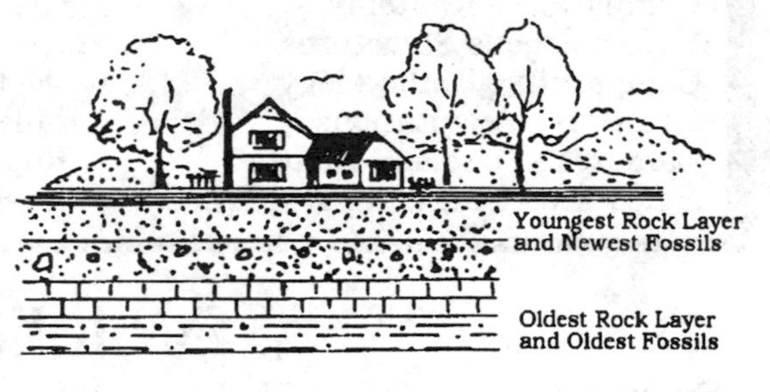

Geologic Record (sedimentary strata)

2. Comparative Biochemistry

Nucleic acids, their structure and function, are similar in living organisms even though they may or may not display structural similarity. Many different organisms have similar proteins and enzymes. Therefore, their DNA must be similar. The greater their biochemical similarity, the closer the relationship among organisms, thus suggesting evolutionary relationships.

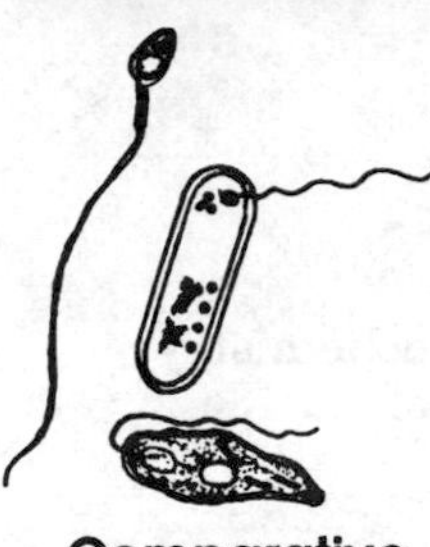

Comparative Cytology

3. Comparative Cytology

According to the cell theory, the cell is the unifying structure for living things. Organelles such as cell membranes, ribosomes, and mitochondria are structurally and functionally similar in most divergent organisms. This suggests that all living things are related to some degree. The fewer the differences in these cell structures, the closer the relationship appears.

4. Comparative Anatomy

A comparative study of certain organisms indicates similarities in anatomical features. This is a basis for the development of the five kingdom system of classification.

Homologous structures are anatomical parts that are similar in their structure and origin although they may function differently. Homologous bones exist in the forelimbs of many different vertebrates such as frogs, birds, horses, bats, whales, and humans.

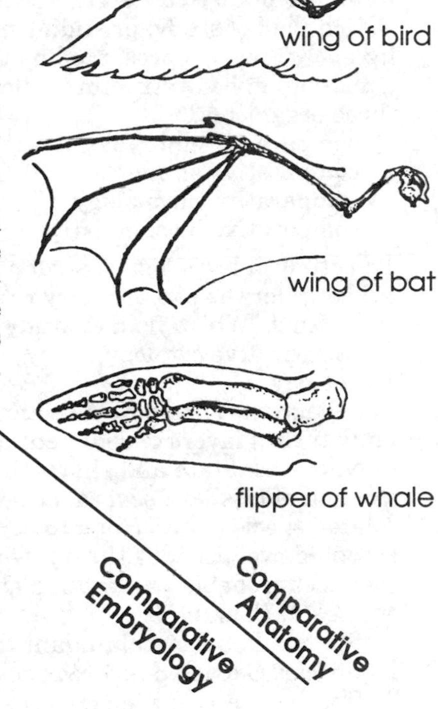

salamander early embryo

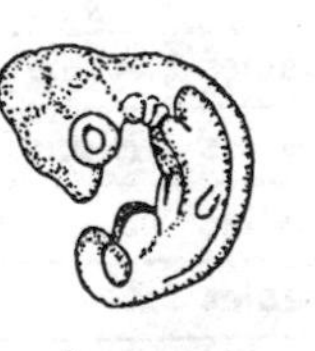

pheasant early embryo

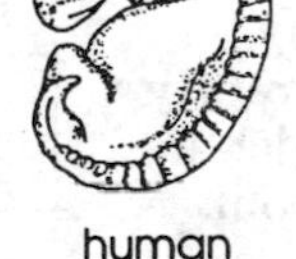

human early embryo

5. Comparative Embryology

Comparison of early embryonic development among groups of organisms reveals similarities which suggest common ancestry. Early vertebrate embryos closely resemble one another. As development proceeds, the distinctive traits of each species become apparent. This is believed by many scientists to be an indicator that species have a common ancestry.

Questions

1 Thyroxin synthesized by cattle has been used in the treatment of certain human thyroid disorders. Cattle also synthesize pepsin that is similar to the pepsin produced by humans. These facts provide evidence of evolution that support the concept of

1 fossil formation
2 genetic dominance
3 reproductive isolation
4 common ancestry

2 Organic evolution is best described as
1 a process of change through time
2 a process by which an organism becomes extinct
3 the movement of large landmasses
4 the spontaneous formation of all species

3 The comparative sciences of anatomy, embryology, and biochemistry provide evidences of evolution which support the concept of
1 genetic dominance
2 geographic isolation
3 fossil formation
4 common ancestry

4 The diagrams at the right represent the forelimbs of three different vertebrates. Which field of study, providing evidence for evolution, is represented by the similarity in bone structure among these three organisms?

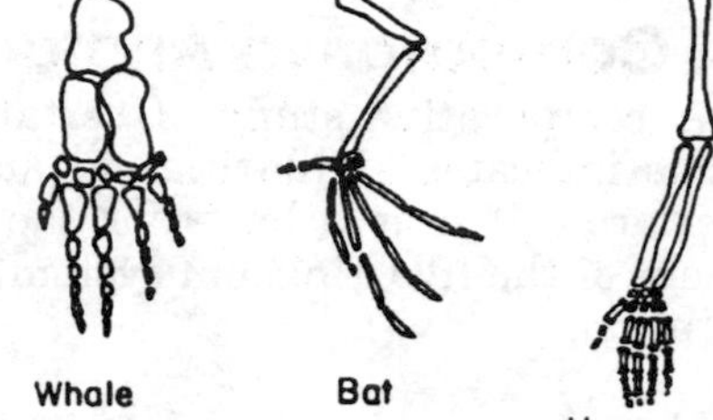

1 comparative embryology
2 comparative anatomy
3 comparative physiology
4 comparative biochemistry

5 The study of homologous structures in mature organisms provides evidence for the evolutionary relationships among certain groups of organisms. Which field of study includes this evidence of evolution?
1 comparative cytology
2 geology
3 biochemistry
4 comparative anatomy

6 The diagram represents a section of undisturbed layers of sedimentary rock in New York State and shows the location of fossils of several closely related species. According to currently accepted evolutionary theory, which is the most probable assumption about species *A*, *B*, and *C*?

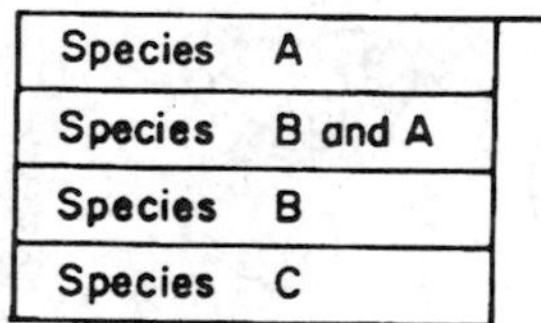

1 Species *B* is more abundant than species *C*.
2 Species *C* existed before species *B*.
3 Species *A* and *B* are genetically identical.
4 Species *B* descended from species *A*.

7 The similarity among the blood proteins of all the mammals may be taken as evidence for evolutionary relationships based upon
1 comparative anatomy
2 geographic isolation
3 comparative embryology
4 comparative biochemistry

8 If a fossil mammoth were discovered frozen in ice, its cells could be analyzed to determine whether its proteins were similar to those of the modern elephant. This type of investigation is known as comparative
1 anatomy
2 embryology
3 biochemistry
4 ecology

9 Fossils are used as evidence for evolution because they
1 may show a pattern of consecutive changes
2 are as old as the oldest rocks on the Earth
3 are composed of mineral substances
4 may be formed in lava rock

10 A geologist finds fossils in each of the undisturbed rock layers represented in the diagram. The fossils are all structurally similar. Which is the most likely conclusion that the geologist would make?

1 All the fossils are of the same age.
2 The relative ages of the fossils cannot be determined.
3 The fossils in rock layer *D* are older than those in layer *A*.
4 The fossils in rock layer *B* are older than those in layer *C*.

11 Structural and physiological changes within a species occur over an extended period of time. These changes appear to be the product of the natural selection of favorable traits within that species. These statements best describe the concept of
1 spontaneous mutation
2 reproductive isolation
3 homeostasis
4 evolution

12 A bird's developmental stages resemble those of a reptile. This observation is often used to illustrate the probable common ancestry of these organisms through the study of
1 comparative biochemistry
2 comparative embryology
3 punctuated equilibrium
4 natural selection

13 Which is an example of evidence of evolution based on comparative biochemistry?
1 Sheep insulin can be substituted for human insulin.
2 The structure of a whale's flipper is similar to that of a human hand.
3 Human embryos have a tail-like structure at one stage in their development.
4 Both birds and bats have wings.

14 The presence of gill-like slits in a human embryo is considered to be evidence for the
1 theory that fish and mammals have a common ancestry
2 theory that the first organisms on Earth were heterotrophs
3 close relationship between fish and mammalian reproductive patterns
4 close relationship between humans and annelids

15 Biochemical analysis has shown that hemoglobin molecules found in monkeys are very similar to those found in humans. Which concept is supported by this analysis?
1 Homologous structures exist in all vertebrates.
2 Embryonic development in humans and monkeys is identical.
3 Monkeys and humans have a common ancestor.
4 Invertebrates and vertebrates have a common ancestor.

B. Theories of Evolution

Theories of evolution are attempts to explain the diversities among species. These explanations are theories and not unchanging. Many attempts to explain evolution have gone through revisions and continue to be modified based on new scientific evidence. **Adaptations** are a major component of these theories. Adaptations are features (including structure, function, or behavior) which make a species more suited to survive and reproduce in its environment.

1. Lamarck

In 1809, a French biologist Jean Baptiste de Lamarck proposed an explanation of the origin of species in his book, *Zoological Philosophy*. Lamarck's theory encompassed two main ideas:

- **Use and Disuse.** New organs arise according to the needs of an organism, and the size of organs is determined by the degree to which they are used. Lamarck suggested that the "needs" of an organism determined the characteristics that evolved in that organism.
- **Transmission of Acquired Characteristics.** Useful characteristics acquired by an individual during its lifetime can be transmitted to its offspring. These acquired characteristics result in species better adapted to their environment. For example, the offspring of an organism that acquired a desired characteristic would also possess that characteristic, passed on from the parents.

Scientists that followed Lamarck conducted experiments that failed to support his theory. **August Weismann** conducted experiments involving the removal of the tails of mice over several generations. Weismann proved that the offspring of "tail-less" mice did not pass to their offspring their "tail-less" characteristic. His work helped to disprove Lamarck's theory of the inheritance of acquired characteristics.

2. Darwin

In the 19th century, English naturalist **Charles Darwin** developed a theory of evolution based on observations of plants and animals that he studied all over the world (during his voyage on the *H.M.S. Beagle*) and on the writings of several geologists.

In his book, *Origin of Species,* Darwin presented evidence that demonstrated that all living things evolved from other living things.

Darwin's theory of evolution was based on **variation** and **natural selection.**

Charles Darwin

It encompassed the following ideas:

- **Overproduction.** Within a population more offspring are born than can possibly survive to reproduce themselves.
- **Competition.** Since the number of individuals in a population tends to remain constant from generation to generation, a struggle for survival is suggested. That is, within a particular species, there is competition for the basics of life, such as food and shelter.
- **Survival of the Fittest.** The individuals who survive are the ones best adapted to exist in their environment due to the possession of **variations** (differences in form) that maximize their fitness.
- **Reproduction.** Individuals that survive would then reproduce and transmit these variations to their offspring.
- **Speciation.** As time and generations continue, many adaptations are perpetuated in individuals until new species evolve in forms different from a common ancestral species.

The major weakness of Darwin's theory of natural selection was that he did not explain the genetic basis for variations. At the time Darwin made his observations and developed his theory, there was very little information on the mechanisms of genetic inheritance. Also, Darwin did not take into consideration the origin of life itself.

Questions

1 According to Darwin's theory of natural selection, individuals who survive are most likely the ones best adapted to exist in their environment. Their survival is due to the
   1 possession of structures developed through use
   2 possession of variations that maximize fitness
   3 lack of competition within the species
   4 ability to change their genotype

2 Which theory maintains that new organs arise according to the needs of an organisms and that the size of the organs is determined by the degree to which they are used?
   1 use and disuse theory
   2 natural selection
   3 geographic isolation
   4 modern evolutionary theory

3 Which concept was not included in Charles Darwin's theory of natural selection?
   1 survival of the fittest
   2 struggle for existence
   3 overproduction of offspring
   4 punctuated equilibrium

4 Which statement most accurately expresses an evolutionary theory of Lamarck?
   1 Variations in organisms occur as a result of both gene mutations and chromosome mutations.
   2 Variations in organisms occur as a result of genetic recombination.
   3 New organs arise according to the needs of an organism.
   4 Since more organisms are produced than can survive, a struggle for existence occurs.

5 Which is a major concept in Lamarck's theory of evolution?
  1 Change is the result of mutations.
  2 New organs arise according to the needs of the organism.
  3 Dominant genes increase the rate of evolution.
  4 Sexual reproduction is the genetic basis for variations.

6 "It is likely that ducks developed webbed feet because ducks need webbed feet for efficient swimming."
  This attempt to explain the development of webbed feet in ducks most nearly matches the theory of evolution proposed by
  1 Jean Lamarck
  2 Charles Darwin
  3 Gregor Mendel
  4 Francis Crick

7 Which statement is in agreement with Darwin's theory of evolution?
  1 More offspring are produced than can possibly survive.
  2 The organisms that are the most fit are always those with the greatest strength.
  3 Mutations are always beneficial.
  4 Acquired characteristics are inherited.

8 An athlete explains that his muscles have become well developed through daily activities of weight lifting. He believes that his offspring will inherit this trait of well developed muscles. This belief would be most in agreement with the theory set forth by
  1 Darwin
  2 Lamarck
  3 Weismann
  4 Mendel

9 Natural selection can best be defined as
  1 survival of the strongest organisms
  2 elimination of the smallest organisms by the largest organisms
  3 survival of those organisms genetically best adapted to the environment
  4 survival and reproduction of the organisms that occupy the largest area in an environment

10 A weakness in Darwin's theory of evolution was that he was not able to
  1 explain selection of favorable traits
  2 account for an increase in population
  3 explain the genetic basis for variation in populations
  4 understand competition among individuals of a species

3. Modern Evolutionary Theory

The modern theory of evolution supports Darwin's concepts of variation and natural selection. Since modern research has provided information on the mechanisms of genetic inheritance, the modern theory of evolution, based on Darwin's theory, incorporates the genetic basis of variation in individual organisms and populations.

Producing Variation

The genetic basis for variation within a species is provided by mutations and sexual reproduction. Mutations are *spontaneous* and provide the raw material for evolution. Most mutations are either harmful or neutral, but a small percentage provide adaptations that help the organism better survive.

These mutations are transmitted to offspring when they occur in the sex cells of the parents. Sexual reproduction involves the sorting out and recombination of genes, thus producing and maintaining variations.

Natural Selection

Natural selection involves the struggle of organisms to survive and reproduce in their environment.

- Traits which are *beneficial to the survival* of an organism in a particular environment tend to be retained and passed on, and therefore, increase in frequency within a population.
- Traits which have *low survival value* to organisms tend to diminish in frequency from generation to generation.

If environmental conditions change, traits that were formerly associated with a low survival value may, in a changed environment, have greater survival value and increase accordingly. Examples include:

- In recent years, roaches, mosquitoes, and houseflies have developed a resistance to insecticides. The pesticides once used with success against certain organisms now have little or no effect on these organisms. *(Note: Resistance is not in response to the insecticide. The insecticide acts as a selecting agent.)*
- The wide-spread use of some antibiotics has produced a similar effect. Penicillin-resistant strains of microorganisms have developed. Again, as in the case of insecticides, the resistance is a result of random mutations and is not the result of changing due to the presence of the antibiotic.

Geographic Isolation

Geographic isolation favors **speciation** by segregating a small group from the main population. Changes in gene frequencies are more likely in small populations than in large populations. In time, this isolated population may evolve into a separate species due to the following factor(s):

- It may have possessed different initial gene frequencies than the main population.
- Different mutations occur within the main population and the isolated population.
- Different environmental factors, and thus, different selection pressures, may be acting on each population.

It is probable that initially the isolated population had a different gene frequency than the original population. This is the **Founder Effect**.

Examples of the Founder Effect include Darwin's finches on the Galapagos Islands and marsupials in Australia.

- During his voyage on the *H.M.S. Beagle*, Darwin discovered that more than a dozen species of finches inhabited the Galapagos Islands. On different islands, each species had distinct adaptations. He hypothesized that when a population is isolated from other populations, the species developed differing characteristics that better adapted that species for survival in that particular environment.
- In Australia, marsupial species have developed. It is believed that these unique organisms have arisen over thousands of years due to their isolation from ancestral species and freedom from preditors.

Reproductive Isolation

These separated groups may become so divergent that, if geographic barriers were removed, interbreeding could not take place. Thus, the two populations have become reproductively isolated and have become two distinct species.

4. Time Frame for Evolution

While the essentials of Darwin's theory of evolution, variation, and natural selection are generally accepted by the scientific community, considerable discussion exists within this community as to the time frame in which evolution occurs.

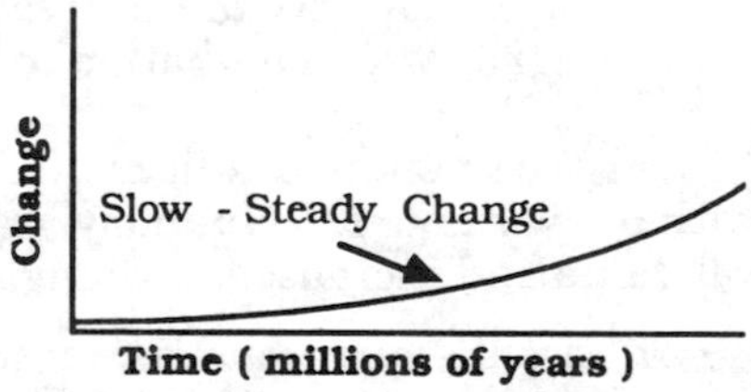

Change
Slow - Steady Change
Time (millions of years)

Gradualistic Changes

Gradualism

Gradualism proposes that evolutionary change is slow, gradual, and continuous. This view point is supported with evidence in geologic fossils which show slight changes in organisms between adjacent layers of sedimentary rocks. Between bottom and top layers, greater and significant differences (divergence) are found between the organisms.

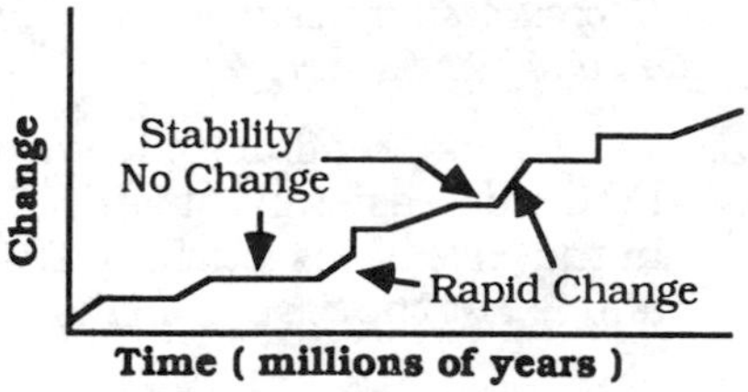

Change
Stability
No Change
Rapid Change
Time (millions of years)

Punctuated Changes

Punctuated Equilibrium

Punctuated Equilibrium proposes that species have long periods of stability (typically several million years) interrupted by geologically brief periods of significant change during which new species may evolve. According to some scientists, geologically brief periods represent approximately one percent of the duration of a species. This view is also supported in the fossil record where little change is observed from layer to layer, but sudden changes ("bursts") are observed between some adjacent layers.

In general, the fossil record indicates that most invertebrate species have an average duration between "bursts" of 5-10 million years.

Questions

1 According to modern evolutionary theory, the mechanism by which new species arise is based on a situation in which
  1 organisms of different phyla interbreed and produce fertile offspring
  2 environmental conditions remain stable for long periods
  3 individual organisms change drastically in a single lifetime
  4 a series of mutations lead to changes in genetic structure

2 The theory of continental drift states that Africa and South America were once a single land mass, but have drifted apart over millions of years. The "Old World" monkeys of Africa, although similar, show several genetic differences from the "New World" monkeys of South America. Which factor is probably the most important for maintaining these differences?
  1 fossil records
  2 use and disuse
  3 comparative anatomy
  4 geographic isolation

3 A possible conclusion based on the modern theory of evolution is that
  1 most species have changed
  2 all living things developed from fish
  3 most plants and animals can interbreed
  4 all dogs are more closely related to fish than to whales

4 Over a period of 28 million years, various genera of hoofed mammals called *titanotheres* showed a continuous change in body and horn size before they eventually became extinct. This slow rate of evolutionary change is an example of
  1 punctuated equilibrium
  2 gradualism
  3 reproductive isolation
  4 overproduction

5 Geographic isolation of organisms increases the likelihood of genetic differentiation. This genetic differentiation occurs because geographic isolation
  1 prevents interbreeding between populations
  2 prevents interbreeding within populations
  3 stimulates the production of different kinds of enzymes
  4 accelerates the production of new mutations

6 Over a long period of time the organisms on an island changed so that they could no longer interbreed with the organisms on a neighboring island. This inability to interbreed is known as
  1 hybridization
  2 reproductive isolation
  3 artificial selection
  4 survival of the fittest

7 In areas of the American Southwest, certain insect species are quickly becoming resistant to continuous applications of chemical insecticides. The increase in the number of insecticide-resistant species is due to
  1 inheritance of acquired traits
  2 asexual reproduction
  3 geographic isolation
  4 natural selection

8 According to the theory of natural selection, genes responsible for new traits that are beneficial to the survival of a species in a particular environment will usually
  1 decrease suddenly in frequency
  2 decrease gradually in frequency
  3 not change in frequency
  4 increase in frequency

9 According to modern evolutionary theory, which factor least influences the pattern of evolution in a population?
1 sexual reproduction
2 environmental change
3 geographic isolation
4 use and disuse

10 In an environment, barriers prevent an organism from entering other environments. This phenomenon illustrates the concept of
1 punctuated equilibrium
2 geographic isolation
3 genetic variation
4 natural selection

III. Heterotroph Hypothesis

The cell theory explains how current organisms are formed: "All cells arise from preexisting cells." However, it does not explain how the "first" cells were developed. Some scientists have proposed the **Heterotroph Hypothesis** as an explanation for how early life forms may have developed on the primitive Earth. Like many scientific explanations of incompletely understood phenomena, the heterotroph hypothesis is based upon logical extensions of certain basic assumptions. According to the hypothesis, the first life forms were not able to synthesize their own organic nutrients from inorganic compounds.

A. Primitive Life Forms

Raw Inorganic Materials

It is assumed that the primitive Earth was an exceptionally hot body consisting of inorganic substances in solid, liquid, and gaseous states, with a rich supply of energy in the environment.

Matter. Water (H_2O), condensing and falling as rain, carried dissolved atmospheric gases (ammonia - NH_3, methane - CH_4, and hydrogen - H_2) and minerals into the seas, forming a "hot, thin soup." Evidence suggests that both gaseous oxygen and carbon dioxide were *not* present at this early stage in the Earth's history.

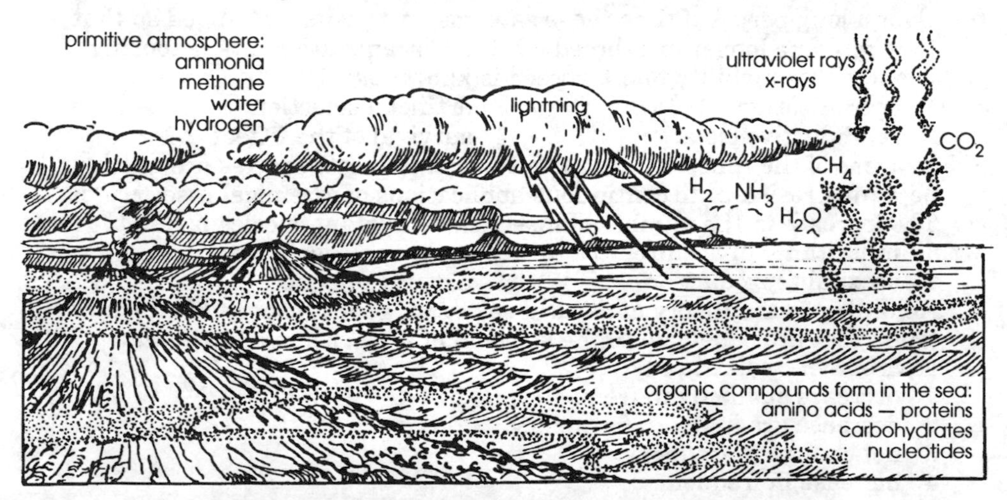

Heterotroph Hypothesis (primitive Earth conditions)

Energy Sources. In addition to heat, energy in the form of lightning, solar radiation (including x-rays and ultraviolet rays), and radioactive materials in the rocks, provided an energy rich environment. Since there was no oxygen in the atmosphere, there was no ozone layer.

Synthesis of Organic Molecules

Energy from the Earth's environment contributed to the formation of chemical bonds among the dissolved particles in the "hot, thin soup" of the seas. Organic molecules were synthesized. This type of synthesis in the seas led to the formation of organic molecules such as simple sugars, amino acids, and nucleotides.

- Experiments by **Stanley Miller** and **Harold Urey** have simulated the primitive environment in the laboratory and have resulted in the production of organic compounds. In time, these organic molecules interacted and formed more complex organic molecules.

- Experiments by **Sidney Fox** have demonstrated interactions among these organic molecules. These experiments joined Miller's organic compounds into aggregates, similar in form to "cell-like" structures.

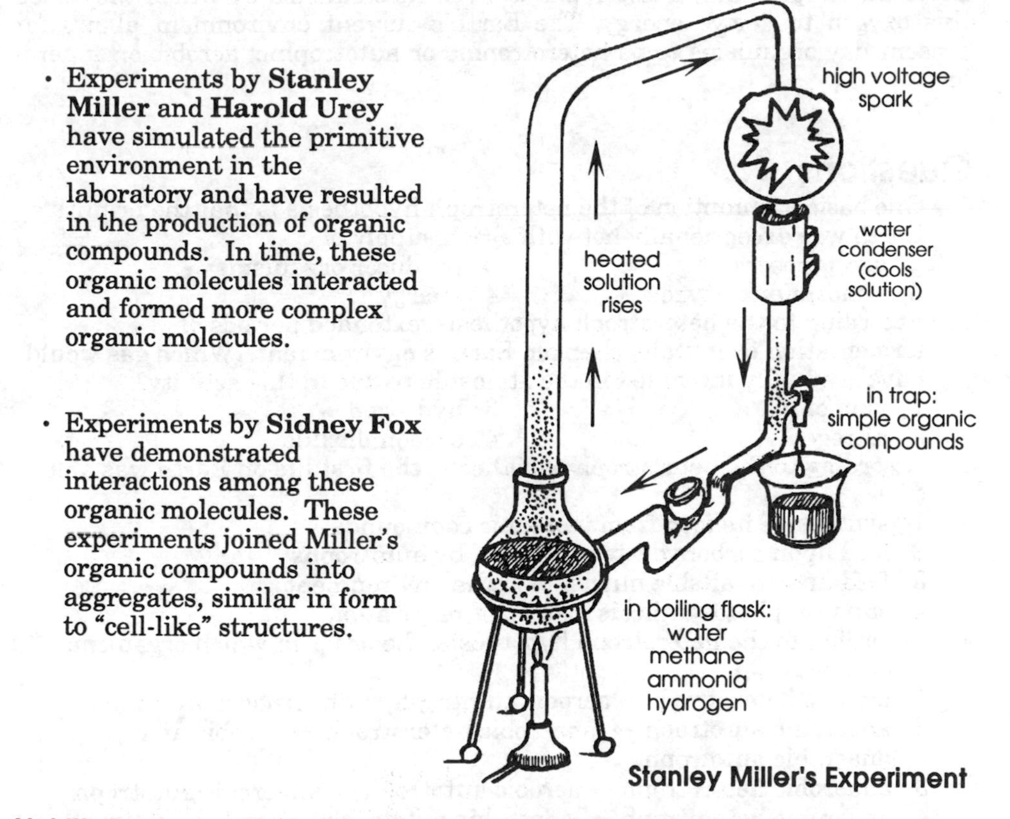

Stanley Miller's Experiment

Nutrition

Some of the large, complex molecules formed **aggregates**. These aggregates probably incorporated molecules from the seas as "food," thus carrying on heterotrophic nutrition.

Reproduction

In time, as these aggregates became increasingly more complex and highly organized, the ability to reproduce evolved. These aggregates are considered to have been "alive" when they developed the ability to reproduce.

B. Heterotroph to Autotroph

It is thought that these heterotrophs evolved a pattern of respiration similar to the anaerobic process of fermentation carried on by present day yeast cells. Extended periods of fermentation activity by these organisms added quantities of carbon dioxide to the environment. Some heterotrophs evolved a means of using carbon dioxide to synthesize organic compounds. These were the **pioneer autotrophs**, such as blue-green algae.

C. Anaerobic to Aerobic

Autotrophic activity added free oxygen to the environment for the first time. Some autotrophs and heterotrophs evolved mechanisms by which they used this oxygen to derive energy. The Earth's current environment allows for present day organisms to be heterotrophic or autotrophic; aerobic or anaerobic.

Questions

1 One basic assumption of the heterotroph hypothesis is that the primitive Earth was exceptionally hot with a rich supply of
1 decomposers
2 atmospheric oxygen
3 producer organisms
4 energy

2 According to the heterotroph hypothesis, extended periods of fermentation eventually changed Earth's environment. Which gas would have gradually increased in the atmosphere due to this activity?
1 ammonia
2 nitrogen
3 hydrogen
4 carbon dioxide

3 According to the heterotroph hypothesis, the first life on Earth was able to
1 synthesize its food from inorganic compounds
2 feed upon carbohydrates produced by autotrophs
3 feed upon available nutrients in the environment
4 carry on photosynthesis instead of respiration

4 According to the heterotroph hypothesis, the order in which organisms evolved was
1 aerobic heterotroph → aerobic autotroph → anaerobic autotroph
2 anaerobic autotroph → anaerobic heterotroph → aerobic and anaerobic autotroph
3 anaerobic heterotroph → aerobic autotroph → anaerobic autotroph
4 anaerobic heterotroph → anaerobic autotroph → aerobic autotroph and heterotroph

5 Which gas became more abundant in the Earth's primitive atmosphere as a result of long periods of fermentation by anaerobic organisms?
1 hydrogen
2 carbon dioxide
3 ammonia
4 methane

6 According to the heterotroph hypothesis, the first forms of life on Earth probably obtained energy by anaerobic respiration. What material is thought to have been added to early Earth's atmosphere by this process?
1 methane
2 carbon dioxide
3 nitrogen
4 hydrogen

7 In an experiment by Stanley Miller, the chemicals methane, hydrogen, ammonia, and water vapor were subjected to a high-energy electrical sparking device at high temperatures. This experiment was an attempt to
 1 produce organic compounds
 2 produce elements
 3 duplicate aerobic respiration
 4 duplicate photosynthesis

8 The heterotroph hypothesis is an attempt to explain
 1 how the Earth was originally formed
 2 why simple organisms usually evolve into complex organisms
 3 why evolution occurs very slowly
 4 how life originated on the Earth

9 According to the heterotroph hypothesis, the earliest heterotrophs carried out what type of energy-releasing process?
 1 protein synthesis
 2 anaerobic respiration
 3 photosynthesis
 4 aerobic respiration

10 According to the heterotroph hypothesis, where were the first cell-like structures probably formed?
 1 in fresh water
 2 in the ocean
 3 on land
 4 within crustal rocks

SELF–HELP: Unit VI *"Core"* Questions

1 Structures having a similar origin but adapted for different purposes, such as the flipper of a whale and arm of a human, are called
 1 homozygous structures
 2 homologous structures
 3 identical structures
 4 embryological structures

2 Which group of organisms is believed to be among the earliest to evolve on Earth?
 1 arthropods
 2 coelenterates
 3 protozoans
 4 reptiles

3 The constant change in the kinds and frequencies of genes in a population over a period of time is a modern definition of
 1 a gene pool
 2 evolution
 3 Mendelian heredity
 4 genetics

4 The fact that humans and sheep have similar hormones suggests that humans and sheep
 1 are members of the same species
 2 share the same environment
 3 have similar nucleic acids
 4 appeared on Earth at the same time

5 The leg structures of many different vertebrates are quite similar in number and location of bones. Most scientists would probably explain this on the basis of
 1 need of the organism
 2 common ancestry
 3 chance occurrence
 4 inheritance of acquired traits

6 Some starfish larvae resemble some primitive chordate larvae. This similarity may be used to suggest that primitive chordates
 1 share a common ancestor with starfish
 2 evolved from modern-day starfish
 3 evolved before starfish
 4 belong to the same population as starfish

7 According to modern biologists, heredity variations are due to the
1 use or lack of use of organs
2 inheritance of characteristics acquired during development
3 need for adapting to a changed environment
4 genetic changes resulting from mutation and recombination

8 Which sequence most probably occurred in the evolution of animals on Earth?
1 mammals → fish → coelenterates → protozoa
2 coelenterates → protozoa → fish → mammals
3 protozoa → fish → mammals → coelenterates
4 protozoa → coelenterates → fish → mammals

9 According to a modern evolution theory, which factor favors speciation?
1 asexual reproduction
2 geographic isolation
3 gene pool stability
4 use and disuse

10 Which factor would have the least effect on changing the gene frequencies in a population?
1 asexual reproduction in the population
2 migration of half the population
3 separating the population into two groups by a geographic barrier
4 a change in the population's environment

11 Favorable adaptations are genetic characteristics that
1 cannot be passed on to the next generation
2 are acquired by increased use of an organ
3 reduce the organisms' chances of survival
4 improve the organisms' chances of survival

12 According to the heterotroph hypothesis, which gas was absent in primitive Earth's atmosphere?
1 methane
2 ammonia
3 molecular oxygen
4 water vapor

Base your answers to questions 13 and 14 on the diagram and information below and on your knowledge of biology.

The diagram represents jars containing all the nutrients necessary for the growth and reproduction of fruit flies. A strip of sticky flypaper was suspended from the top of the experimental jar. Some fruit flies with wings and some fruit flies that lacked wings were placed in both jars. After a week, only the wingless flies were alive in the experimental jar, while in the control jar, both varieties of flies were still alive, as indicated in the diagram.

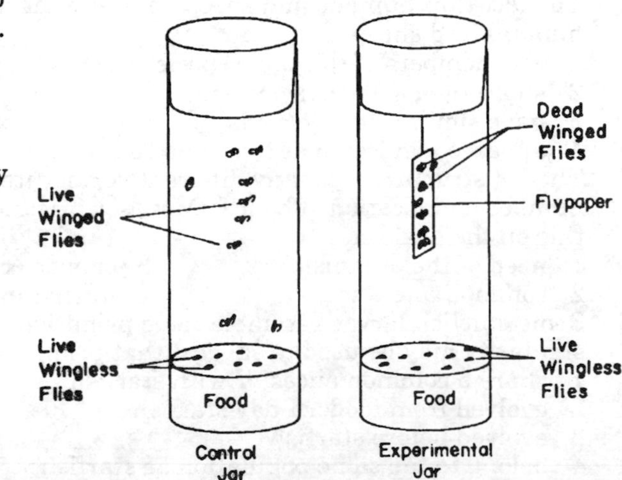

13 Darwin would have been most likely to agree with which statement describing the results of this investigation?
  1 Mutation is responsible for the survival of the control jar flies.
  2 The type of food provided in the experimental jar did not promote survival of winged flies.
  3 In all environments, wingless flies are better adapted for survival than winged flies.
  4 The flypaper is a selecting agent against the winged flies.

14 Which is the best conclusion to be drawn from this investigation?
  1 Winglessness is an advantage in the experimental jar.
  2 Winglessness is a disadvantage in the control jar.
  3 The winged trait is an advantage in the experimental jar.
  4 The winged trait is a disadvantage in the control jar.

15 The undisturbed upper layers of rocks usually contain fossils of organisms that are
  1 more complex than those found the lower layers
  2 less complex than those found in the lower layers
  3 identical to those found in the lower layers
  4 different from any other fossils found in the same layer

16 Based on Lamarck's theory, what is an explanation for the evolution of a tailless species of monkey from an ancestor that had a tail?
  1 A gene for a tailless trait is dominant over a gene for a tailed trait.
  2 A mutation occurred in the tailed species.
  3 The tailless monkey did not overpopulate the inhabited area.
  4 The tail was no longer needed to help the monkey escape predators.

17 In future generations, the frequency of a mutant gene may tend to increase. This would occur in an environment if the new trait has
  1 the ability to form a fossil
  2 a high survival value
  3 a common ancestry
  4 no advantage in competition

18 Which theory is best illustrated by the flow chart below?

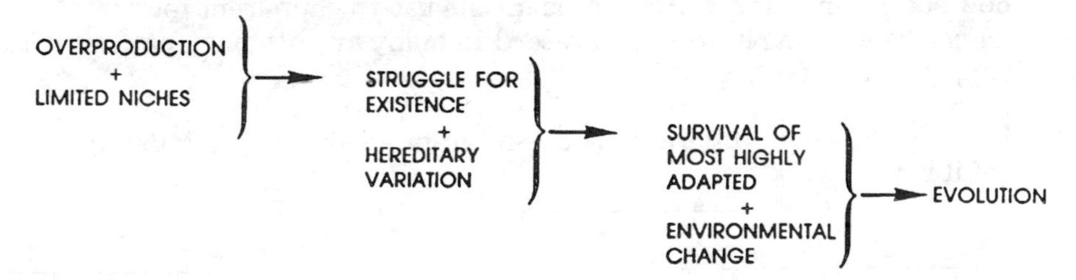

  1 cell theory
  2 theory of acquired characteristics
  3 use and disuse theory
  4 theory of natural selection

19 A supporter of the evolutionary theory set forth by Lamarck would probably theorize that the giraffe evolved a long neck due to
  1 need and inheritance of acquired traits
  2 mutations and genetic recombination
  3 variations and survival of the fittest
  4 overproduction and struggle for survival

20 Which is one basic assumption of the heterotroph hypothesis?
   1 More complex organisms appeared before less complex organisms.
   2 Living organisms did not appear until there was oxygen in the atmosphere.
   3 Large heterotrophic organisms appeared before small photosynthesizing organisms.
   4 Autotrophic activity added carbon dioxide to the environment.

SELF–HELP: Unit VI *"Extended"* Questions

There is *no* extended area in Evolution.

SELF–HELP: Unit VI *"Skill"* Questions

Base your answers to questions 1 through 3 on the reading passage below.

Time Frame for Speciation

Evolution is the process of change through time. Theories of evolution attempt to explain the diversification of species existing today. The essentials of Darwin's theory of natural selection serve as a basis for our present understanding of the evolution of species. Recently, some scientists have suggested two possible explanations for the time frame in which the evolution of species occurs.
Gradualism proposes that evolutionary change is continuous and slow, occurring over many millions of years. New species evolve through the accumulation of many small changes. Gradualism is supported in the fossil record by the presence of transitional forms in some evolutionary pathways.

Punctuated equilibrium is another possible explanation for the diversity of species. This theory proposes that species exist unchanged for long geological periods of stability, typically several million years. Then, during geologically brief periods of time, significant changes occur and new species may evolve. Some scientists use the apparent lack of transitional forms in the fossil record in many evolutionary pathways to support punctuated equilibrium.

1 Identify one major difference between gradualism and punctuated equilibrium. ______________________________

____

2 According to the theory of gradualism, what may result from the accumulation of small variations? ______________________________

____

3 What fossil evidence indicates that evolutionary change may have occurred within a time frame known as gradualism?

____

____

Base your answers to questions 4 through 6 on the information below and on your knowledge of biology.

Man has modified some animal species by breeding only those that have certain desirable traits. As a result, we have race-horses and greyhounds that are faster than their predecessors.

In a similar way many animals have been modified naturally. The giraffe has long forelegs and a long neck, head, and tongue which make it well adapted for browsing in the higher branches of trees. Therefore the giraffe can obtain food that is beyond the reach of other animals, especially during droughts. Ancient populations of giraffes varied in the relative length of their body parts. Those giraffes that were able to browse the highest were more likely to survive. They mated and their offspring often inherited the structural characteristics suitable for high browsing. The giraffes that could not reach the food supply most likely died of starvation and therefore did not produce as many offspring as those that could reach higher.

4 The variations to which the author refers are the direct result of
1 asexual reproduction
2 regenerative ability
3 inherent need
4 gene recombination

5 The modification of some animal species by humans, as described in the reading passage, results from the process known as
1 natural selection
2 artificial selection
3 vegetative propagation
4 chromosomal mutation

6 Which idea included in Darwin's theory of evolution is not found in the reading passage?
1 variation
2 overproduction
3 struggle for existence
4 survival of the fittest

Base your answers to questions 7 through 10 on the reading passage below.

The baleen whales belong to the order Mysticeti. Instead of teeth, baleen whales possess plates on each side of the jaw that filter food from the water. These whales collect food while they swim. Baleen whales feed on strained plankton* and krill. Krill is a crustacean that measures 7.5-12.5 centimeters in length and grows most plentifully in the cold oceanic water near the North and South Poles.

Even though they feed on such small animals, many species of baleen whales grow to enormous sizes. The blue whale (*Balaenoptera musculus*) reaches lengths of 35 meters with recorded masses of 100,000 kilograms. Heavy whaling of this valuable species has put it in danger of extinction. Among the other baleen whales are the grey whale (*Eschrictus glaucos*) which grows to 15 meters, the minke whale (*Balaenoptera acutorostrata*) which grows to 10 meters, the right whale (*Balaena mysticetue*) which grows to 20 meters, and the humpback whale (*Megaptera novaengliae*) which grows to 13 meters.

-National Geographic (adapted)

*Plankton: aquatic organisms that live suspended in open water

7 Blue whales and grey whales are marine organisms that possess
1 jaws for chewing food
2 plates for straining food
3 spouts for collecting food
4 teeth for grinding food

8 The major food supply for baleen whales comes from the
1 continent of Antarctica
2 tidal pools along seashores
3 plant life on the ocean floors
4 open waters of the oceans

9 The humpback whale is classified as belonging to the order
1 *Balaena*
2 *Megaptera*
3 *Mysticeti*
4 *novaengliae*

10 Even though they feed on such small animals, the grey whales grow to a length of
1 10 meters
2 15 meters
3 35 meters
4 20 meters

Unit 7 Ecology

Objectives

The student should be able to:

- Describe the interdependence of organisms on each other and on their environment.
- Identify and define the ecological levels of organization of the living world.
- Identify and describe the components that form and maintain an ecosystem.
- Explain how interactions of living organisms with each other and their environment result in succession.
- Assess human influence on the balance of nature.

Ecology – Terms & Concepts

Ecology
- Population - Community
- Biosphere

Ecosystems
- Abiotic (non-living)
- Biotic (living)

Heterotrophs
- Saprophytes - Herbivores
- Carnivores - Omnivores

Symbiotic Relationships
- Commensalism - Mutualism
- Parasitism

I. Ecology

Ecology is the study of the interactions among organisms and their interrelationships with the physical environment. No organism exists as an entity, separate and distinct from its environment. All living organisms are dependent upon other living things as well as dependent on the nonliving environment.

II. Ecological Organization

Population. A population is all the members of a species inhabiting a given location at a specific time.

Community. All the interacting populations in a given area represent a community and are dependent upon each other.

Ecosystem. An ecosystem is the living community and the physical environment functioning together as an interdependent, self-sufficient, and relatively stable system.

Biosphere. The biosphere is the portion of the Earth in which life exists. It is composed of numerous and varied complex ecosystems.

III. Ecosystems

The **ecosystem** is the structural and functional unit studied in ecology. It is the lowest level of organization in ecology in which all living and nonliving environmental factors exist and interact.

A. Ecosystem Structure and Function

An ecosystem involves interactions between **abiotic** (nonliving) and **biotic** (living) factors. An ecosystem is a self-sustaining unit if the following requirements are met:

- It must have a **constant source of energy** and a living system capable of incorporating this energy into organic compounds.
- It must have mechanisms to **cycle and recycle materials** between the organisms and their environment.

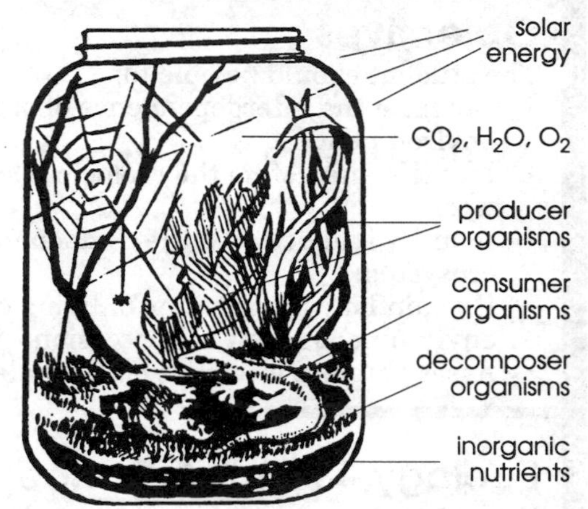

Terrarium: Balanced Ecosystem

1. Components of an Ecosystem

Components of the ecosystem involve interacting abiotic and biotic factors.

Abiotic Factors

The **abiotic environment** includes physical and chemical factors which affect the ability of organisms to live and reproduce.

The abiotic factors include:

- Intensity of **light**
- Range of **temperatures**
- Amount of **moisture**
- Type of **substratum**
- Availability of **inorganic substances** such as minerals
- Supply of **gases** such as O_2, CO_2, N_2
- **pH**

Each of these factors varies in the environment and, as such, may act as a **limiting factor**, determining the types of organisms which may exist in that environment. Examples of limiting factors include:

- Some plants live well on a forest floor under tall trees, but would not do well in an open field due to the increased intensity of **light**.
- A low annual **temperature** common to the northern latitudes determines in part what species of plants can exist in that area because enzymes of different species work best at different temperatures.

- The amount of **oxygen** dissolved in a body of water will help determine which species will exist there. Fish that need high oxygen levels would suffocate and die with a severe drop in the oxygen level in the water.
- The salt-laden **air** and **water** of coastal areas limit what species can exist in those regions. Some species of fish, shellfish, and other marine species would die in freshwater due to an imbalance of water pressure in their tissues.
- More than 200 lakes in the Adirondacks have virtually no life in them because of low pH caused by acid rain.

Biotic Factors

Biotic factors are all the living things that directly, or indirectly, affect the environment. Thus, the organisms, their presence, parts, interaction, and wastes, all act as biotic factors. Biotic factors interact in many ways such as in **nutritional relationships** and **symbiotic relationships**.

Nutritional Relationships

Nutritional relationships involve the transfer of nutrients from one organism to another within an ecosystem.

Autotrophs. These organisms can synthesize their own food (organic nutrients, such as carbohydrates, proteins, lipids, and nucleic acids) from inorganic compounds and a usable energy source.

Heterotrophs. These organisms can *not* synthesize their own food and are dependent upon other organisms for food. On the basis of this dependency, organisms are classified as either saprophytes, herbivores, carnivores, or omnivores.

- **Saprophytes** include those heterotrophic (nongreen) plants, fungi, and bacteria which live on dead matter. Saprophytes function to recycle materials in the environment. Examples include mushrooms, bread mold, and bacteria of decay.
- **Herbivores** are those animals which consume plants. Herbivores include the "grazing" animals such as cows, rabbits, and deer.
- **Carnivores** are those animals which consume other animals. These include: **predators** which are animals which kill and consume their prey (such as wolves and eagles); and, **scavengers** which are animals which feed on other animals they have not killed (such as buzzards and crabs).
- **Omnivores** are those animals that consume both plants and animals. The human is an example of an omnivore.

Extended Area: **Ecology**

Symbiotic Relationships

Different organisms live together in a close association (dependency) which may include: nutritional, reproductive, and protective relationships. This living together in close association is known as **symbiosis**. Symbiotic relationships may or may not be beneficial to the organisms involved. Types of symbiosis include:

Commensalism. In this relationship one organism is benefited and the other is not adversely affected **(+,0)**. Examples include: barnacles on whales, orchids on large tropical trees, and the remora fish and shark.

Mutualism. In this relationship both organisms benefit from the association **(+,+)**. Examples include: nitrogen-fixing bacteria within the nodules of legumes, certain protozoa within termites, and a flower and bee.

Parasitism. In this relationship, the parasite benefits at the expense of the host **(+,-)**. Examples include: athlete's foot fungus on humans and tapeworm, heartworm, and fleas in dogs.

Recent research indicates that lichens may represent a *controlled parasitic relationship* of the fungus on an algae host.

Questions

1 The study of the interrelationships of plants and animals and their interaction with the physical environment is known as
  1 evolution
  2 ecology
  3 anatomy
  4 taxonomy

2 Because of the foods they eat, most humans are classified as
  1 autotrophs
  2 carnivores
  3 herbivores
  4 omnivores

3 All the plant and animal life present in one cubic foot of soil makes up a
  1 community
  2 population
  3 species
  4 biosphere

4 Which term describes all the individuals of any one species present in a particular environment?
  1 a community
  2 an ecosystem
  3 a biosphere
  4 a population

5 Knowledge of ecology would be used most directly in studying the
1 production of hormones and neurotransmitters in two related organisms
2 current decline of bighorn sheep in the Rocky Mountains
3 structure of subcellular organelles
4 biochemical nature of genetic transmission

6 A natural community interacting with its abiotic environment is a description of
1 a population
2 an organ system
3 an organism
4 an ecosystem

7 All the brook trout in a stream constitute
1 an ecological succession
2 a population
3 a habitat
4 a food chain

8 The timber wolves, rabbits, bacteria, insects, and vegetation in a particular region of northern Utah all together constitute part of a
1 population 2 community 3 genus 4 species

9 Which represents a community?
1 all the *Paramecium caudatum* in a pond
2 the abiotic factors in Lake Michigan
3 all the interacting populations in a forest
4 the concentration of minerals in soil

10 Which sequence shows increasing complexity of levels of ecological organization?
1 biosphere, ecosystem, community
2 biosphere, community, ecosystem
3 community, ecosystem, biosphere
4 ecosystem, biosphere, community

11 In a community, the most severe competition develops among those organisms which
1 are active only during the night
2 belong to two different genera
3 depend upon autotrophs for food
4 occupy the same ecological niche

12 Different species of animals in a community would most likely be similar in
1 physical structure
2 abiotic requirements
3 size
4 number of offspring produced

13 Which is an example of a biotic factor that would limit the size of a deer herd?
1 populations of predators
2 severe summer drought
3 lack of oxygen at high altitudes
4 heavy winter snowfalls

14 Fungi and bacteria that depend on dead organic material for their existence are classified as
1 decomposers
2 predators
3 omnivores
4 herbivores

15 Which is a biotic factor that affects the size of a population in a specific ecosystem?
1 the average temperature of the ecosystem
2 the number and kinds of soil minerals in the ecosystem
3 the number and kinds of predators in the ecosystem
4 the concentration of oxygen in the ecosystem

16 The graph at the right represents a predator-prey relationship. What is the most probable reason for the increasing predator population from day 5 to day 6?
  1 an increasing food supply from day 5 to day 6
  2 a predator population equal in size to the prey population from day 5 to day 6
  3 the decreasing prey population from day 1 to day 2
  4 the extinction of the yeast on day 3 and day 10

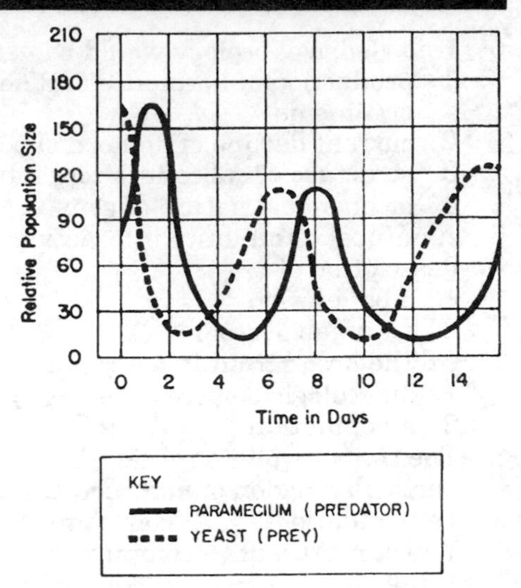

17 The elements stored in living cells of organisms in a community will eventually be returned to the soil for use by other living organisms. The organisms which carry out this process are
  1 producers 2 carnivores 3 herbivores 4 decomposers

18 An ecosystem, (such as a balanced aquarium), is self-sustaining if it involves the interaction between organisms, a flow of energy, and the presence of
  1 equal numbers of plants and animals
  2 more animals than plants
  3 material cycles
  4 pioneer organisms

19 Which sequence of organisms best represents the flow of energy in an ecosystem?
  1 autotrophs → herbivores → carnivores
  2 secondary consumers → producers → heterotrophs
  3 carnivores → decomposers → producers
  4 consumers → heterotrophs → saprophytes

20 Which condition would most likely upset the dynamic equilibrium of an ecosystem?
  1 a constant source of energy entering the environment
  2 a cycling of elements between organisms and the environment
  3 organisms capable of using radiant energy to make organic compounds
  4 a greater number of heterotrophs than autotrophs

2. Energy Flow Relationships

Ecology – Terms & Concepts

Energy Flow
- Food Chains (direct link)
- Food Webs (complex)

Organism Relationships
- Producers
- Consumers
- Decomposers

Material Cycles
- Carbon - Hydrogen - Oxygen
- Water - Nitrogen Cycles

Succession
- Pioneer Organisms
- Limiting Factors - Competition
- Climax Community

If an ecosystem is to be self-sustaining it must contain a constant supply of energy which is available to all the organisms within the ecosystem. The energy must flow from organism to organism.

Energy Flow

Those life activities which are characteristic of living organisms require the expenditure of energy. The pathways of energy through the living components of an ecosystem are represented by food chains and food webs.

Example of a Food Chain

Food Chain

Green plants convert radiant energy from the Sun into chemical energy (food). A **food chain** involves the transfer of energy from green plants through a series of organisms with repeated stages of eating and being eaten. For example, green grass obtains its energy directly from Sunlight; in turn, a frog may obtain its energy from the plant; a snake could use the frog as its source of energy; and finally, when the snake dies, its remains may be consumed by bacteria and fungi providing them with an energy source and recycling materials back into the ecosystem.

Food Web

In a natural community, the flow of energy and materials is much more complicated than is illustrated by one food chain. Since practically all organisms may be consumed by more than one species, many interactions occur among the food chains of any community. These interactions are described as a **food web**. Interactions in a food web involve:

- **Producers**. The energy for a community is derived from the organic compounds synthesized by green plants. *Autotrophs* are therefore considered the *primary producers* in all ecosystems.

- **Consumers**. Organisms that feed directly upon green plants are *primary consumers* or herbivores. *Secondary consumers*, or carnivores, feed upon other consumers. Omnivores may be either primary or secondary consumers.

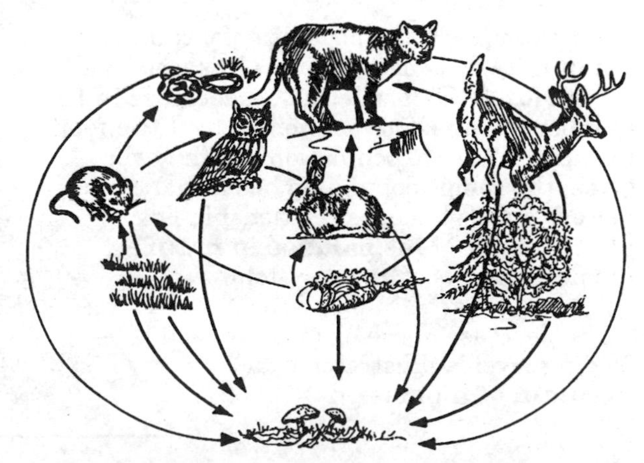

Example of a Complex Food Web

- **Decomposers**. Organic wastes and dead organisms are eventually broken down to simpler substances by **decomposers**, such as the bacteria of decay. Through this action, chemical substances are returned to the environment where they can be used by other living organisms.

Pyramid of Energy

There must be much more energy at the producer level in a food web than at the consumer levels. In turn, there is more energy at the primary consumer level than at the secondary consumer level. A pyramid of energy can be used to illustrate the *loss of usable energy* at each feeding level.

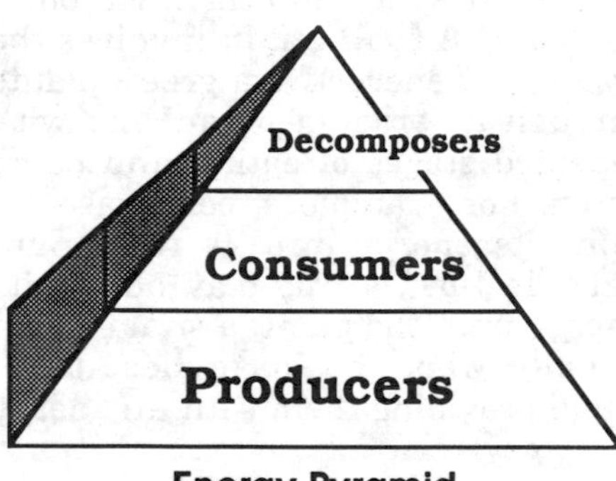

Energy Pyramid Relationships

Each consumer level of the food pyramid utilizes approximately 10% of its ingested nutrients to build new tissue. This new tissue represents the food for the next feeding level. The remaining energy is lost in the form of heat and unavailable chemical energy.

Eventually, the energy in an ecosystem is lost and is radiated from the Earth's system. Thus, an ecosystem *cannot* sustain itself without the *constant input of energy* from the Sun.

Extended Area: ***Ecology***

Biomass Pyramid

In general, the decrease of energy at each successive feeding level means that *less* biomass (amount of organic matter) can be supported at each level. Thus, the total mass of carnivores in a particular ecosystem is less than the total mass of the producers.

For example, if the population of rabbits in a community decreases, there is less food available for the foxes. This will cause a decrease in the number of foxes born as there is too little food to support a large population. With fewer foxes, the rabbit population has a chance to increase. The biomass relationship is a good example of the **balance in nature**, the *homeostasis* of an ecosystem.

The decrease of the biomass at each feeding level is illustrated by a **pyramid of biomass**.

Questions

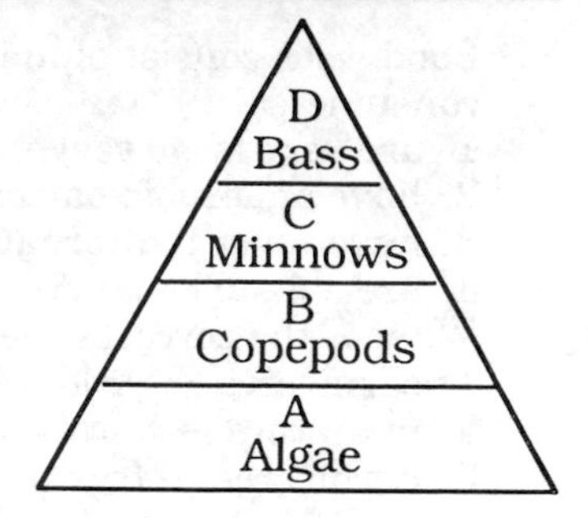

Base your answers to questions 1 through 3 on the diagram at the right which represents a food pyramid for organisms inhabiting a pond.

1 At which level of the food pyramid is the smallest percentage of total stored energy found?
 1 *A*
 2 *B*
 3 *C*
 4 *D*

2 Which organisms in the food pyramid function as primary consumers?
 1 bass
 2 minnows
 3 copepods
 4 algae

3 What is the original source of energy for all organisms in this food pyramid?
 1 water
 2 sunlight
 3 the substratum
 4 carbon dioxide

Base your answers to questions 4 through 7 on the diagram at the right which represents a food web. The numerals I, II, III, and IV represent four nutritional levels within the community in which different species compete.

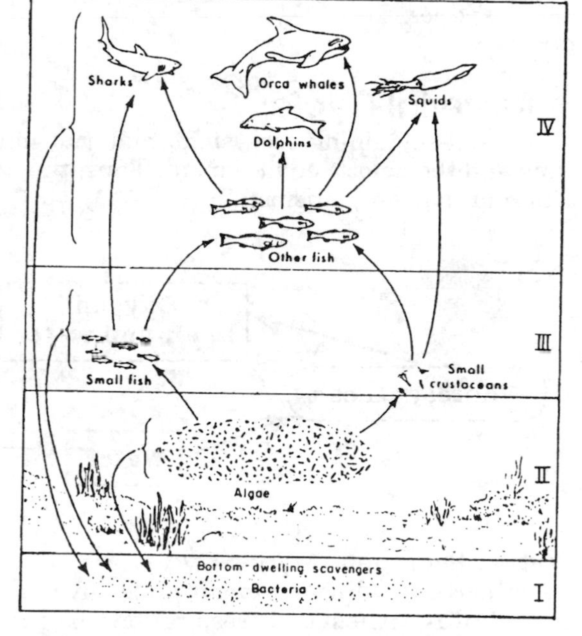

4 Which is an abiotic factor that is needed to make this marine community a self-sustaining ecosystem?
 1 sunlight
 2 primary consumers
 3 producers
 4 decomposers

5 Which organisms are producers?
 1 sharks
 2 algae
 3 small crustaceans
 4 bottom-dwelling scavengers

6 When an organism in this web dies, the organic molecules in its body are finally broken down and made available for recycling by the action of organisms in level
 1 I
 2 II
 3 III
 4 IV

7 Which level in this marine community contains the greatest total amount of energy?
 1 I
 2 II
 3 III
 4 IV

8 Food webs consist of many predator-prey relationships because many consumers
   1 are very large scavengers
   2 have anaerobic patterns of respiration
   3 have several alternate nutrient supplies
   4 are able to carry on photosynthesis

9 What is the correct order of organisms in a food chain?
   1 carnivores → producers → herbivores
   2 producers → herbivores → carnivores
   3 herbivores → producers → carnivores
   4 producers → carnivores → herbivores

10 Which statement best explains why a food web is a more realistic representation of nutritional patterns than a food chain?
   1 Energy is consumed in metabolic activities and gained at every feeding level.
   2 Decomposers return materials from the top-order predators to inorganic form.
   3 Energy always flows through the consumers to the producers.
   4 Practically all species are consumed by or feed on more than one species.

3. Material Cycles

In a self-sustaining ecosystem, material must be cycled among the organisms and the abiotic environment. Thus, the same materials can be reused by different living organisms.

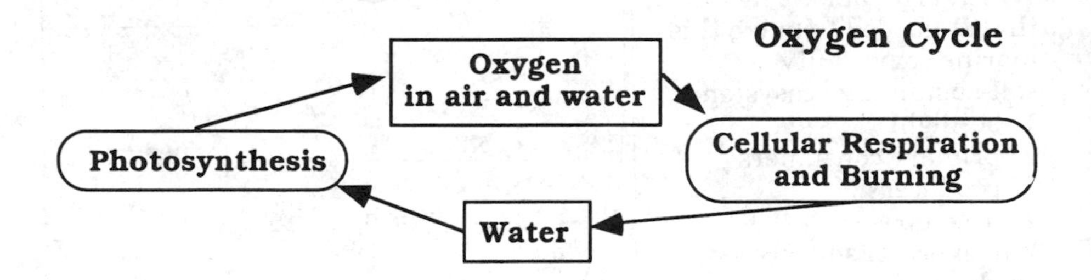

Carbon-Hydrogen-Oxygen Cycle

The carbon-hydrogen-oxygen cycle involves the processes of **respiration** and **photosynthesis**. In respiration, oxygen and glucose are combined releasing energy and producing water and carbon dioxide. In photosynthesis, water and carbon dioxide with the energy from the Sun are combined to produce glucose (containing the energy) and oxygen. Each process compliments the other, and the ecosystem maintains its balanced communities.

Water Cycle

Water is vital to all living organisms and is a primary limiting factor within any ecosystem. The water cycle involves the processes of **photosynthesis**, **transpiration**, **evaporation** and **condensation**, **respiration**, and **excretion**.

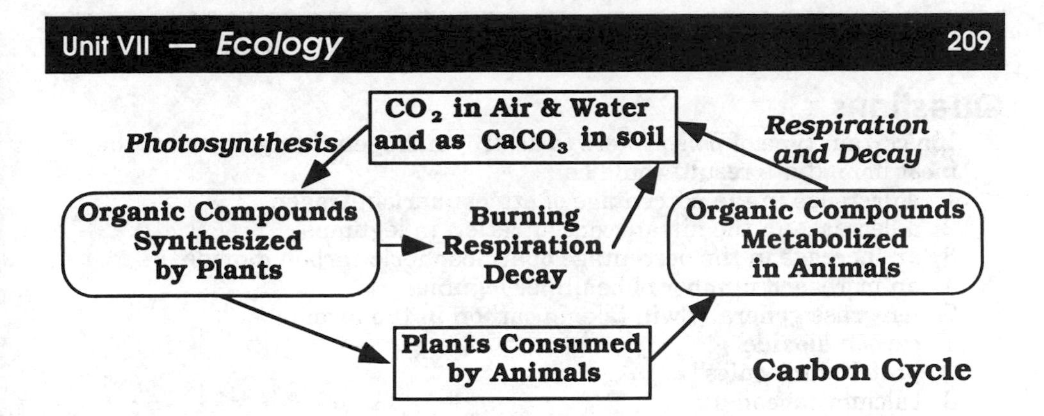

Nitrogen Cycle

The nitrogen cycle "recycles" the nitrogen necessary for the production of proteins, essential to all living things. It is an example of a material cycle involving decomposers and other soil bacteria which, in part, break down and convert nitrogenous wastes and the remains of dead organisms into materials usable by autotrophs.

Extended Area: ***Ecology***

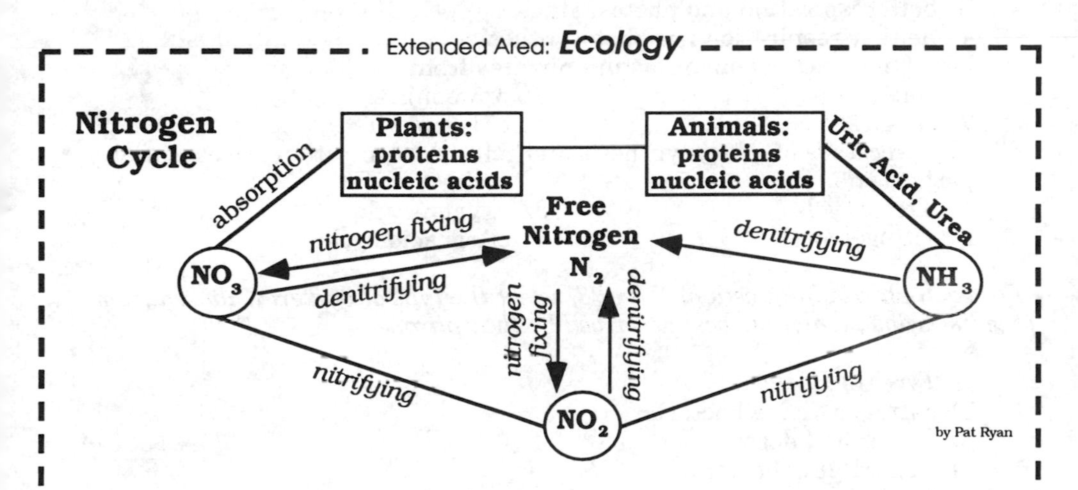

The essential parts of the nitrogen cycle include:

- Atmospheric nitrogen is converted into nitrates by **nitrogen-fixing bacteria.**
- Plants use **nitrates** for protein synthesis.
- Animals which eat plants convert the plant protein into animal protein.
- Nitrogenous wastes and the bodies of dead plants and animals are broken down by decomposers (**bacteria of decomposition**) and ammonia is released.
- Ammonia may be converted into nitrates by **nitrifying bacteria**.
- Nitrogen containing compounds may also be broken down by **denitrifying bacteria**, resulting in the release of nitrogen into the atmosphere.

Questions

1 If a certain type of poison were to destroy nitrogen-fixing bacteria, the most immediate result would be
1 a decrease in the percentage of atmospheric nitrogen
2 a decrease in the nitrate concentration in legumes
3 an increase in the percentage of atmospheric carbon dioxide
4 an increased number of healthier legumes

2 Green grass generally will take in carbon in the form of
1 carbon dioxide
2 protein molecules
3 calcium carbonate
4 carbon monoxide

3 Most plants can use nitrogen in the form of
1 gaseous nitrogen
2 uric acid
3 ammonia
4 nitrates

4 The carbon cycle and the oxygen cycle involve
1 respiration only
2 photosynthesis only
3 both respiration and photosynthesis
4 neither respiration nor photosynthesis

5 Nitrifying bacteria manufacture nitrates from
1 water
2 cellulose
3 ammonia
4 chlorophyll

6 The processes of photosynthesis and cellular respiration are essential parts of the
1 Kreb's cycle
2 nitrogen cycle
3 oxygen cycle
4 citric acid cycle

For each phrase in questions 7 and 8, select the type of bacteria, chosen from the list below, which is best described by that phrase.

Types of Bacteria
1) Nitrogen-fixing bacteria
2) Bacteria of decay
3) Denitrifying bacteria
4) Nitrifying bacteria

7 Returns molecular nitrogen to the atmosphere

8 Converts gaseous nitrogen into nitrates

9 Bacteria capable of fixing the nitrogen in the air into nitrates are found in the roots of
1 corn plants
2 tomato plants
3 legume plants
4 lettuce plants

10 The processes of photosynthesis, transpiration, evaporation and condensation, respiration, and excretion are all part of the
1 nitrogen cycle
2 carbon cycle
3 energy cycle
4 water cycle

B. Ecosystem Formation

Ecosystems tend to go through dynamic change with time until a stable system (climax community in a state of equilibrium) is attained. The type of ecosystem that is formed depends on the climatic limitations of a particular geographical area.

1. Succession

The replacement of one community by another until a stable stage (**climax community**) is reached is called **ecological succession**.

Extended Area: ***Ecology***

Pioneer Organisms

Succession may be said to begin with **pioneer organisms**, since these are the first living things to populate a given location. For example, lichens (a symbiotic association between fungus-alga) are the pioneer organisms on bare rock. Pioneer organisms modify their environment. Seasonal die-back and erosion, for example, would create pockets of **"soil"** in the crevices in bare rock.

Changes in Succession

Each community modifies the environment, often making it more unfavorable for itself and, apparently, more favorable for the following community which infiltrates the first community over a period of years. For example, as lichens grow and reproduce, they add organic matter and moisture to their substratum. After a period of time, humus is made and is too rich and moist for the lichen to survive. The lichens die but produce a richer substratum that will support seeds for the development of grasses and herbs, the next stage of succession.

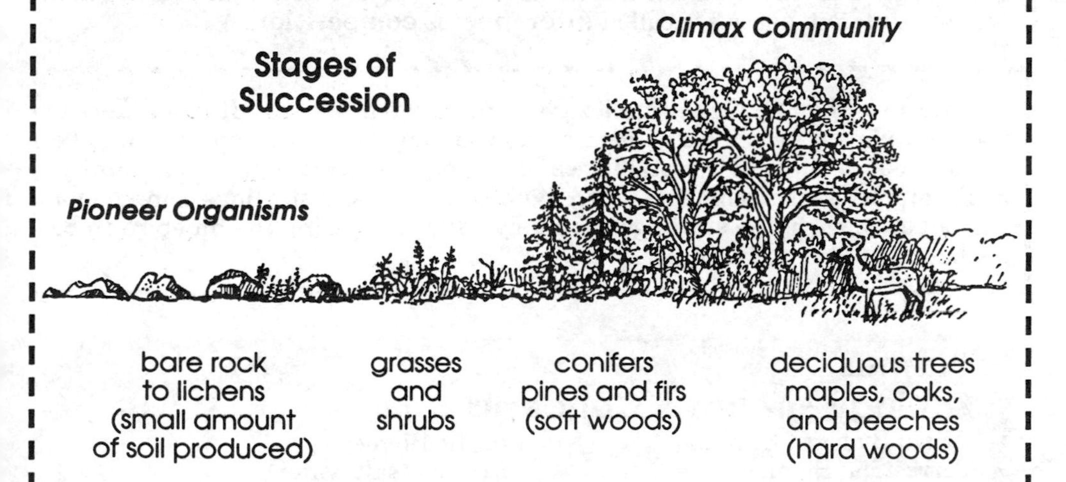

A typical successional sequence might be:

lichen (pioneer), grass, shrub, conifer, and deciduous woodland (climax)

Plant species (**flora**) dominate in the sense that they are the most abundant food sources. **Plant succession** is a major limiting factor for animal (**fauna**) succession. Communities are composed of populations able to exist under the prevailing conditions and are identified by their *dominant* plant species — the one that exerts the most influence over the other species present. Some examples include: Pine Barrens and a Sphagnum Bog.

Climax Community

A climax community is a self-perpetuating community in which populations remain stable and exist in balance with each other and the environment. The oak-hickory and the hemlock-beech-maple associations represent two climax communities found in New York State. A **climax community** continues until a catastrophe or a change in a major biotic or abiotic factor alters or destroys it, thus producing "non-climax" conditions. Some examples of natural and man-caused factors that affect a climax community include:

- forest fires
- abandoned farmlands
- areas where topsoil has been removed

Thereafter, succession once again occurs leading to another climax community. The original climax community may be reestablished or a new climax community may be established if the abiotic environment has been permanently altered.

2. Competition

Competition occurs when different species of organisms living in the same environment (**habitat**) utilize the same limited resources, such as food, space, water, light, oxygen, and minerals. The more similar the requirements of the organisms involved, the more intense the competition. This competition *between* different species is called **interspecies competition**.

A **niche** is the role that a species plays in its environment. If two different species compete for the same food or reproductive sites, one species may be eliminated. This usually establishes one species per niche in a community. For example, in environments where both bluebirds and starlings compete for reproductive sites, the starlings are likely to "win" causing the bluebird to be pushed out of that community.

3. Biomes

Ecology – Terms & Concepts

World Biomes - Climate	Aquatic Biomes
Terrestrial Biomes	Marine (salt water)
Tundra - Taiga	Coastal Zone
Temperate - Deciduous	Freshwater
Tropical - Grassland - Desert	Latitude - Altitude Compared

The term **biome** refers to the most common climax ecosystem that will form in geographic regions of similar climatic conditions. Biomes are **terrestrial** or **aquatic**. The temperature deciduous forest of the U.S. is a terrestrial biome. The ocean is an aquatic biome.

Extended Area: ***Ecology***

Terrestrial Biomes

The major plant and animal associations on land are determined by the major climatic zones of the world, which are modified by local land and water conditions. Climates will vary as to **temperature**, **solar radiation**, and **precipitation**. The presence or absence of water is a major limiting factor for terrestrial biomes.

Characteristics. Land biomes are characterized and sometimes named by the climax vegetation in the region. The major land biomes, and their characteristic flora, and fauna are listed in the following chart.

World Biomes

BIOME	CLIMAX FLORA	CLIMAX FAUNA	CHARACTERISTICS
Tundra	lichens, mosses, grasses	caribou, snowy owl	permanently frozen subsoil
Taiga	conifers	moose, black bear	long, severe winters, summers with thawing subsoil
Temperate-Deciduous Forest	trees that shed leaves (deciduous trees)	grey squirrel, fox, deer	moderate precipitation, cold winters, warm summers
Tropical Forest	many species of broad-leaved plants	snake, monkey, leopard	heavy rainfall, constant warmth
Grassland	grasses	pronghorn antelope, prairie dog, bison	rainfall & temperature vary greatly, strong prevailing winds
Desert	drought-resistant shrubs and succulent plants	kangaroo rat, lizard	Sparse rainfall, extreme daily temperature changes

Geographic Factors. Climatic conditions change with latitude and altitude. Earth latitude and altitude are similar in that as both increase, the limiting factors change in a similar manner, and the organisms change as well. (Note: see chart on the next page)

Aquatic Biomes

Aquatic biomes represent the largest ecosystems on Earth. More than 70 percent of the Earth's surface is covered by water and most of the life on this planet exists under conditions where water is the principal external medium.

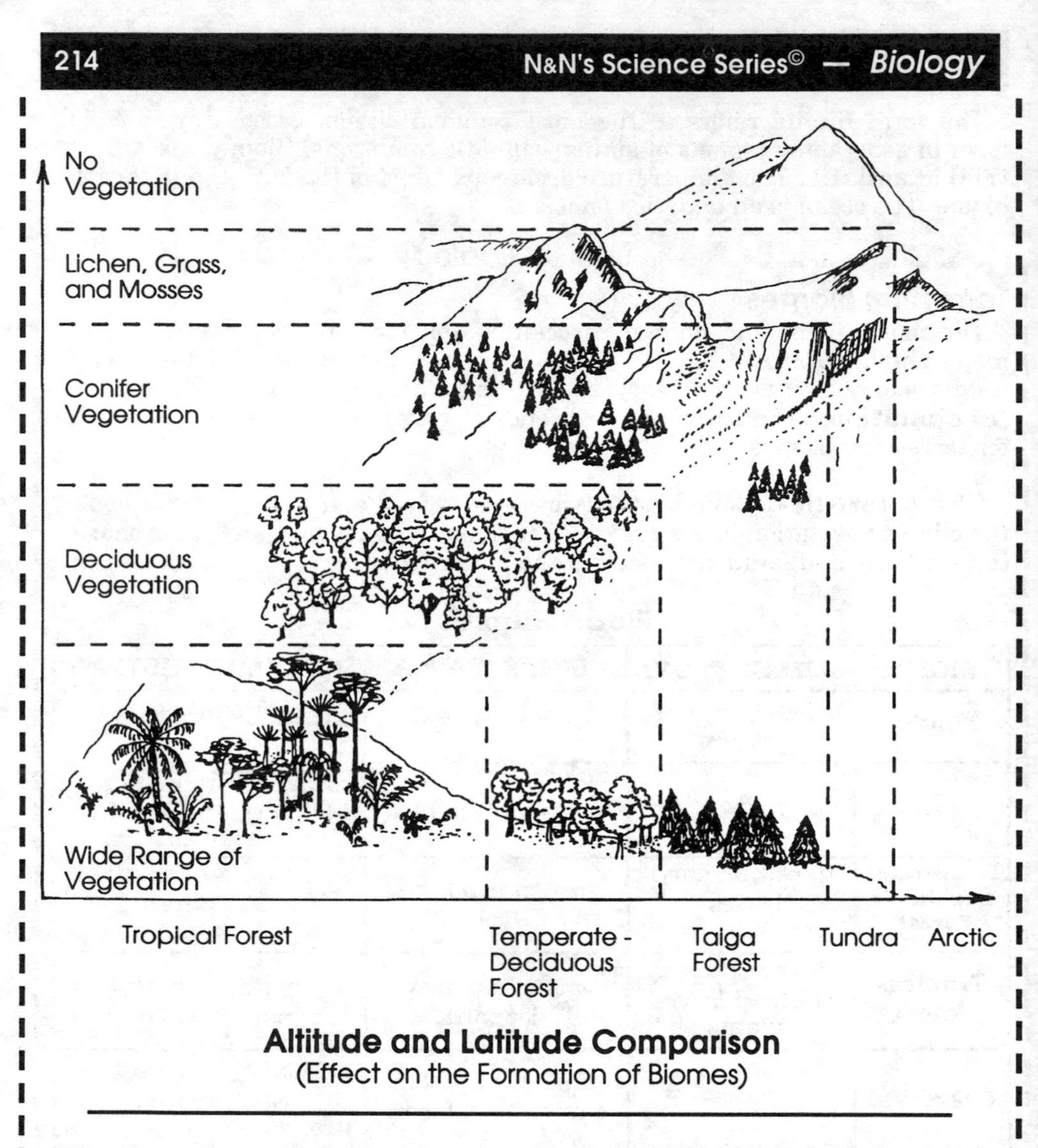

Altitude and Latitude Comparison
(Effect on the Formation of Biomes)

The temperature variation in aquatic biomes is not as great as in the terrestrial biomes, due to the ability of water to absorb and hold heat. In addition, moisture is not a limiting factor. Thus, aquatic biomes are typically more stable than terrestrial biomes.

Such factors as the quantity of available oxygen and carbon dioxide, temperature, light, dissolved minerals and suspended particles are the major factors affecting the kinds and numbers of organisms in an aquatic biome.

Aquatic organisms are well adapted to remove oxygen which is dissolved in water. They must also maintain a proper water balance. This water balance is affected by the concentration of dissolved salts in the water. Because light is most intense near the surface of the aquatic biome, plants exist there in large numbers and most photosynthesis takes place in the first one hundred feet of water.

Marine. The oceans of the world are a continuous body of water that

- Provides the most stable aquatic environment.
- Absorbs and holds large quantities of solar heat and helps to stabilize the Earth's atmosphere.
- Contains a relatively constant supply of nutrient materials and dissolved salts.
- Serves as a habitat for a large number of diverse organisms.

A great amount of food production in the world occurs in the oceans along the edges of the land masses (coastal waters), the deeper regions being too dark.

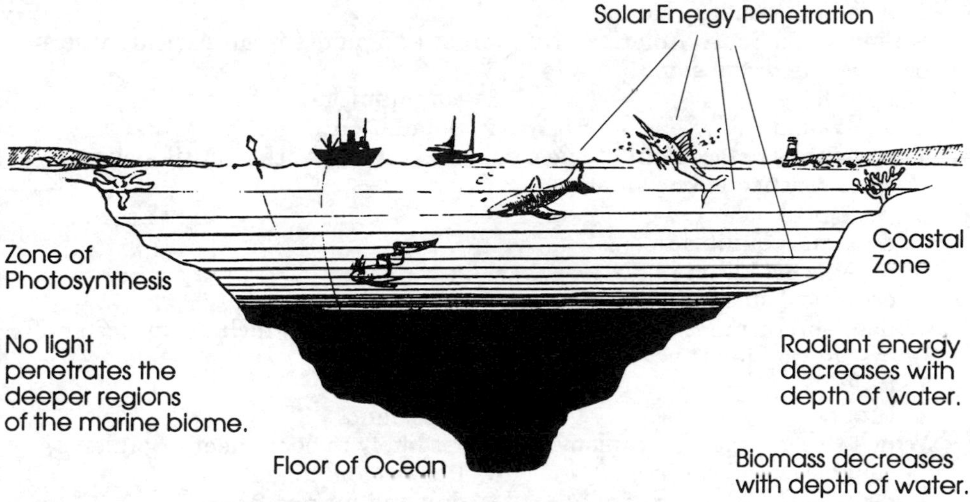

Solar Energy Penetration

Zone of Photosynthesis

Coastal Zone

No light penetrates the deeper regions of the marine biome.

Radiant energy decreases with depth of water.

Floor of Ocean

Biomass decreases with depth of water.

Marine Biome — Coastal Zone

Fresh Water. The fresh water biome includes ponds, lakes, and rivers. The areas which make up a fresh water biome show considerable variation in

- size
- current velocity
- temperature
- concentration of dissolved gases
- suspended particles
- rate of succession

Ponds and small lakes, for example, fill in due to the seasonal die-back of aquatic vegetation and the erosion of their banks. Eventually a small body of water enters into terrestrial succession terminating in a terrestrial climax community.

Questions

1 In a forest ecosystem, the abiotic factors include
1 light, temperature, and plants
2 animals, water, and soil
3 minerals, oxygen, and protists
4 water, soil, and temperature

2 In a given location, the replacement of one community by another until a climax stage is reached is referred to as
1 ecological succession
2 modern evolution
3 energy cycling
4 a food pyramid

3 Which condition is not necessary for an ecosystem to be self-sustaining?
1 a greater number of consumers than producers
2 the presence of decomposers
3 the presence of autotrophic organisms
4 a constant energy source

4 Animals in a South American rain forest and plants in an African desert both belong to the same
1 biosphere
2 ecosystem
3 community
4 population

5 The natural replacement of one community with another until a climax stage is reached is known as
1 ecological balance
2 dynamic equilibrium
3 organic evolution
4 ecological succession

6 Which land biome is characterized by conifers, which include spruce and fir, as the dominant vegetation?
1 taiga
2 tundra
3 desert
4 grassland

7 Which two groups of organisms are most likely to be pioneer organisms?
1 songbirds and squirrels
2 deer and black bears
3 lichens and algae
4 oak and hickory trees

8 Drastic changes in air temperature would be least likely to affect which biome?
1 tundra
2 marine
3 temperate deciduous forest
4 taiga

9 A wooded area which has undergone succession for 200 years consists of a mixture of mature beech and maple trees. Which is the most plausible assumption concerning this wooded area?
1 The trees are primary organisms.
2 The grasses will eventually take over.
3 The climax stage has occurred.
4 The bryophytes will eventually take over.

10 This biome receives the least amount of solar radiation. The ground is permanently frozen (permafrost) throughout the year. During the summer season, plants quickly grow, reproduce, and form seeds during their short life cycle. Lichens and mosses grow abundantly on the surface of the rocks.
1 grassland
2 desert
3 taiga
4 tundra

11 Following a major forest fire, an area that was once wooded is converted to barren soil. Which of the following schemes describes the most likely sequence of changes in vegetation in the area following the fire?
   1 shrubs → maples → pines → grasses
   2 maples → pines → grasses → shrubs
   3 pines → shrubs → maples → grasses
   4 grasses → shrubs → pines → maples

12 Which marine biome zone usually has the highest food production per unit volume?
   1 ocean floor zone
   2 deep ocean zone
   3 open ocean zone
   4 coastal ocean zone

13 In which of the following biomes does most of the photosynthesis taking place on the Earth occur?
   1 deciduous forests
   2 oceans
   3 deserts
   4 coniferous forests

14 This biome is found in the foothills of the Adirondack and Catskill Mountains of New York State and supports the growth of dominant vegetation including maples, oaks, and beeches.
   1 grassland
   2 temperate deciduous forest
   3 tundra
   4 taiga

15 This biome receives less than 10 inches of rainfall per year. Extreme temperature variations exist throughout the area over a 24-hour period. Water-conserving plants such as cacti, sagebrush, and mesquite are found.
   1 grassland
   2 temperate deciduous forest
   3 desert
   4 taiga

16 Generally, an increase in altitude has the same effect on the habitat of organisms as
   1 an increase in latitude
   2 a decrease in available light
   3 an increase in moisture
   4 a decrease in longitude

17 An abiotic factor which affects the ability of pioneer organisms such as lichens to survive is
   1 the type of climax vegetation
   2 the type of substratum
   3 the species of algae
   4 the species of bacteria

18 Which world biome has the greatest numbers of organisms?
   1 tundra
   2 temperate deciduous forest
   3 tropical forest
   4 marine

19 Which abiotic factor is most important in determining the type of land biome which usually develops in a particular region?
   1 type of vegetation
   2 rate of photosynthesis
   3 annual rainfall
   4 species of animals

20 Which is true of the major land biomes?
   1 They are characterized by the animals living in the region.
   2 They are unaffected by major climatic changes.
   3 They are named by the climax vegetation in the region.
   4 They are located predominantly at lower latitudes and altitudes.

IV. Biosphere and Humans

Ecology – Terms & Concepts

Environmental Modifications
- Population - Malthus Theory
- Resource Management
  - Game Animals
  - Importation of Organisms
  - Poor Land Use
- Pollution of Water, Air, Land
- Pest Control
  - Biocide - Herbicide

Future Technology

Waste Disposal
- Solid Waste - Land Fills
- Chemical and Nuclear
- Toxic Waste in Water, Land

Conservation of Natural Resources
- Population - Pollution Controls
- Flora and Fauna Preservation
- Biological Controls

Environmental Laws
- Freshwater Wetlands Acts

A. Past and Present

As is true with all living organisms, humans are dependent on a balanced environment. However, humans, in exercising a unique and powerful influence on the physical and living world, have modified their environment more than any other living thing.

1. Negative Aspects

Natural systems have been upset because humans have not realized that they not only influence other individuals, other species, and the nonliving world, but are, in turn, influenced by them. Although most ecosystems are capable of recovering from the impact of minor disruptions, human activities have sometimes increased the magnitude of such disruptions so as to bring about a more lasting and less desirable change in the environment upon which all life depends. *Such disruptions directly affect at least one of the components of an ecosystem and this, in turn, affect the remaining components.*

Human Population Growth

The total human population of the world has risen at a rapid rate, partly because of the removal of natural checks on the population, such as disease. This continued increase in the human population has far exceeded the food-producing capacities of many ecosystems of the world. The change in the world population of humans is illustrated in the population graph.

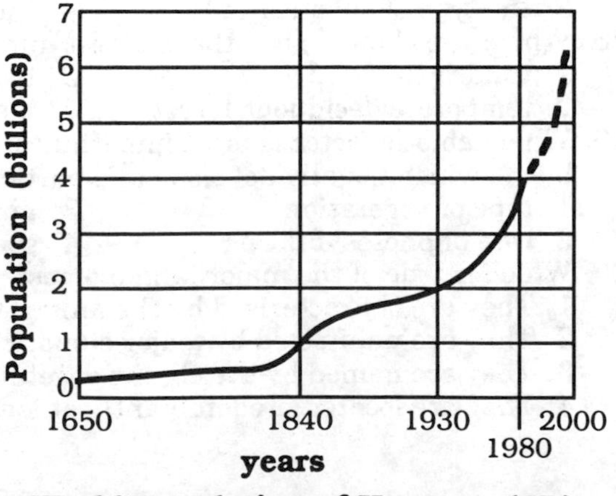

World Population of Humans (AD)

Human Activities

Some human activities have led to the extinction or endangerment of numerous species of plants and animals as well as producing less favorable living conditions for many species, including humans. Such activities include:

Overhunting. Uncontrolled hunting, trapping, and fishing still occur in many parts of the world. The extinction of the dodo bird and the passenger pigeon resulted from such activities. Several countries, including Japan, continue to hunt whales and other marine mammals in large numbers. This unchecked overhunting has endangered certain marine species including the blue whale.

Importation of Organisms. Humans have accidentally and/or intentionally imported organisms to areas where they have no natural enemies leading to the disruption of existing stable ecosystems. Examples include: the Japanese beetle, Gypsy Moth and disease-causing organisms such as those that cause Dutch Elm Disease.

Exploitation. The exploitation of wildlife, both flora and fauna, for their products and the pet trade has led to threatened populations and ecosystem disruptions. The African elephant and the Pacific walrus have been killed for just their ivory tusks. The desire of Americans to have exotic birds as pets has led to endangerment of the Colombian parrot. It is estimated that only one in 50 parrots survives the transition from the wild to private collectors. The Japanese plywood industry has depleted Southeast Asia and the Amazon region of Brazil of their tropical rain forests.

Poor Land Use Management. Increased urbanization and suburbanization claims increasing amounts of agricultural lands, modifies watersheds, disrupts natural habitats (including wetlands), and threatens the existence of wildlife species. Poor land use management practices have led to overcropping, overgrazing, and failure to use cover crops. This has resulted in the loss of valuable soil nutrients and topsoil. In regions of the world, such as northern Africa, the land has been so depleted of soil nutrients as to make it useless for farming. This has resulted in widespread starvation.

Technological Oversights

Technological oversights have led to unplanned consequences which have contributed to the pollution of the water, air, and land.

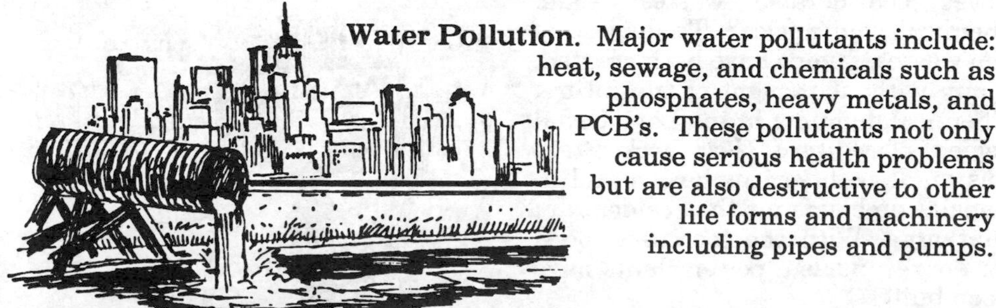

Water Pollution. Major water pollutants include: heat, sewage, and chemicals such as phosphates, heavy metals, and PCB's. These pollutants not only cause serious health problems but are also destructive to other life forms and machinery including pipes and pumps.

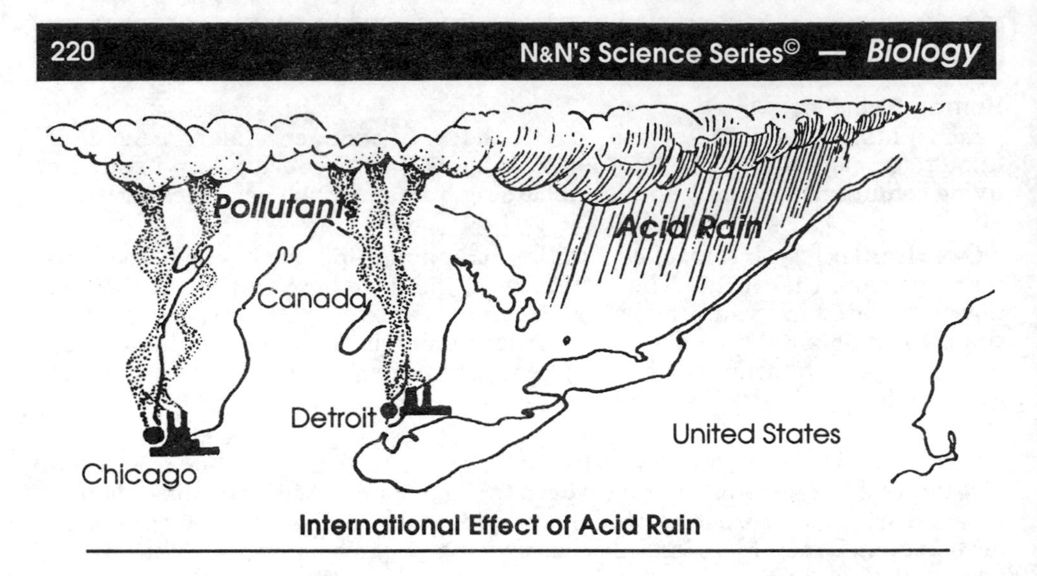

International Effect of Acid Rain

Air Pollution. Major air pollutants include: carbon monoxide, hydrocarbons, and particulates. Nitrogen oxides and sulfur dioxide combine with water vapor creating acid rain problems. Acid rain has been shown to be responsible for the destruction of populations of plants and animals in forests and bodies of water in both North America and Europe. In addition to the destruction of living things, air pollution causes damage to buildings, historic monuments, and machines. In Rome, the exhausts from cars has produced severe damage to many of the centuries old statues, monuments, and buildings that have even endured through wars.

Biocide Use. The use of some biocides (such as pesticide and herbicide) without a complete assessment of their environmental impact has contaminated the soil, atmosphere, water supply, and has disrupted food webs. For example, the widespread use of the pesticide DDT has produced harmful effects on the bald eagle and peregrine falcon. The DDT disrupts the embryological processes of many birds, causing them to become endangered. Other examples include Dioxin (Agent Orange) and the use of Temick on Long Island.

Disposal Problems. The affluent lifestyle of humans currently requires increasing supplies of products and energy, the production of which produces considerable wastes: solid, chemical, and nuclear. To dispose of the wastes, dumps have been created, many without thought of the future. Chemical dumping in the Love Canal region of western New York State has made both serious health and financial problems for the residents of that area. With the increased need for energy, nuclear power plants have been built.

Although the cost of energy produced by nuclear plants is relatively low as compared to the conventional fossil fuel power plants, nuclear plants produce "permanent" dangerous radioactive wastes. The safe disposal of these long-lasting nuclear wastes has become a global problem.

2. Positive Aspects

Through increased awareness of ecological interactions humans have attempted to prevent continued disruption of the environment and to counteract the results of many of our past negative practices. Many changes in how humans view their environment are occurring.

Population Control

Methods of controlling the human reproductive rate have been, and continue to be, developed. Family planning has become a responsible approach to the population problems of the whole world.

Conservation of Resources

In an attempt to reduce the loss of valuable agricultural land, soil cover plantings (reforestation and covercropping) are serving as erosion controls. Water and energy conserving measures are currently being implemented. The economic significance of recycling is now being realized. In many states, the recycling of metals, glass, and paper are now mandated by law. Along with maintaining valuable resources, the recycling of materials is reducing the need for land-fills.

Pollution Controls

Attempts are being made to control air and water pollution by laws and by the development of new techniques of sanitation. Due to technological improvements in the treatment of wastes and the proper disposal of industrial hazardous materials, the quality of water in the Hudson River is improving.

Species Preservation

Some efforts to sustain endangered species have included habitat protection (wildlife refuges and national parks) and wildlife management (game laws and fisheries). Some animals which were once endangered are presently successfully reproducing and increasing their numbers.

In the Northeast U.S., efforts to reintroduce peregrine falcons into the environment have been successful. In the South and West, the wild populations of whooping cranes, bald eagles, egrets, and the American bison have been improved. However, the future of some other species is still in doubt.

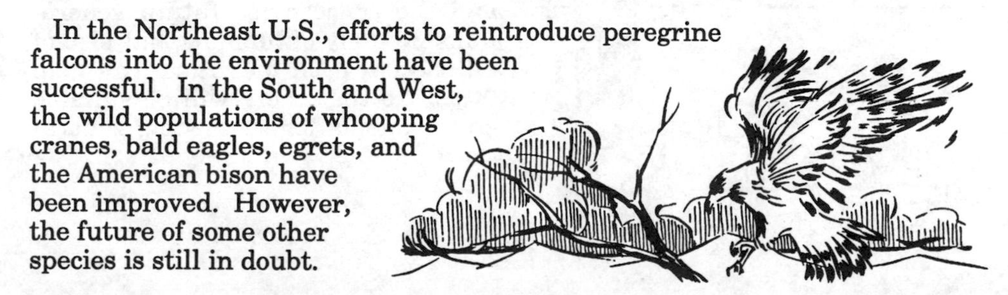

Biological Control

The biological control of insect pests continues to be encouraged, rather than the use and overuse of man-made pesticides. This method is less likely to: affect those species which are beneficial to humans, disrupt food webs, and contaminate the land. Examples include: the use of sex hormones and natural parasites.

Laws

In recent years, new and tougher laws which regulate and guide the use of natural habitats have been enacted throughout the country. Some examples of these laws include:

- **1985 Farm Act** is a Federal law which helps in the efforts to preserve natural wetlands. Other Federal laws were passed in the 1980's such as the **Clean Air Act** giving the EPA more power to regulate industrial air pollutants, the **Clean Water Act**, and the **Safe Drinking Water Act**.

- **SEQR** is a New York State law designed to provide the opportunity for citizen and community review and comment of the environmental impact of any proposed development that has been determined to have significant effect on the environment.

- **Freshwater Wetlands Acts** are laws designed to regulate the use of large or unique freshwater wetlands both publicly and privately owned so as to prevent their destruction and thus maintain valuable wetlands for all life forms.

C. The Future

While human technological advances have led to a higher standard of living for many, the environment has often suffered. Through a greater awareness of ecological principles, the wise use of our energy resources, and the concern for future generations not only of humans, but also of all species, each individual can help to assure that there will be suitable environments for succeeding generations.

Questions

1 Humans have been responsible for some of the negative changes that occur in nature because they
  1 have controlled the use of chemical biocides
  2 have passed laws to preserve the environment
  3 are able to conserve scarce resources
  4 are able to modify their physical environment

2 The resistance of some of the mushrooms to the fungicide was
  1 caused by the existence of mutations
  2 transmitted to the mushrooms from the fungicide
  3 transferred through the food web
  4 developed in response to the fungicide

3 An example of a biological control against insects is the use of
  1 herbicides   3 pesticides
  2 wildlife refuges   4 sex hormones

Base your answers to questions 4 and 5 on the paragraph below and on your knowledge of biology.

A fungicide was used to kill the mushrooms in a lawn. Some mushrooms were not affected by the fungicide. The resistant mushrooms reproduced.

4 The fungicide acted as a
  1 neurotransmitter   3 selecting agent
  2 saprophyte   4 biological control agent

5 A poor land use practice that usually leads to the loss of soil nutrients is
  1 reforestation   3 overcropping
  2 recycling   4 sewage control

6 A farmer abandons one of his fields, and over the years he notices that one community is replaced by another community. This replacement represents part of
  1 a food chain   3 a pyramid of energy
  2 an abiotic community   4 an ecological succession

7 Which pollutant is produced by the burning of coal and oil and can result in the production of acid rain?
  1 phosphate   3 lead
  2 sulfur dioxide   4 hydrogen chloride

8 Increasing populations of gypsy moths have caused the defoliating of many forested areas. This illustrates the kind of environmental disruption which may result from
  1 human population-control measures
  2 efforts to sustain endangered species
  3 increased urbanization and concentration of food crops
  4 introduction of organisms into areas where they have no natural enemies

9 A scorpion stalks, kills, and then eats a spider. Based on its behavior, which ecological terms describe the scorpion?
  1 producer, herbivore, decomposer
  2 producer, carnivore, heterotroph
  3 predator, carnivore, consumer
  4 predator, autotroph, herbivore

10 If excessive amounts of hot water are discharged into a lake, the immediate result will most likely be
 1 an increase in the sewage content of the lake
 2 a decrease in the amount of dissolved oxygen in the lake
 3 an increase in the amount of PCB pollution in the lake
 4 a decrease in the amount of phosphates in the lake

11 An abiotic factor that could be studied by an ecologist in Africa is the
 1 amount of rainfall in northern Africa
 2 birth rate in central Africa
 3 type of plants that grow in southern Africa
 4 species of mosquitoes in western Africa

12 Which activity helps scientists control harmful pests and also protects the environment from pollution?
 1 feeding toxic substances to organisms
 2 feeding herbicides to organisms
 3 using sex hormones as biological controls
 4 using pesticides ten times a year

13 Which human activity has probably contributed most to lake acidification in the Adirondack region?
 1 passage of environmental protection laws
 2 reforestation projects in lumbered areas
 3 production of sulfur and nitrogen air pollutants
 4 use of biological insect control

14 Recent evidence indicates that lakes in large areas of the Northeast U.S. and Eastern Canada are being affected by acid rain. The major effect of acid rain in the lakes is
 1 an increase in game fish population levels
 2 the stimulation of a rapid rate of evolution
 3 the elimination of many species of aquatic life
 4 an increase in agricultural productivity

15 A major problem with the disposal of wastes from nuclear power plants is that radioactive wastes
 1 take up space needed for the disposal of other inorganic wastes
 2 cannot be properly and safely packaged to prevent environmental damage
 3 often require many years before their radioactive decay makes them harmless
 4 make good landfills, so they are in demand by many communities

SELF–HELP: Unit VII *"Core"* Questions

1 Which would be considered a biotic factor in a pond ecosystem?
 1 snails 2 water 3 oxygen 4 sunlight

2 Cattails in freshwater swamps in New York State are being replaced by purple loosestrife plants. The two species have very similar environmental requirements. This observation best illustrates
 1 variations within a species
 2 competition between species
 3 isolation of species populations
 4 random recombination

3 Crows frequently are observed feeding on dead animals they have not killed. On this basis, the crow is classified as a
 1 predator 2 scavenger 3 decomposer 4 herbivore

4 A sudden increase in the number of producers in an ecosystem would first affect the population of
1 carnivores 2 herbivores 3 saprophytes 4 decomposers

5 A snapping turtle will kill animals for food, as well as feed on dead organisms. The snapping turtle is considered both a
1 saprophyte and a herbivore
2 scavenger and a herbivore
3 predator and a scavenger
4 predator and an omnivore

6 Nitrogen-fixing bacteria living on the roots of legumes are examples of a nutritional relationship known as
1 parasitism
2 mutualism
3 commensalism
4 saprophytism

7 An adult frog feeds on insects and represents a type of consumer known as a
1 producer 2 carnivore 3 saprophyte 4 parasite

8 Which organisms are classified as herbivores?
1 algae, tadpole, raccoon
2 tadpole, worm, grasshopper
3 worm, snake, bacteria
4 grasshopper, bacteria, frog

9 Which statement about the producers, algae, and grass, is true?
1 They are classified as omnivores.
2 They parasitize the animals that consume them.
3 They contain the greatest amount of stored energy.
4 They decompose nutrients from dead organisms.

10 In the biosphere, what are some of the major abiotic factors which determine the distribution and types of plant communities?
1 temperature, sunlight, rainfall
2 soil type, bacteria, water
3 humidity, location, humans
4 insects, carbon dioxide, nitrogen

11 In an ecosystem, the ultimate source of all energy is
1 photosynthesis
2 oxygen
3 fermentation
4 sunlight

12 The diagram represents some of the food relationships between several organisms in a marine community.

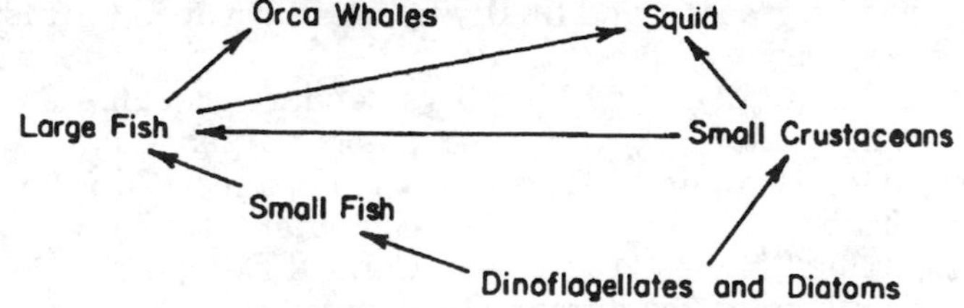

Which organisms would normally be the least numerous in this marine community?
1 diatoms
2 small fish
3 small crustaceans
4 orca whales

13 Successive transfers of energy from green plants through a series of organisms is referred to as
1 a saprophytic relationship
2 a water cycle
3 a food chain
4 an ecological succession

14 In the food web shown at the right, which organisms contain the greatest amount of stored energy?
1 snakes and hawks
2 weeds and shrubs
3 rodents and insects
4 insects and hawks

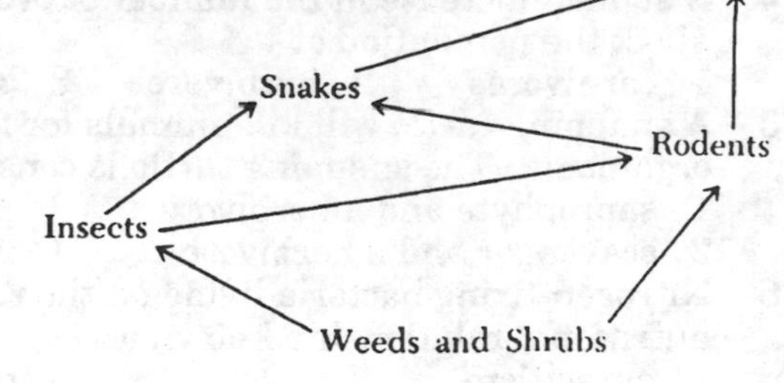

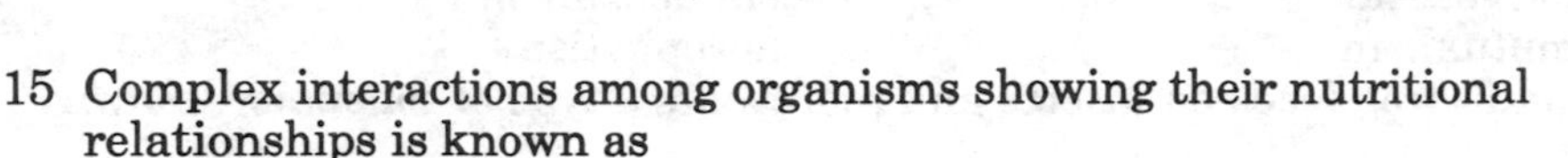

15 Complex interactions among organisms showing their nutritional relationships is known as
1 a food web
2 geographic isolation
3 abiotic factors
4 organic evolution

16 *Bacillus popilliae* is a bacterium which causes "milky disease" in the Japanese beetle. Using *Bacillus popilliae* to decrease a Japanese beetle population is an example of the
1 abiotic control of insect pests
2 use of biological control of insect pests
3 use of artificial insecticides
4 destruction of the abiotic environment

17 In a climax community, the establishment of a single species per niche is most directly the result of
1 mutualism
2 interbreeding
3 competition
4 overproduction

18 In a certain area, over 80 inches of fairly evenly distributed rainfall occurs. As a result, there is no well-defined dry season. This area is known as a
1 tundra
2 temperate deciduous forest
3 taiga
4 tropical rain forest

19 A slab of bare rock is covered with lichens. In time, mosses cover the rock, followed by grasses, and finally by small shrubs. In this example, the lichens represent
1 a climax community
2 a dominant species
3 secondary consumers
4 pioneer organisms

20 A coniferous forest would be least likely to appear within the
1 United States
2 Arctic Circle
3 Canadian Provinces
4 Central Europe

21 Which accomplishment by humans has made the most positive ecological impact on the environment?
1 the importation of organisms such as the starling and the Japanese beetle into the United States
2 the reforestation and soil cover planting measures to prevent soil erosion
3 the extinction or near extinction of many predators to prevent the death of prey animals
4 the use of pesticides and other similar crop-improvement chemicals to regulate the insect population

22 Compared to large land area, large bodies of water, such as oceans, absorb heat
 1 much faster and release it much faster
 2 faster and release it more slowly
 3 more slowly and release it faster
 4 slowly and release it at a slower rate

23 Which is an example of a biological control of a pest species?
 1 DDT was used to destroy the red mite.
 2 Most of the predators of a deer population were destroyed by humans.
 3 Gypsy moth larvae (a tree defoliator) are destroyed by beetle predators which were cultured and released.
 4 Drugs are used in the control of certain pathogenic bacteria.

24 The number of African elephants has been drastically reduced by poachers who kill the animals for the ivory in their tusks. This negative aspect of human involvement in the ecosystem could best be described as
 1 poor land use management
 2 poor agricultural practices
 3 importation of organisms
 4 exploitation of wildlife

25 In a community, the most severe competition develops among those organisms which
 1 are active only during the night
 2 belong to two different genera
 3 depend upon autotrophs for food
 4 occupy the same ecological niche

SELF–HELP: Unit VII *"Extended"* Questions

1 In the diagram at the right, regions *A*, *B*, and *C* represent areas of varying altitude on the Earth, and regions *D*, *E*, and *F* represent areas of varying latitude.

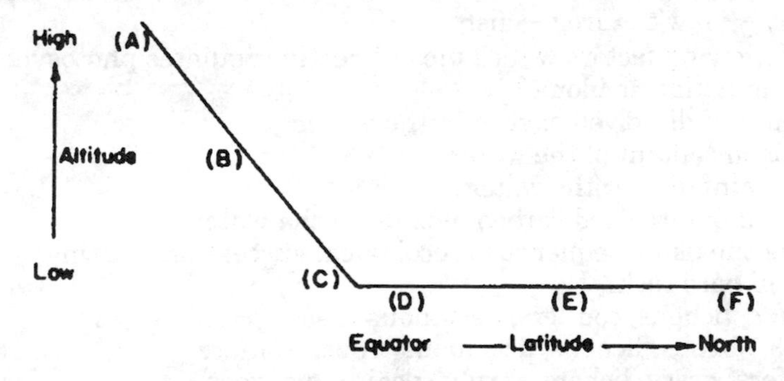

Which two life zones exhibit the most similar climatic conditions?
 1 *A* and *F* 2 *B* and *D* 3 *C* and *E* 4 *D* and *E*

2 Which biome is characterized by warm temperatures, an abundance of snakes, monkeys, and large cats, over 80 inches of annual rainfall evenly distributed with no well-defined dry season?
 1 taiga
 2 tundra
 3 tropical rain forest
 4 temperate deciduous forest

Base your answers to questions 3 through 5 on the diagram of a lake ecosystem. The diagram shows a cross section of a deep lake. The dashed line which separates level A from level B indicates the depth beyond which light cannot penetrate.

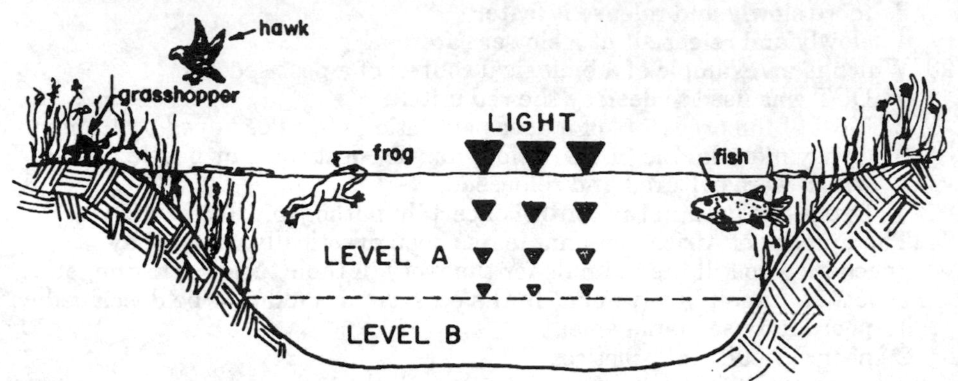

3 Minerals entering this system from the surrounding soil would reach all parts of the lake by the process of
1 active transport
2 ecological succession
3 diffusion
4 transpiration

4 Which type of organism that ordinarily inhabits a lake ecosystem would not be found in level *B* because of the lack of light penetration?
1 decomposers
2 scavengers
3 carnivores
4 producers

5 A possible food chain represented by the diagram would be
1 plant → grasshopper → frog → fish
2 hawk → plant → grasshopper → frog
3 grasshopper → fish → frog → plant
4 plant → hawk → frog → fish

6 Of the following factors, which most directly regulates photosynthetic activity in a marine biome?
1 amount of dissolved oxygen in the water
2 nitrogen content of the water
3 total rainfall over the water
4 amount of dissolved carbon dioxide in the water

7 Which is the usual sequence for ecological succession starting on the surface of bare rock?
1 shrubs, lichens, conifers, deciduous trees, grass
2 grass, shrubs, lichens, deciduous trees, conifers
3 conifers, grass, lichens shrubs, deciduous trees
4 lichens, grass, shrubs, conifers, deciduous trees

8 Some hydras have green algae living symbiotically inside their bodies. The algae produce food for the hydra and receive carbon dioxide and shelter from the animal. What type of relationship exists between the two organisms?
1 parasitism
2 commensalism
3 mutualism
4 saprophytism

Base your answers to questions 9 and 10 on the key below and on your knowledge of biology. The symbols in the key represent possible effects on an organism of some nutritional relationships.

Key
(+) Organism benefits
(-) Organism is harmed.
(0) Organism neither benefits nor is harmed.

9 Which represents the effects of the relationship between nitrogen-fixing bacteria and leguminous plants such as clover?
1 nitrogen-fixing bacteria (-); clover (-)
2 nitrogen-fixing bacteria (-); clover (+)
3 nitrogen-fixing bacteria (0); clover (-)
4 nitrogen-fixing bacteria (+); clover (+)

10 Which relationship would be illustrated by (+) for one organism and (0) for the other organism?
1 mutualism
2 parasitism
3 commensalism
4 autotrophism

Base your answers to questions 11 through 15 on the chart shown below and on your knowledge of biology.

A	Characteristics	Climax Flora	Climax Fauna
B	Long, severe winters	*D*	Moose, Black Bear
Tropical Rain Forest	Heavy rainfall	Many species of broadleaf plants	*E*
Desert	*C*	Succulent plants	Lizards

11 Which heading belongs in box *A*?
1 Land Biome
2 Aquatic Ecosystem
3 The Biosphere
4 Succession Stage

12 Which name belongs in box *B*?
1 Tundra
2 Taiga
3 Grassland
4 Temperate Forest

13 Which characteristic belongs in box *C*?
1 extreme daily temperature fluctuations
2 constant rainfall
3 seasonal animal migrations
4 strong prevailing winds

14 Which organisms belongs in box *D*?
1 maple trees
2 cactus plants
3 lichens
4 conifers

15 The climax fauna in box *E* would probably include
1 lizards and caribou
2 red fox and whitetail deer
3 bison and antelope
4 monkeys and snakes

Base your answers to questions 16 through 20 on the diagram below and on your knowledge of biology. The diagrams show the types of plants which grew in a farm field in the 200 years after it was abandoned. Different types of plants appeared and disappeared during this time.

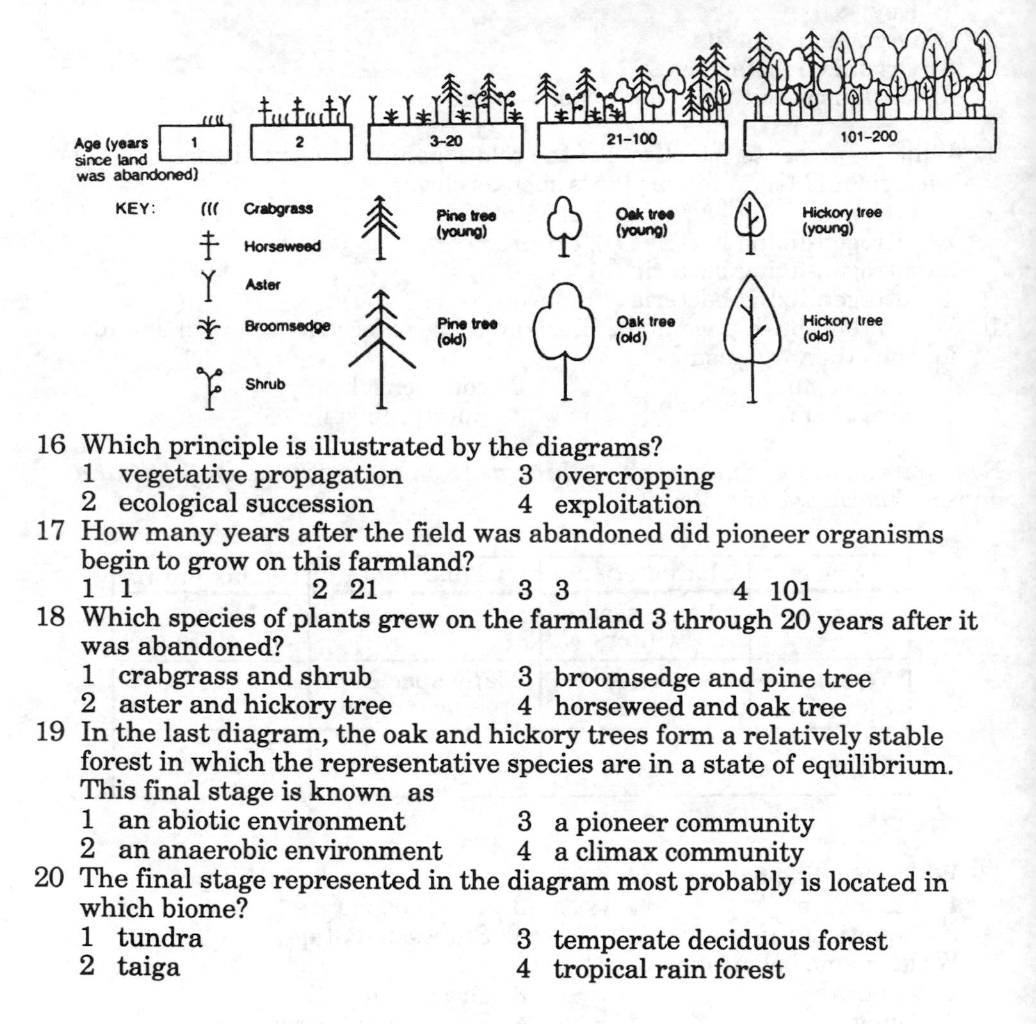

16 Which principle is illustrated by the diagrams?
1 vegetative propagation
2 ecological succession
3 overcropping
4 exploitation

17 How many years after the field was abandoned did pioneer organisms begin to grow on this farmland?
1 1
2 21
3 3
4 101

18 Which species of plants grew on the farmland 3 through 20 years after it was abandoned?
1 crabgrass and shrub
2 aster and hickory tree
3 broomsedge and pine tree
4 horseweed and oak tree

19 In the last diagram, the oak and hickory trees form a relatively stable forest in which the representative species are in a state of equilibrium. This final stage is known as
1 an abiotic environment
2 an anaerobic environment
3 a pioneer community
4 a climax community

20 The final stage represented in the diagram most probably is located in which biome?
1 tundra
2 taiga
3 temperate deciduous forest
4 tropical rain forest

SELF–HELP: Unit VII *"Skill"* Questions

1 In 1986, an accident at a nuclear plant in Chernobyl, Ukraine., resulted in widespread radioactive contamination of the surrounding environment. Using one or more complete sentences, describe one possible effect of this contamination on the humans who were living in that area and who survived the accident. ______________________________

____

____

____

Base your answers to questions 2 through 4 on the graph below and on your knowledge of biology. The graph illustrates a comparison between pH conditions and species survival rates in certain Adirondack lakes.

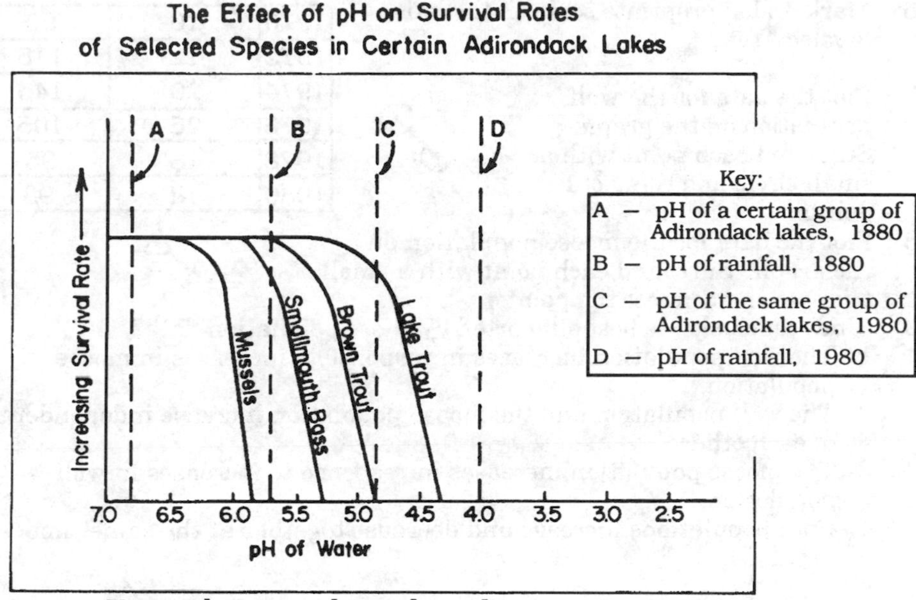

—*National Geographic* (adapted)

2 Which species can tolerate the highest level of acidity in its water environment?
1 mussels
2 smallmouth bass
3 brown trout
4 lake trout

3 In the years between 1880 and 1980, which species would most likely have been eliminated first due to the gradual acidification of Adirondack lakes?
1 mussels
2 smallmouth bass
3 brown trout
4 lake trout

4 What is the total change in the pH value of rainwater from 1880 to 1980?
1 1.3
2 1.7
3 5.3
4 9.7

Base your answers to questions 5 through 9 (on the next page) on the information below and data table on the next page and on your knowledge of biology.

The data table shows the wolf and moose populations recorded at the end of June from 1970 to 1980 on an isolated island national park where no hunting by humans is allowed. Prior to the arrival of wolves on the island (1965), the moose population had increased to over 300 members. Wolves have been observed many times on this island hunting cooperatively to kill moose.

Using the information in the data table, construct a line graph, following the directions on the next page.

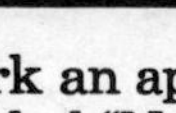

5 Mark an appropriate scale on the axis labeled "Number of Members of Each Population."

6 Mark and appropriate scale on the axis labeled "Year."

Data Table

Year	Number of Members	
	Wolf Population	Moose Population
1970	10	90
1972	12	115
1974	20	145
1976	25	105
1978	18	95
1980	18	98

7 Plot the data for the wolf population on the graph. Surround each point with a small circle and connect the points.

8 Plot the data for the moose population on the graph. Surround each point with a small triangle and connect the points.

9 What conclusion is best supported by these population data?
 1 The wolf population increases in response to increases in moose population.
 2 The wolf population and the moose population increase independently of each other.
 3 The moose population increases in response to increases in wolf population.
 4 Both populations increase and decrease together at the same time.

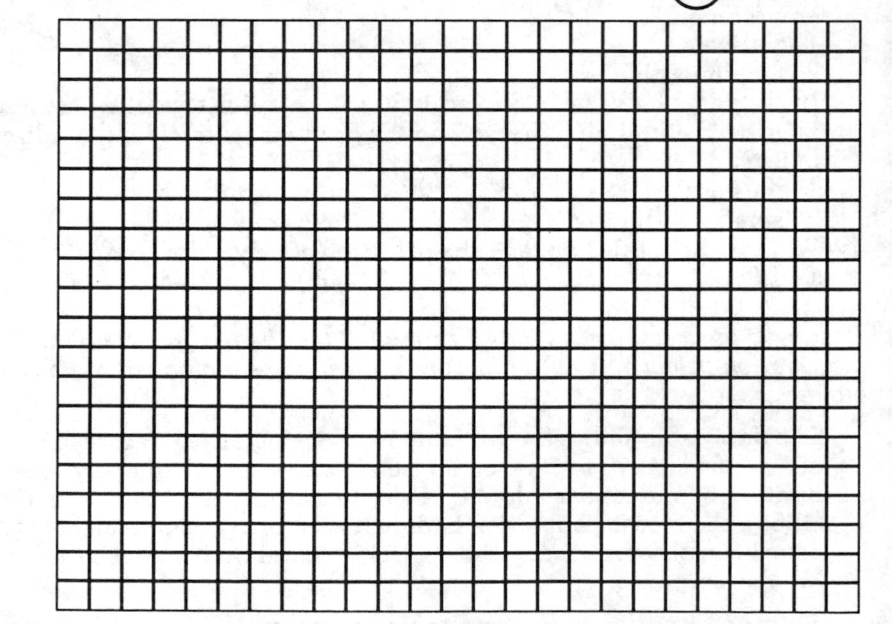

Laboratory Skills, Safety, and Concepts

Recommended Lab Skills

The biology student should become proficient in understanding and the performance of the following laboratory skills:

- **Formulate a question or define a problem and develop a hypothesis** to be tested in an investigation.
- Given a laboratory problem, **select suitable lab materials, safety equipment, and appropriate observation methods.**
- **Distinguish between controls and variables** in an experiment.
- **Identify parts of a light microscope** and their functions and focus in low and high power.
- **Determine the size of microscopic specimens** in micrometers.
- **Prepare wet mounts** of plant and animal cells and apply staining techniques using iodine or methylene blue.
- **Identify cell parts** under the compound microscope, such as the nucleus, cytoplasm, chloroplast, and cell wall.
- **Use and interpret indicators** such as pH paper, Benedict's (Fehling's) solution, iodine (Lugol's) solution, and bromthymol blue.
- **Use and read measurement instruments**, such as metric rulers, Centigrade (Celsius) thermometer, and graduate cylinders.
- **Dissect plant and animal specimens** for the purpose of exposing major structures for suitable examination. Suggestions of specimens include seeds, flowers, earthworms, grasshoppers, etc.
- **Demonstrate safety skills** involved in heating materials in test tubes or beakers, use of chemicals, and handling of dissection instruments.
- **Collect, organize, and graph data**
- **Make inferences and predictions** based on data collected and observed.
- **Formulate generalizations or conclusions** of the investigation.
- **Assess the limitations and assumptions** of the experiment.
- **Determine the accuracy and repeatability** of the experimental data and observations.

I. Lab Safety

The following "Safety Suggestions" should be explained by your teacher according to the special conditions of your classroom and your school's laboratory equipment.

- 1st **Always follow your teacher's instructions.**
- 2nd **In the event that you are not under the direction of your teacher, follow the safety precautions outlined here.**

Safety Suggestions

- **Do not handle chemicals or equipment in the laboratory until you have been given specific instructions.**
- Report at once any equipment in the laboratory that appears to be unusual or improper such as broken, cracked, or jagged apparatus, or any reactions that appear to be proceeding in an abnormal fashion.
- **Report any personal injury** or damage to clothing to the teacher immediately no matter how trivial it may appear.
- Prevent loose clothing and hair from coming in contact with any science apparatus, chemicals, or sources of heat or flame.
- **Do not transport laboratory materials** through hallways unsupervised or during the passing of classes.
- **Understand all instructions** for the proper use of dissection instruments before beginning the lab procedure.
- Do not taste or inhale directly any unknown chemicals.
- Do not pour reagents back into stock bottles or exchange stoppers.
- Be aware of dangers involved in the handling of hot glassware or other equipment. **Know the proper devices for handling these items.**
- Do not use any apparatus with frayed wiring, loose connections, or exposed electrical wires.
- **Know the location and use of the available safety equipment,** including the eye baths, fire extinguisher, etc.
- **Know the specific safety instructions** before beginning the lab procedures which are applicable to particular experiments.

Questions

1 Which would be the proper laboratory procedure to follow if some laboratory chemical splashed into a student's eyes?
   1 Send someone to find the school nurse.
   2 Rinse the eyes with water and do not tell the teacher because he or she might become upset.
   3 Rinse the eyes with water, then notify the teacher and ask farther advice.
   4 Assume that the chemical is not harmful and no action is required.

2 A student performing an experiment noticed that the beaker of water being heated had a slight crack in the glass, but was not leaking. What should the student do?
   1 Discontinue heating and attempt to seal the crack.
   2 Discontinue heating and report the defect to the instructor.
   3 Discontinue heating and immediately take the beaker to the instructor.
   4 Continue heating as long as fluid does not seep from the crack.

3 Using one or more complete sentences, describe two safety procedures that should be followed when heating a liquid in a test tube.

____

____

____

____

4 Using one or more complete sentences, describe one safety procedure that should have been used when a student smelled an unknown chemical substance. ______________________________

____

____

5 Using one or more complete sentences, describe one safety procedure that should have been used when a student was to begin a dissection procedure. ______________________________

____

____

II. Measurement

Metric Measurement

Using a millimeter ruler to determine actual object size. *(Note that ruler shown is not drawn to scale.)*

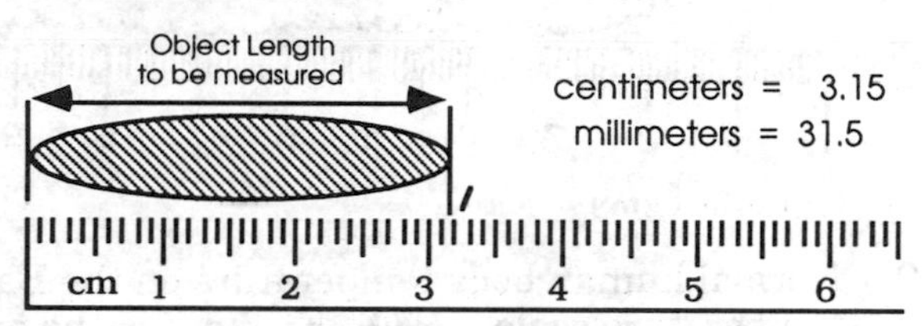

Metric (SI) Units

Some commonly used units:

Volume:		
1 liter (l)	=	0.26 gal
1 liter (l)	=	1,000 ml
1 milliliter (ml)	=	0.001 liter

Mass:		
1 gram (g)	=	0.035 ounce
1 kilogram (kg)	=	1000 g
1 milligram (mg)	=	0.001 g

Length:		
1 meter (m)	=	39 inches
1 kilometer (km)	=	1000 m
1 centimeter (cm)	=	0.01 m
1 millimeter (mm)	=	0.001 m
1 micrometer (μm)	=	0.001 mm

Meniscus Reading

How to read the meniscus on a graduated cylinder correctly.

Note the correct eye level. It is most important that the reading is taken from the lowest top surface of the solution in the graduated cylinder (right).

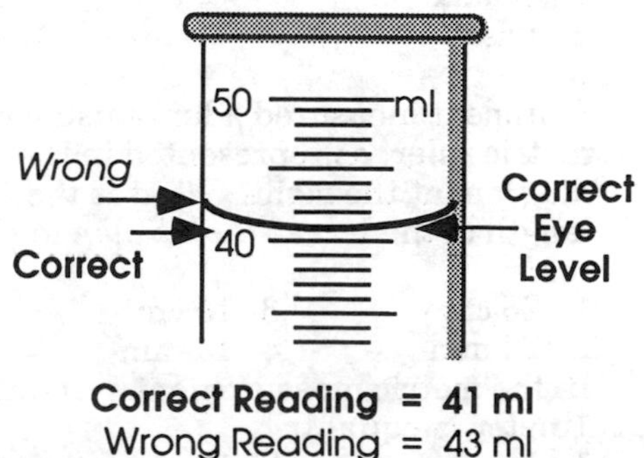

Temperature Measurement

Celsius is the most commonly used temperature unit in the biology lab. The scales at the right give a general comparison of Celsius and Fahrenheit temperatures.

°C: -40, -20, 0, 20, 37, 40, 60, 80, 100
°F: -40, 0, 32, 98.6, 160, 212
water freezes | body temp | water boils

Celsius and Fahrenheit Temperature Scales

Questions

1 The diagram below represents a segment of a metric ruler and part of an earthworm. What is the length of the part of the earthworm shown? Your answer must include the correct units. *(not drawn to scale)*

____

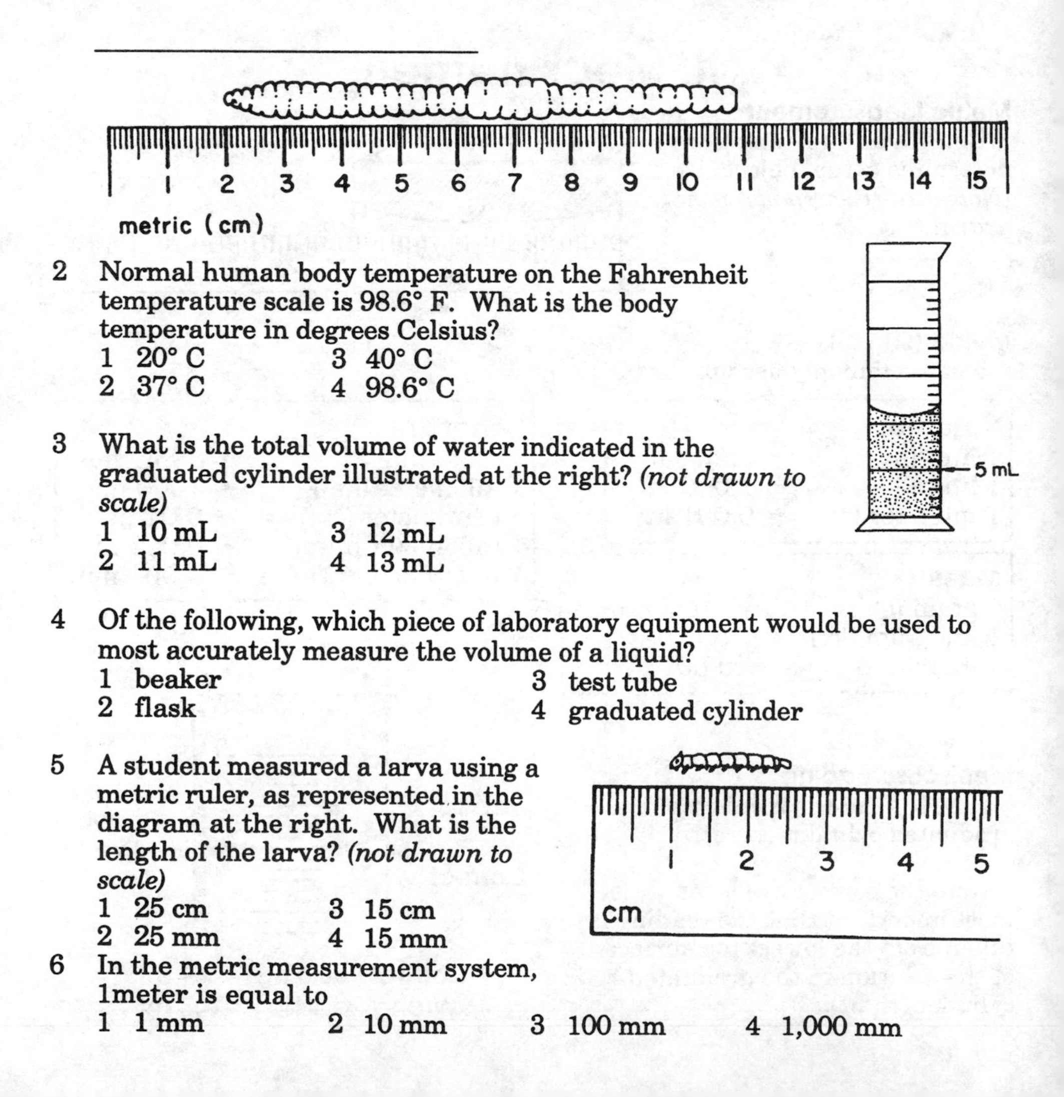

2 Normal human body temperature on the Fahrenheit temperature scale is 98.6° F. What is the body temperature in degrees Celsius?

1 20° C
2 37° C
3 40° C
4 98.6° C

3 What is the total volume of water indicated in the graduated cylinder illustrated at the right? *(not drawn to scale)*

1 10 mL
2 11 mL
3 12 mL
4 13 mL

4 Of the following, which piece of laboratory equipment would be used to most accurately measure the volume of a liquid?

1 beaker
2 flask
3 test tube
4 graduated cylinder

5 A student measured a larva using a metric ruler, as represented in the diagram at the right. What is the length of the larva? *(not drawn to scale)*

1 25 cm
2 25 mm
3 15 cm
4 15 mm

6 In the metric measurement system, 1meter is equal to

1 1 mm
2 10 mm
3 100 mm
4 1,000 mm

7 In the metric system, what is the unit meaning "1,000 times" the unit size?
1 milli 2 kilo 3 centi 4 micro

8 Celsius and Fahrenheit temperature readings for the freezing point of water are
1 0°C and 32°F
2 100°C and 212°F
3 10° C and 100°F
4 32°C and 0°F

9 Which unit of measurement should be used when determining the size of an organelle of a cell?
1 meter 2 millimeter 3 micrometer 4 kilometer

10 When correctly reading the volume of a liquid in a graduated cylinder, the student should look at the
1 top of the fluid where it touches the sides of the cylinder
2 bottom of the fluid where it touches the sides of the cylinder
3 top of the fluid at the meniscus
4 bottom of the fluid in the middle of the cylinder

III. Microscope

Refer to Unit I, pages 10 through 12, for a description of the use and parts of the compound light microscope. In addition the following information concerning measurements in the field of view and magnification will be useful for answering *"Skill"* questions.

Microscope Parts

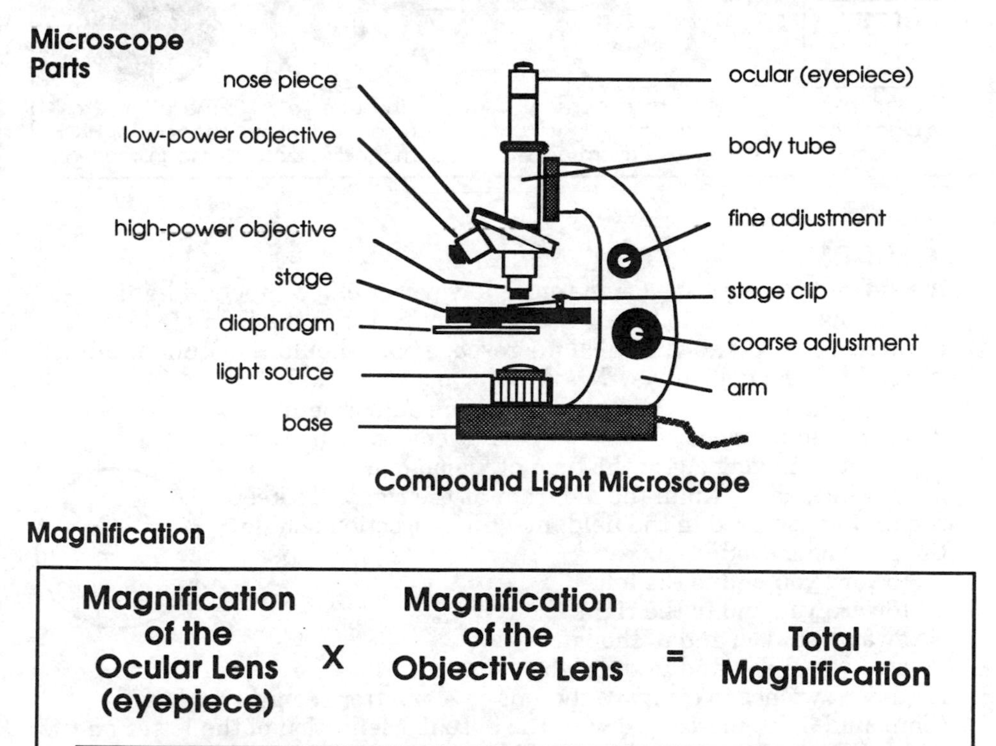

Compound Light Microscope

Magnification

Magnification of the Ocular Lens (eyepiece)	X	Magnification of the Objective Lens	=	Total Magnification
10X	x	30X	=	300X

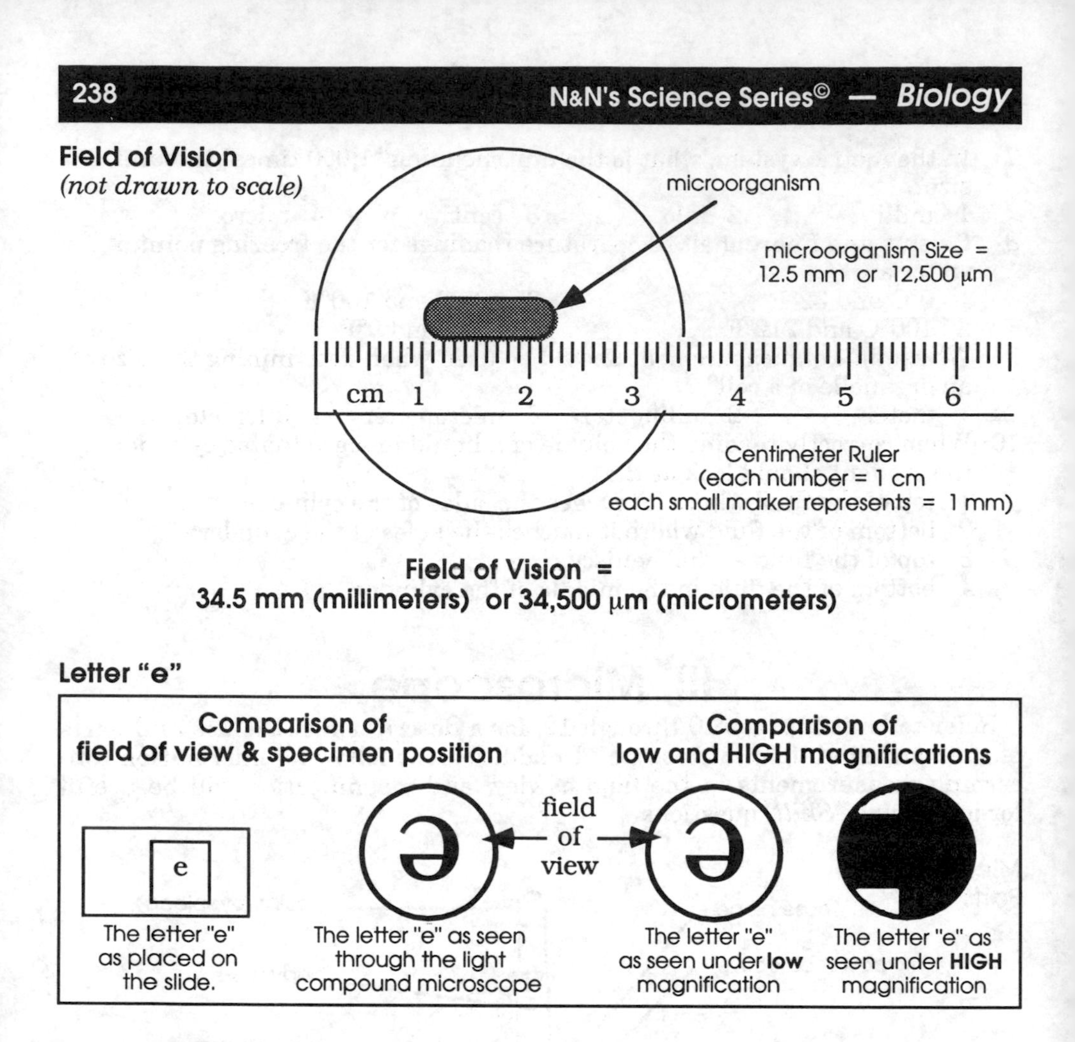

Questions

1 A student viewing a specimen under low power of a compound light microscope switched to high power and noticed that the field of view darkened considerably. Which microscope part should the student adjust to brighten the field of view?
 1 diaphragm
 2 coarse adjustment
 3 fine adjustment
 4 eyepiece

2 A protist is moving out of the field of vision of a microscope, as shown in the diagram at the right. To keep the protist centered in the field, in which direction should the slide be moved?
 1 toward you and to the left
 2 toward you and to the right
 3 away from you and to the left
 4 away from you and to the right

3 A student wishes to compare the image of a letter seen through a compound light microscope with the actual orientation of the letter on the slide. Which of the following would be the most suitable letter to use?
 1 F 2 I 3 O 4 Z

4 A student calculated the diameter of the high-power field of a microscope to be 0.3 millimeter. If ten cheek cells fit across the diameter of the field, what is the average diameter of one of the cheek cells?
1 0.3 millimeter
2 3.0 millimeter
3 3.0 micrometers
4 30 micrometers

5 A student observed an ameba under the low-power objective (10x) of a compound microscope and noted that the organism occupied one-fourth of the field of view after it was centered. When the student changed to the high-powered objective (40x), how much of the field of view would the ameba most likely occupy?
1 0%
2 50%
3 75%
4 100%

Base your answers to questions 6 and 7 on the information below and on your knowledge of biology.

A student was using a microscope with an 10x eyepiece and 10x and 40x objective lenses. He viewed the edge of a millimeter ruler under low power and observed the following field of vision.

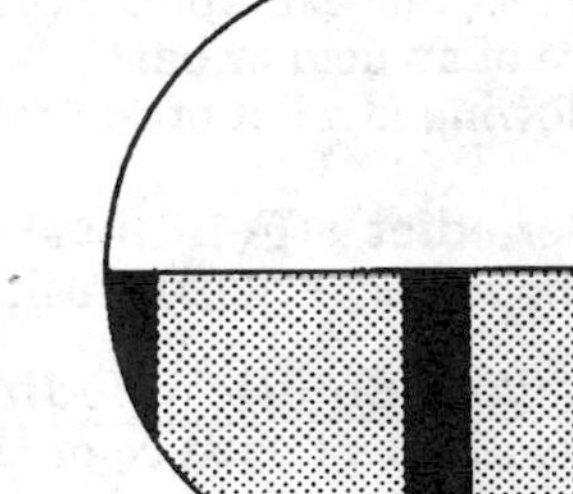

6 What is the diameter of the low-power field of vision in micrometers?
1 1 μm
2 2 μm
3 1,000 μm
4 2,000 μm

7 The diameter of the high-power field of vision of the same microscope would be closest to
1 0.05 μm
2 0.5 μm
3 5 μm
4 500 μm

8 The diagram at the right represents the view of a stained wet mount of human cheek cells prepared by a student and observed under low power of a compound light microscope. What do the dark-rimmed circles labeled A represent?

cheek cells
A

1 nuclei
2 red blood cells
3 air bubbles
4 chloroplasts

9 While focusing a microscope on high power, a student crushed the cover slip. The student probably
1 closed the diaphragm
2 rotated the eyepiece
3 turned up the light intensity
4 used the coarse adjustment

10 The low-power diameter of the field of a microscope is determined to be 1.5 millimeters. A slide of onion skin cells is put on the stage, and exactly 3 cells can be counted lengthwise across the diameter of the field. The average length of each cell is
1 100 μm
2 500 μm
3 750 μm
4 1,500 μm

IV. Stains and Indicators

Unstained cells are translucent. Therefore, in order to study tissues and organisms in greater detail and more clearly under the light microscope, it is necessary to use stain.

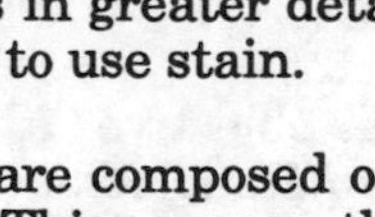

Stains are composed of dyes which bind to certain proteins and provide a contrast. This causes the structure or organelle containing that specific protein to become more easily observed. Examples include methylene blue and Lugol's iodine.

Indicator Solutions are substances which are added in small amounts to a test solution or material in order to indicate (show) certain changes or conditions. For example, a pH indicator is added to a solution to test for the presence of an acid or base.
The following is a list of commonly used stains and indicators:

- **Benedict's (Fehling's) Solution**: Test for simple sugars (monosaccharides). Positive test: blue color changes to orange.
- **Lugol's Solution (Iodine)**: Test for a polysaccharide (starch). Positive test: rust color changes to purple.
- **Bromthymol blue**: Test for presence of gas. Blue color indicates O_2. Yellow color indicates CO_2.
- **pH paper (litmus)**: Test for presence of an acid or base. Most commonly a change to a blue color indicates a basic solution, and a change to a red color indicates an acid solution. Other pH indicators include methyl orange, phenolphthalein, and bromthymol blue.

Questions

Base your answers to questions 1 and 2 on the diagram at the right and on your knowledge of biology.

Glass Container
Tight Seal
Selectively Permeable Membrane Pouch
Water and Iodine Solution
Glucose, Starch and Water Solution

1 A student tested a sample of the fluid in the glass container for glucose 30 minutes after the apparatus had been set up. Which indicator should be used for this test?
 1 iodine solution
 2 bromthymol blue
 3 Benedict's solution
 4 pH paper

2 If glucose is detected in the fluid in the glass container after 30 minutes, its presence is probably due to the process of
 1 starch hydrolysis
 2 passive transport
 3 dehydration synthesis
 4 transpiration pull

3 Using a complete sentence, state why iodine (Lugol's) solution cannot be used as a stain for observing the activity of living cells.

____

____

____

4 A student was testing the composition of exhaled air by exhaling through a straw into a solution of bromthymol blue. The presence of carbon dioxide in the exhaled air would be indicated by
1 a color change in the solution
2 a change in atmospheric pressure
3 the formation of a precipitate in the solution
4 the release of bubbles from the solution

5 Which substance, when added to a wet mount containing starch grains, would react with the starch grains and make them more visible?
1 litmus solution
2 iodine solution
3 distilled water
4 bromthymol blue

6 Each test tube at the right contains an equal amount of a food sample and Benedict's solution. Each test tube has been heated to the same temperature. Which test tube contains the greatest amount of simple sugar?
1 *A*
2 *B*
3 *C*
4 *D*

brick red
bright blue
green
yellow
A B C D

For each statement in questions 7 through 9, select the substance, chosen from the list below, that is best described by the statement and record its number.

Substances:
1) Distilled water
2) Lugol's iodine solution
3) Concentrated salt solution
4) Benedict's solution

____7 When added to a wet mount of elodea cells which are then observed with a microscope, this colorless substance decreases the volume of water in those cells.

____8 When added to a glucose solution and heated, this substance turns orange-red.

____9 When used to prepare a wet mount of plant cells which are then observed with a microscope, this substance makes the nucleus more visible.

10 Write the name of one indicator a student could use in a test tube to detect the presence of glucose.

____

V. Tools of the Scientist

Refer to Unit I, pages 10 through 12, for a description of some of the tools that a biologist uses in the laboratory. Review your laboratory investigations and the equipment that you have used this year for the information to answer the other questions.

Questions

Base your answers to questions 1 through 3 on the four sets of laboratory materials listed below and on your knowledge of biology.

Laboratory Materials

Set *A*
Anti-A and Anti-B sera
Microscope slides
Sterile absorbent cotton
Alcohol
Sterile disposable lancet
Wright's stain
Compound microscope
Paper toweling
Medicine droppers

Set *B*
Human saliva
Test tubes
Starch solution
Lugol's iodine solution
Graduated cylinder
Benedict's solution
Glucose solution
Bunsen burner
Test tube holder
Safety goggles

Set *C*
Filter paper
Test tubes
Rubber stoppers to fit test tubes
Alcohol
Mortar and pestle
Test tube rack
Scissors
Metric ruler
Safety goggles
Solvent mixture
Spinach

Set *D*
Toothpick
Microscope slides
Cover slips
Medicine dropper
Lugol's iodine solution
Compound microscope
Methylene blue
Onions
Water

1 Which set should be used by a student to study the effect of an enzyme on a polysaccharide?
1 *A* 2 *B* 3 *C* 4 *D*

2 Which set should a student select to study differences between plant and animal cells?
1 *A* 2 *B* 3 *C* 4 *D*

3 Which set should be used to extract and separate chlorophyll and other pigments from a leaf?
1 *A* 2 *B* 3 *C* 4 *D*

4 An instrument used to collect ribosomes for chemical analysis is
1 a compound microscope
2 an ultracentrifuge
3 a scalpel
4 an electron microscope

5 The diagram represents the result of spinning a suspension of broken cells in an ultracentrifuge. Which is a correct conclusion?

1 Ribosomes are more dense than mitochondria.
2 Nuclei are more dense than mitochondria.
3 Mitochondria and ribosomes are equal in density.
4 The cell consists of only solid components

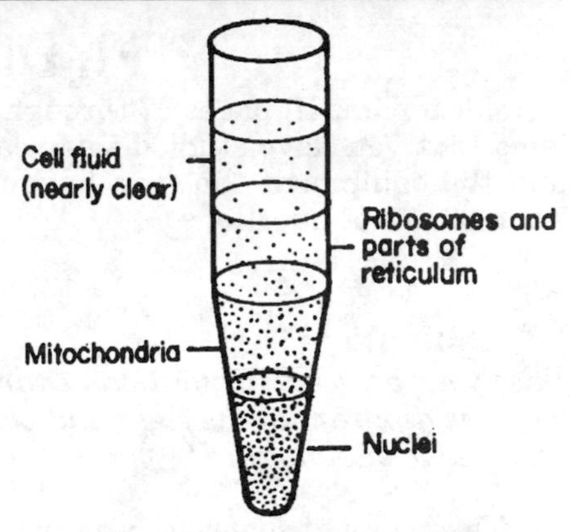

Ultracentrifuge tube showing various layers.

Base your answers to questions 6 and 7 on the diagrams of the laboratory equipment below and on your knowledge of biology.

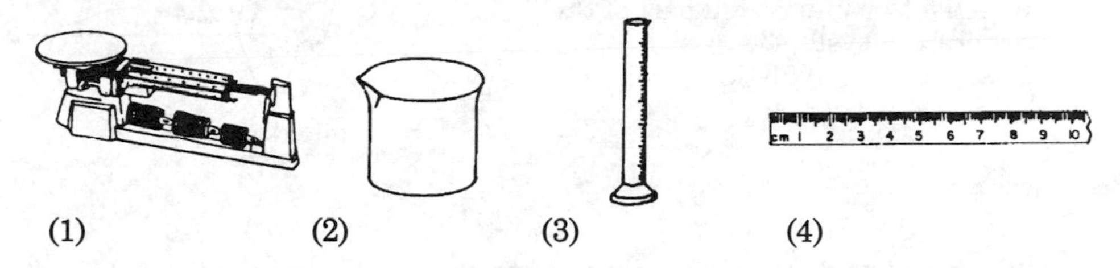

(1) (2) (3) (4)

6 Which piece of laboratory equipment would be used to obtain the most accurate measure of the volume of a glucose solution?
1 1 3 3
2 2 4 4

7 Which piece of equipment could be used to determine the mass of an earthworm?
1 1 3 3
2 2 4 4

8 An instrument used to separate cell parts according to density is the
1 compound microscope 3 microdissector
2 electron microscope 4 ultracentrifuge

9 Which instrument would provide the most detailed information about the internal structure of a chloroplast?
1 a compound light microscope
2 an electron microscope
3 a phase contrast microscope
4 an interference microscope

10 To transplant a nucleus from one cell to another cell, a scientist would
1 use an electron microscope
2 use staining techniques
3 use an ultracentrifuge
4 use microdissection instruments

VI. Dissection

Refer to Unit II, pages 37 through 80, for a description of some of the organisms that you have studied this year. Review your dissection investigations and the equipment that you have used this year for the information to answer the other questions.

Questions

Base your answers to questions 1 through 3 on the diagram at the right and on your knowledge of biology.

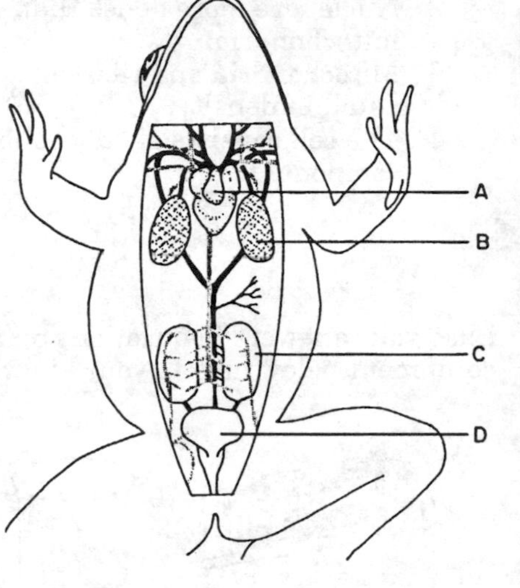

1 Through which surface was the initial incision for this dissection made?
1 ventral
2 dorsal
3 lateral
4 anterior

2 Which letter indicates a part of the respiratory system?
1 *A*
2 *B*
3 *C*
4 *D*

3 Which materials probably were used to prepare the specimen as shown?

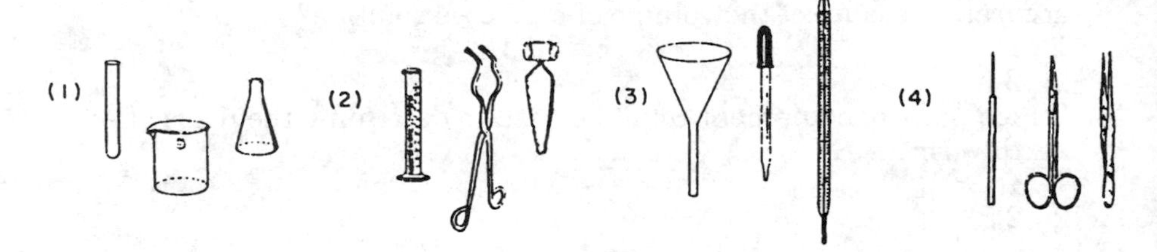

4 Which group of equipment would a student use to dissect a preserved earthworm?
1 dissecting pan, dissecting pins, electron microscope, and safety goggles
2 dissecting pan, methylene blue, eyedropper, and safety goggles
3 dissecting pan, litmus paper, scissors, and safety goggles
4 safety goggles, dissecting pan, scissors, and dissecting pins

5 The diagram below represents a portion of a dissected earthworm. What is the name of the organ labeled *X*?
1 crop
2 gizzard
3 esophagus
4 intestine

POSTERIOR END
X
ANTERIOR END
DIRECTION OF FOOD MOVEMENT

VII. Graphing and Scientific Method

Scientific Method

The scientific method is a way of thinking about problems and solutions. The general rules were worked out by many scientists over a very long period of time. However, many scientists find it difficult to tell in what order they actually use the steps of the scientific method. The following are the major parts in the formal scientific method:

- **State the problem**
- **Form a hypothesis** (best possible explanation prior to experimentation)
- **Observe and experiment**
- **Organize the data collected**
- **Interpret the data** (inference)
- **Determine conclusions**

A **control** is that part of an experiment that does not change. It is the standard for comparison since its results are known prior to the experiment.

A **variable** is the one factor allowed in an experiment that changes. It is the one unknown factor in the experiment which is being tested for its results.

An **inference** is a conclusion based on the collected and organized data.

Developing Data Tables and Graphs

Data tables and graphs are used to organize and visualize the observations of an experiment. The following is an example of an experiment and the resulting data table and graph:

A student performs an experiment to determine the relative effects of light intensity during photosynthesis. Oxygen (O_2) is a waste product of photosynthesis. Therefore, by comparing the amount of O_2 produced during the experiment by photosynthesis, the effect of light may be determined.

With all other conditions, such as soil, water, and available carbon dioxide being the same, only the light intensity (variable) is different for each plant. Plant A is given only a low level of light, and plant B is given a high level of light. Over a five day period, the following observations are recorded.

Day 1 - Plant A produces 10 ml of O_2, and plant B produces 20 ml of O_2.
Day 2 - Plant A produces 12 ml of O_2, and plant B produces 22 ml of O_2.
Day 3 - Plant A produces 08 ml of O_2, and plant B produces 18 ml of O_2.
Day 4 - Plant A produces 10 ml of O_2, and plant B produces 22 ml of O_2.
Day 5 - Plant A produces 12 ml of O_2, and plant B produces 24 ml of O_2.

A data table for the experiment on the previous page may look like this (right):

Data Table

Amount of Oxygen Released

	Plant **A**		Plant **B**	
Day	Daily	Total	Daily	Total
1	10 ml	10 ml	20 ml	20 ml
2	12 ml	22 ml	22 ml	42 ml
3	08 ml	30 ml	18 ml	60 ml
4	10 ml	40 ml	22 ml	82 ml
5	12 ml	53 ml	24 ml	106 ml

Note: Organizing data in a table form puts the observations in a clear and concise order, easy to read and graph. A graph of the data in the table (right) is shown below (left).

Note: Plotting information from the data table allows the student to make clear and accurate conclusions based on the data comparison. The student concludes that higher light intensity increases the rate of photosynthesis, because of the increased O_2 produced. A graph of the data with connecting lines is shown below (right).

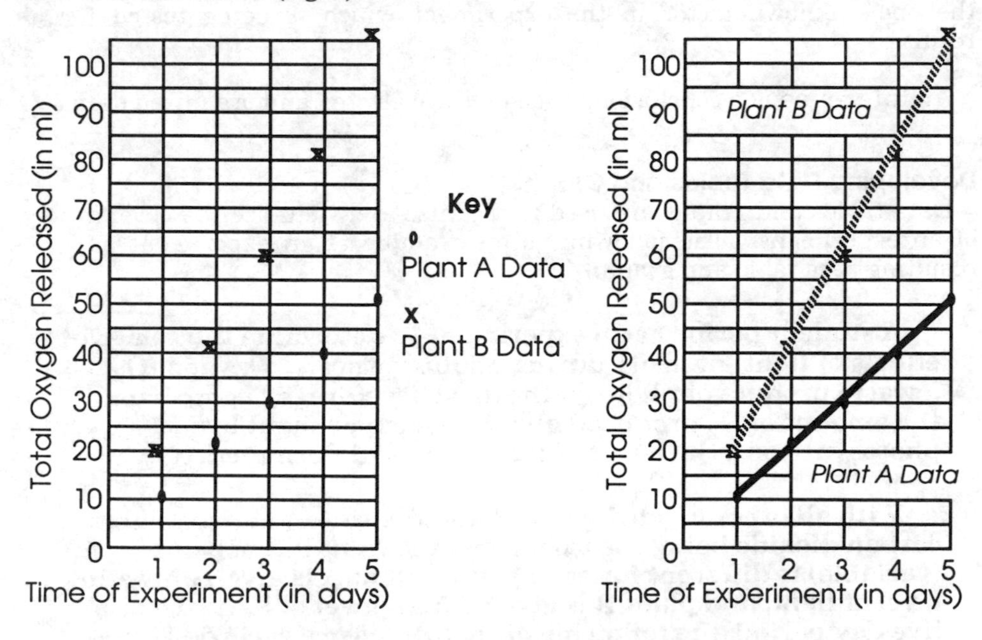

Questions

1 A drug company tested a new medication before putting it on the commercial market. Pills without medication were given to 500 test subjects in group *A* and pills containing medication were given to 500 test subjects in group *B*. In this experiment, the individuals in group *A* served as the

1 host 2 variable 3 control 4 hypothesis

Base your answers to questions 2 through 6 on the information and data table at the right and on your knowledge of biology.

An investigation was conducted to study the effect of two different fertilizers on the growth of corn seedlings. Two hundred individual corn seedlings were used in this study. Half of the seedlings were given fertilizer X and the other half were given fertilizer Z. All other growth conditions such as temperature, water, and light were kept constant for both groups. At the end of each day, the increase in height of each plant was recorded. The results of the investigation are shown in the data table at the right.

Data Table
The Effect of Fertilizers X and Z on the Growth of Corn Seedlings

End of Day	Average Increase in Plant Height (Millimeters)	
	Fertilizer X	Fertilizer Z
1	2	1
2	4	2
3	6	3
4	8	4
5	10	5

Using the information in the data table, construct a line graph on the grid, following the directions below.

2 Mark an appropriate scale on the axis labeled "Time (days)."

3 Mark and appropriate scale on the axis labeled "Average Increase in Plant Height (millimeters)."

4 Plot the data for fertilizer *X* on the grid. Surround each point with a small triangle and connect the points.

5 Plot the data for fertilizer *Z* on the grid. Surround each point with a small circle and connect the points.

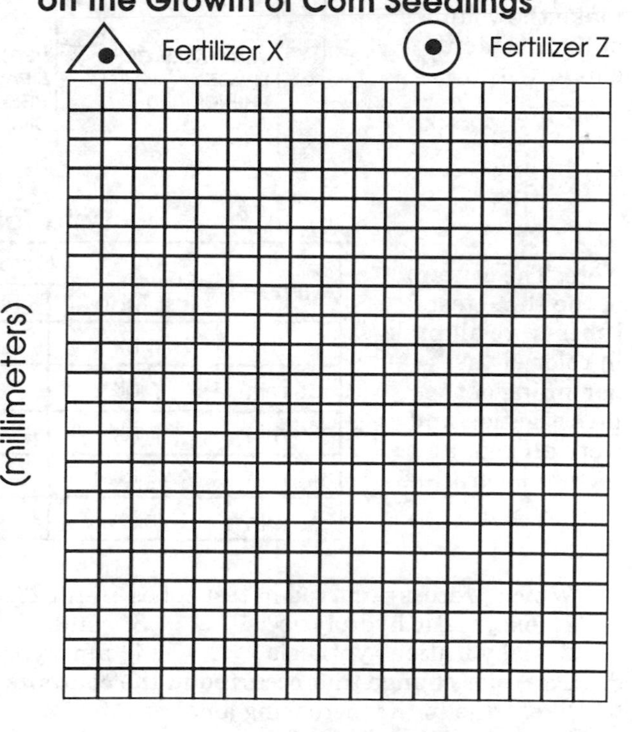

6 Using a complete sentence, state a conclusion that can be drawn from this investigation. ____

____

Base your answers to questions 7 through 10 on the information below and on your knowledge of biology. The information is given in three parts: the procedures for an investigation, diagrams, and a data table showing the results.

Procedures:

1) A student mixed appropriate amounts of vegetable oil (fat source), lipase, and phenol red (an indicator which is pink in the presence of a base and yellow in the presence of an acid) in test tube number 1.
2) In test tube number 2, the student mixed the same amounts of vegetable oil, lipase, and phenol red as in test tube number 1 and then added some bile salts (a fat emulsifier).
3) In test tube number 3, the student mixed the same amounts of vegetable oil, phenol red, and bile salts as in test tube 2.

Distilled water was added to all three test tubes to make the volumes of their contents equal.

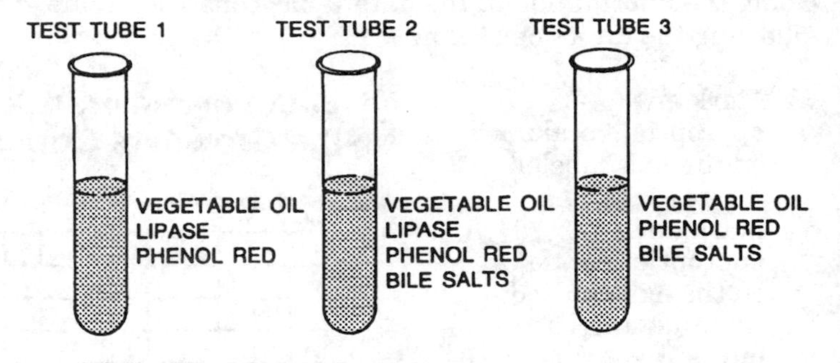

Note: The contents of the three test tubes were all pink in color at the beginning of the investigation and were all kept at room temperature.

Data Table

Time	Color Changes		
	Test Tube 1	Test Tube 2	Test Tube 3
0	pink	pink	pink
15 min	pink	yellow	pink
30 min	pink	yellow	pink
45 min	yellow	yellow	pink
60 min	yellow	yellow	pink

7 Which process occurred in test tubes 1 and 2?
1 enzymatic hydrolysis
2 dehydration synthesis
3 osmosis
4 pinocytosis

8 The color change that occurred in the contents of test tube 2 was most likely due to the increasing amounts of
1 glucose molecules
2 glycerol molecules
3 fatty acid molecules
4 amino acid molecules

9 After 60 minutes, the pH of the contents of test tube number 2 would be closest to
1 14
2 11
3 10
4 6

10 What would be the color of the contents of test tube 3 after 120 minutes?
1 white
2 yellow
3 blue
4 pink

The Biology Achievement Test ©

N & N Publishing expresses our appreciation to the College Board and the Educational Testing Service for providing their permission to use materials from The College Board Achievement Tests in Science.

Also, for students who need a more "in depth" explanation of The Biology Achievement Test, we recommend purchasing the official publication of The College Board:

The College Board Achievement Tests (14 Tests in 13 Subjects) ISBN: 0-87447-162-1

*It is our hope that the addition of these materials in N&N's **Science Series – Biology** will further enhance the education of our students.*

The Biology Achievement Test©

(Excerpts from: The College Board Achievement Tests in Science©)

The Biology Achievement Tests consists of 95 multiple choice questions. The test is based on the assumption that you have had a one-year introductory course in biology at a level suitable for college preparation. The test covers the topics listed in the chart below. Different aspects of each topic are stressed from year to year. Because high school courses differ, both in the percentage of time devoted to each major topic and in the specific subtopics covered, you may encounter questions on topics unfamiliar to you. However, in any typical high school biology course, more topics are usually covered - and in more detail - than there are questions in the Biology Achievement Test.

No one instructional approach is better than another in helping you prepare for the test, provided that you are able to recall and understand the major concepts of biology and to apply the principles you have learned to solve specific scientific problems in biology. In addition, you should be able to organize and interpret results obtained by observation and experimentation and to draw conclusions or make inferences from experimental data. Laboratory experience is a significant factor in developing reasoning and problem-solving skills. Although testing laboratory skills in a standardized test is necessarily limited, reasonable experience in the laboratory is an asset to you in preparing for the Biology Achievement Test.

You will *not* be allowed to use electronic calculators during the test. Numerical calculations are limited to simple arithmetic. In this test, the metric system of units is used.

Content of the Test

Topics Covered	Approximate Percentage of Test
Cellular and Molecular Biology	25 %
Ecology	15 %
Classical Genetics	10 %
Organismal Structure and Function	30 %
Evolution and Systematics	15 %
Behavior	5 %

Skills Specifications	Approximate Percentage of Test
Level I: *Essentially* Recall: (remembering specific facts; demonstrating straight-forward knowledge of information and familiarity with terminology)	**50 %**
Level II: *Essentially* Application: (understanding concepts and reformulating information into other equivalent terms; applying knowledge to unfamiliar and/or practical situations; solving mathematical problems)	**30 %**
Level III: *Essentially* Interpretation: (inferring and deducing from data available and integrating information to form conclusions; recognizing unstated assumptions)	**20 %**

Questions Used in the Test

Classification Questions

Each set of classification questions has five lettered choices in the heading that are used in answering all of the questions in the set. The choices may be statements that refer to concepts, principles, organisms, substances, or observable phenomena; or they may be graphs, pictures, equations, formulas, or experimental settings or situations. The questions themselves may be presented in one of these formats or in the question format directly. To answer each question first consider all of the choices. The directions for this type of question specifically state that you should not eliminate a choice simply because it is the correct answer to a previous question.

Five-Choice Completion Questions

The five-choice completion question is written either as an incomplete statement or as a question. In its simplest application, this type of question poses a problem that intrinsically has a unique solution. The five-choice completion question is also used with other introductory material (such as a summary of an experiment, an outline of a problem, a graph, a chart) to assess your ability to use learned concepts and to apply them to unfamiliar laboratory or experimental situations.

Sets of questions describing laboratory or experimental situations test your understanding of a problem in greater depth than is generally possible with some other formats. In particular, sets test your ability to: (1) identify a problem, (2) evaluate experimental situations, (3) suggest hypotheses, (4) interpret data, (5) make inferences and draw conclusions, (6) check the logical consistency of hypotheses based on relevant observations, and (7) select appropriate procedures for further investigation of the problem described. In the Biology Achievement Test, questions pertaining to laboratory or experimental situations are grouped together in the latter part of the test.

A special type of five-choice completion question allows for the possibility of multiple correct answers. Unlike many quantitative problems that have a unique solution, some questions have more than one correct response. For these questions, you must evaluate each response independently of the others in order to select the most appropriate combination. In questions of this type, several (usually three to five) statements labeled by Roman numerals are given with the question. One or more of these statements may answer the question correctly. The statements are followed by five lettered choices, each consisting of some combination of the Roman numerals that label the statements. You must select from among the five lettered choices the one combination of statements that best answers the question. In the test, questions of this type are intermixed among the more standard five-choice completion questions.

Biology Achievement Supplementary Information

*The following supplementary information is either NOT included or has only limited reference in the previous sections of **N&N's Science Series – Biology**. However, The Biology Achievement Test bases about five percent of its questions on Animal Behavior and may include questions on the other topics explained in this Supplementary Information section.*

Animal Behavior

Basic Definitions:

Innate Behavior is a behavior that is passed from parent to offspring through hereditary. Subdivisions of innate behavior include:

A **reflex** is a response pattern resulting from fixed pathways in the autonomic nervous system. A reflex has a specific and "predetermined" response to certain stimuli (such as the blinking of the eye and the jerking of the knee).

Instinct is a behavior involving "unlearned" complicated responses to specific stimuli (such as a female salmon returning to her hatching stream to deposit eggs and also an animal's courtship behavior).

Learned Behavior can be changed and is developed through trial and error, conditioned responses, and conceptualization. Learned behavior occurs through experience, insight, and/or reasoning.

Additional Explanations of Behavior:

Memory is the process by which information is stored and recalled in the brain of animal organisms. The process involves millions of different nerve cells in the cerebral cortex interacting together. In humans, there are two separate and independent memory systems:

- **Short term memory** is used for the quick storage of information (for example, a telephone number). However, it may result in the rapid loss of that information.

- **Long term memory** is dependent upon short term information being repeated several times and associated with previously stored information (such as a favorite book or a piece of sheet music).

Learning is the structural and biochemical change which occurs in the brain due to experience. Learning results in a change in behavior and a modification in the thinking process. For example, one of the most important activities that you learned in elementary school was to read. The process of reading allowed you to study and understand the world around you.

Habits are learned responses which become "automatic" because of frequent repeating of the behavior (such as a manner of speaking and tying a shoelace). **Habitation** is learning *not* to respond to a certain stimulus. Animals learn to "sort out" and "ignore" stimuli that are not essential to their survival. For example, the sound produced by a rabbit moving through the bushes does not frighten a deer, but the sound produced by a human alerts the deer.

Imprinting is a simple, rapid, and irreversible type of learning which is not a part of the organism's innate behavior. An example of imprinting may be observed in the process by which newly hatched birds follow whatever moving object that they first see. The organism may be its biological parent, another animal, or in rare cases, a human who happened to be near the chick at its birth.

Thomas Malthus

Thomas Malthus (1766-1834) was an English economist best known for his essay on the *Principle of Population,* published in 1798. His main idea in this essay was that population tended to increase more rapidly than food supplies, and that a lack of food inevitably led to war and disease. He believed that war and disease would have to reduce the extra population in order to reestablish a balance with the food supply. He suggested, as an alternative means of population control, that people limit the number of their children.

Bioluminescence

Bioluminescence is the process by which stored energy is converted into light energy by certain bacteria, protozoa, fungi, worms, crustacea, and fireflies. For example, in fireflies, stored energy in the form of ATP is combined by enzymes with oxygen to release energy in the form of light. This energy change is used by the firefly as a mating signal.

Radioactive Isotopes and the Age of Fossils

Isotopes are atoms of the same element that contain extra neutrons in their nuclei. Most atoms have isotopes which are stable, but some decay and release particles (radiation). These radioactive isotopes change into atoms with lighter atomic masses.

The **half-life** of a radioactive isotope is specific and refers to the amount of time necessary for 50% of the radioactive isotopes in a sample to decay.

Carbon 14 is the radioactive isotope of normal Carbon, and Potassium 40 is the radioactive isotope of normal Potassium. Carbon 14 and Potassium 40 are naturally occurring isotopes found in all living organisms. During the lifetime of the organism, the ratio of normal atoms compared to their isotopes is constant and remains fixed at the time of the organism's death. Scientists are able to measure the amount of radioactive decay, and by using the specific half-life of that radioactive isotope, to establish the age of the fossil.

Primates

Primates are a group of animals which include New and Old World monkeys, Great Apes, and humans. New World monkeys, Old World monkeys, and Great Apes share similar characteristics such as binocular vision, long arms for climbing trees, and brachiation (the ability to raise the arms above the head). These animals are also similar biochemically and anatomically. The comparison of these similarities is important in studying evolution, since the similarities suggest a common ancestor for man.

This chart compares the New World monkey to the Great Ape:

Item	New World Monkey	Great Ape
Name	Squirrel Monkey, Spider Monkey, and Howler Monkey	Chimpanzee, Gorilla, and Orangutang
Range	Central and South America	Africa and Asia
Vertical Placement	Middle to upper tree branches	Ground to middle level branches of trees
Weight	20 to 30 pounds	100 to 400 pounds
Shape of the Face	Dog-like	More flattened face
Nostrils of the Nose	Wide, separated by broad septum	Narrow, pinched together
Dental Formula	Upper: 2 : 1 : 3 : 3 Lower: 2 : 1 : 3 : 3	Upper: 2 : 1 : 2 : 3 Lower: 2 : 1 : 2 : 3
Facial Muscles	Essentially one large sheet of muscle	Several different muscles, each with its own nerve control
Tails	Prehensil tail, half again as long as the head and body combined	Little or no tail
Digits	Claws on all digits; thumb has little power of adduction	Nails on all digits; well developed thumb has good adductive power
Uterus	Designed for twin fetuses	Designed for single fetus

Sporulation

Sporulation is the process by which fungi, such as the breadmold *Rhizopus*, reproduce by spores. The mold is composed of individual thread-like structures called *hypae*. The hypae join together to form a large interconnected mass called *mycelium*. Mycelium carry on extracellular digestion, absorb "predigested" nutrients, and produce vertical hypae called *sporangiophores*. At the end of the sporangiophore, a spherical reproductive case called a *sporangium* is formed. Spores are produced within the sporangium.

Alternation of Generations

Alternation of generations describes the diploid (*2n*) and haploid (*n*) stages found in the life cycle of some algae, fungi, all Bryophytes (non-vascular plants), and Tracheophytes (vascular plants).

The diploid or sporophyte state reproduces as a result of meiotic cell division, forming haploid spores. These haploid spores live and grow in the environment and make up the *gametophyte* stage. The haploid cells reproduce by mitotic cell division, eventually forming gametes. These gametes fuse to form the diploid zygote or *sporophyte* stage.

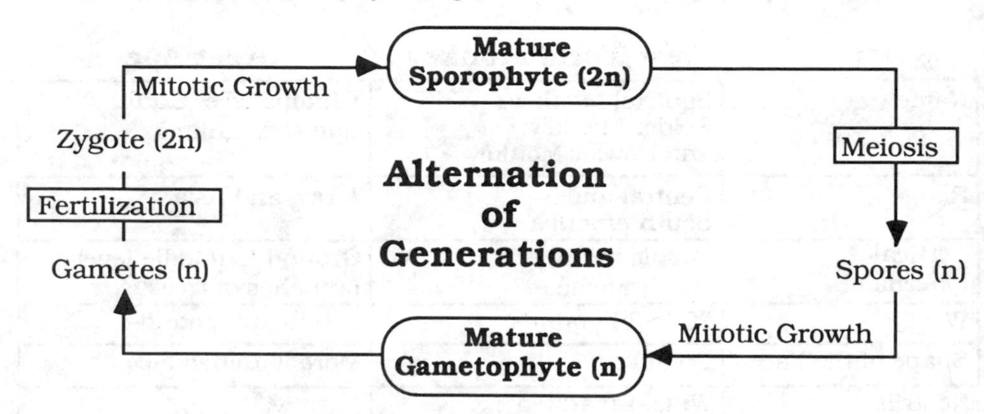

Koch's Postulates

Robert Koch (1843-1910), a German bacteriologist, discovered the cause of tuberculosis and made important contributions to medicine. In recognition of his achievements to science, he was awarded the Nobel prize for medicine. In order to prove that a particular microorganism is the cause of a specific infectious disease, Koch stated the following:

The pathogen
- must be found in every case of the disease
- must be isolated and grown in a pure culture
- must produce the disease in suitable laboratory animals
- must be recovered from the laboratory animal, re-isolated, and grown in a pure culture

Infectious Disease

Microscopic organisms such as viruses, bacteria, yeast, molds, protozoa, and flat and round worms are responsible for infectious diseases which adversely affect man, other animals, and plants. These disease producing organisms (pathogens) exist in a parasitic relationship with a host organism, and because of their activity, cause damage and death to their host.

Growth Curve

The growth curve is a visual method used to illustrate the relationship between the number of organisms in a population and time. The curve (illustrated on the left) is composed of three stages seen on the opposite page:

- First stage (I) shows that more organisms are produced than die.
- Second stage (II) shows a stable population where the number of new organisms is equal to the number of organisms that die.
- Third third stage (III) shows that more organisms die than are produced.

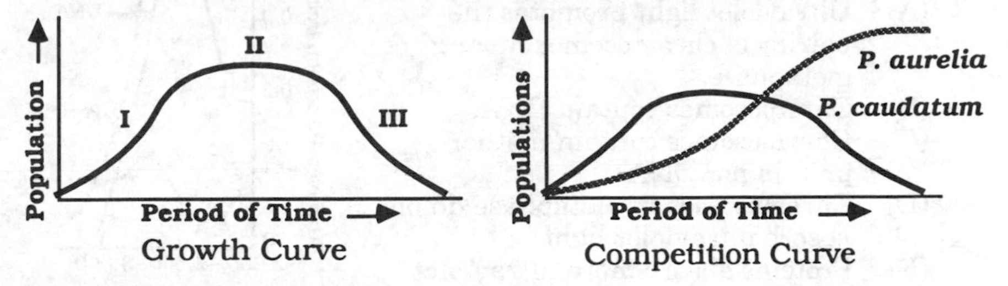

Growth Curve

Competition Curve

Competition Curve

The competition curve is used to compare the population of two different species of organisms under the same set of physical conditions. The graph (illustrated above right) is composed of two separate growth curves which illustrate the population of each species over a given period of time.

Biology Achievement Test Questions

The following questions have been selected (with permission from the Educational Testing Service) from a published Biology Achievement Test. These questions were chosen because we feel that the questions are

(1) "typical" of the style and format of the actual achievement test questions that you will be expected to answer correctly.

(2) "typical" in respect to their difficulty and usage of graphs, illustrations, and diagrams.

(3) "typical" as to the broad range of material that will be covered on the actual achievement test.

(4) "typical" of the most often missed questions (generally, over 50%).

Achievement Test Questions For Practice

1. Animals that possess a closed circulatory system include which of the following?

   I. Earthworm II. Frog III. Grasshopper

   (A) II only
   (B) III only
   (C) I and II only
   (D) I and III only
   (E) I, II, and III

2. Which of the following is NOT found in DNA molecules?

   (A) Adenine
   (B) Deoxyribose
   (C) Phosphorus
   (D) Uracil
   (E) Thymine

3. All of the following are characteristics of enzymes EXCEPT:

   (A) They are proteins.
   (B) They are inactivated by high temperature.
   (C) They are organic catalysts.
   (D) Each binds temporarily with its substrate.
   (E) Each is active within a wide range of pH.

4. From the graph showing the degree of ultraviolet absorption by metaphase chromosomes, pure DNA, and protein, which of the following can be inferred?
   (A) Ultraviolet light promotes the pairing of chromosomes during metaphase.
   (B) Chromosomes contain DNA.
   (C) Chromosomes contain neither protein nor DNA.
   (D) Chromosomes in metaphase do not absorb ultraviolet light.
   (E) Proteins absorb more ultraviolet light than does DNA.

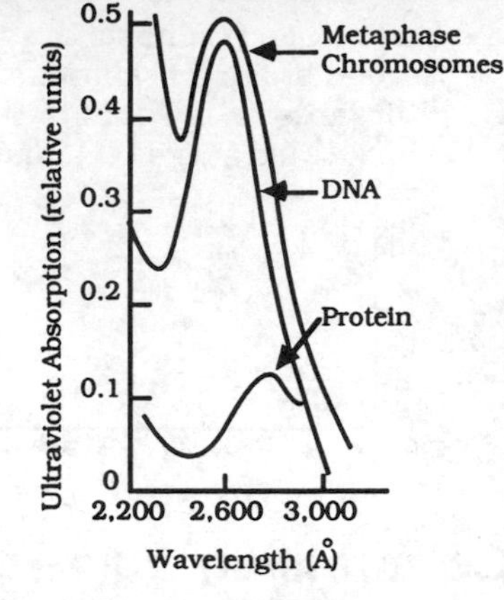

5. During embryonic development, the germ layer from which the inner lining of the digestive tract develops is the
   (A) ectoderm
   (B) epidermis
   (C) mesenchyme
   (D) endoderm
   (E) mesoderm

6. Which of the following invertebrates is considered to be the most primitive?
   (A) Arthropods
   (B) Coelenterates (Cnidaria)
   (C) Mollusks
   (D) Annelids (segmented worms)
   (E) Echinoderms

7. The crab, spider, ant, and lobster are classified in the same phylum. Which of the following sets of characteristics do they share?
   (A) Segmented body and six appendages
   (B) Jointed appendages and a digestive system with a single opening
   (C) Jointed appendages and a chitinous exoskeleton
   (D) A dorsal nerve cord and a chitinous exoskeleton
   (E) A dorsal nerve cord and jointed appendages

8. Which of the following contributes most to genetic variability?
   (A) Mitosis
   (B) Regeneration
   (C) Meiosis
   (D) Linkage
   (E) Vegetative propagation

9. All of the following are endocrine glands EXCEPT the
   (A) salivary
   (B) adrenal
   (C) pituitary
   (D) testis
   (E) thyroid

10. Which of the following is true about the flow of energy in a food chain?
   (A) There is more energy available to consumers than to producers.
   (B) There is more energy available to secondary consumers than to primary consumers.
   (C) There is more energy available to primary consumers than to secondary consumers.
   (D) The energy available to producers is determined by their interactions with primary and secondary consumers.
   (E) There is more energy available to decomposers than to producers.

11. The study of crossing-over between homologous chromosomes is most useful for determining which of the following?
    (A) Number of mutations on a chromosome
    (B) Length of a chromosome
    (C) Position of genes on a chromosome
    (D) Presence of dominant genes
    (E) Occurrence of polyploidy

12. The primary function of NAD in the Kreb's cycle is to serve as
    (A) an oxygen acceptor
    (B) an oxygen donor
    (C) a source of phosphate ions
    (D) a hydrogen ion and electron acceptor
    (E) a photosynthetic pigment

13. The oxygen given off by plants is a product of
    (A) aerobic respiration
    (B) anaerobic respiration
    (C) the light phase of photosynthesis
    (D) the dark phase of photosynthesis
    (E) oxidation of carbohydrates

14. All of the following processes are involved in translocation in plants EXCEPT
    (A) root pressure
    (B) cohesion
    (C) transpiration
    (D) capillarity
    (E) phototropism

15. As one travels northward from the geographical center of the U.S., the correct sequence of major biomes encountered is
    (A) grassland, taiga, arctic tundra
    (B) deciduous forest, alpine tundra, arctic tundra
    (C) grassland, desert, alpine tundra
    (D) desert, deciduous forest, taiga
    (E) desert, grassland, taiga

16. Which of the following is NOT characteristic of monocots?
    (A) Flower parts are usually in threes or multiples of three.
    (B) Leaves have parallel veins.
    (C) Stems contain cambium for production of secondary xylem.
    (D) Seeds contain a single cotyledon.
    (E) Seeds at maturity usually contain a large endosperm.

17. Scientists who study evolution look for homologous structures when determining similarities among species. These structures are said to have been inherited from common ancestors. All of the following are examples of homologous structures EXCEPT the
    (A) wings of a robin and the wings of an owl
    (B) wings of a blue jay and the wings of a butterfly
    (C) wings of an ostrich and the front legs of a dog
    (D) front legs of a horse and the arms of a human
    (E) legs of a chicken and the hind legs of a lizard

18. Which of the following occurs in meiosis but <u>not</u> in mitosis?
    (A) Production of diploid cells
    (B) Synapsis of homologous chromosomes
    (C) Duplication of DNA
    (D) Duplication of the centrioles
    (E) Appearance of spindle fibers

19. If a cell uses 100 amino acid molecules to produce a particular polypeptide chain in a protein, the number of water molecules formed during the process is
    (A) 1 (B) 50 (C) 99 (D) 100 (E) 101

20. The primitive Earth atmosphere before life began is believed to have contained all of the following gases EXCEPT
    (A) hydrogen (C) ammonia (E) methane
    (B) water vapor (D) oxygen

21. Which of the following represents the normal sequence of animal development?
    (A) Gastrula formation → blastula formation → mesoderm formation → cleavage → somite formation
    (B) Cleavage → blastula formation → somite formation → gastrula formation → mesoderm formation
    (C) Cleavage → blastula formation → mesoderm formation → gastrula formation → somite formation
    (D) Cleavage → blastula formation → gastrula formation → mesoderm formation → somite formation
    (E) Blastula formation → gastrula formation → mesoderm formation → cleavage → somite formation

Questions 22-23 refer to the two graphs. Graph I shows the growth curve of Paramecium aurelia and Paramecium caudatum cultured in separate dishes. Graph II shows the growth curve of these two species cultured in the same dish. Similar conditions of food and other requirements were provided in each case.

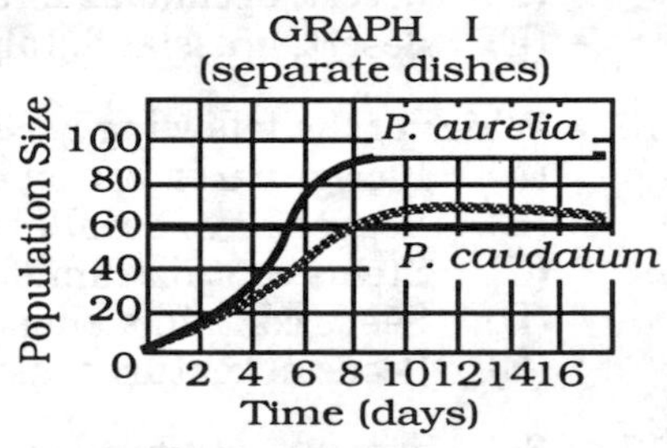

22. In graph I, the most likely explanation for the fact that the population curves of both *P. aurelia* and *P. caudatum* flatten out after an initial period of growth is that
    (A) organisms of both populations are no longer reproducing
    (B) population sizes have reached the maximum capacity that their environment can support
    (C) organisms of both populations have used up all the food
    (D) organisms of both populations are dying faster than reproducing
    (E) the oxygen required by the organisms has been all used up

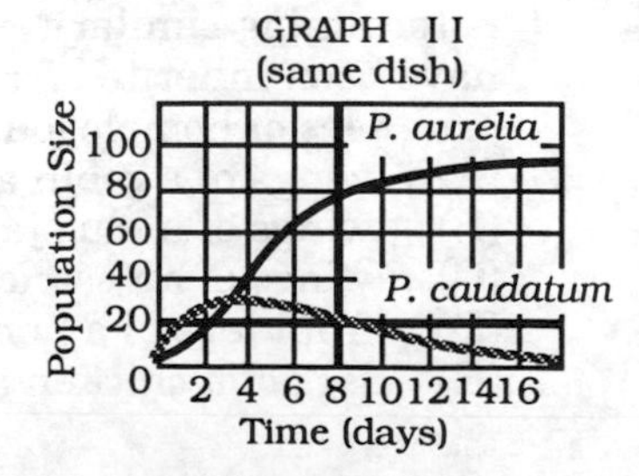

23. The effects shown in graph II are most likely the result of
    (A) mutualism between two species
    (B) parasitism of one species by another
    (C) competition between members of the same species
    (D) competition between members of different species
    (E) toxic poisoning of one species after the first day

Questions 24-25
(A) Xylem vessel
(B) Companion cell
(C) Guard cell
(D) Apical meristem
(E) Sieve-tube element

24. Is usually dead in a mature plant; is cut when a finished board of lumber is sawed

25. Transports manufactured food from leaves to roots in green plants

Questions 26-27
(A) Amnion
(B) Allantois
(C) Yolk sac
(D) Chorion
(E) Eggshell

26. Provides for storage of the nitrogenous wastes of a developing chick embryo and functions in respiratory gas exchange

27. Forms a fluid-filled chamber in which a terrestrial animal may develop in an aquatic medium

Questions 28-30 pertain to the muscle twitch of the gastrocnemius muscle of a frog in response to a stimulus. The diagram below illustrates a muscle twitch recorded on a kymograph.

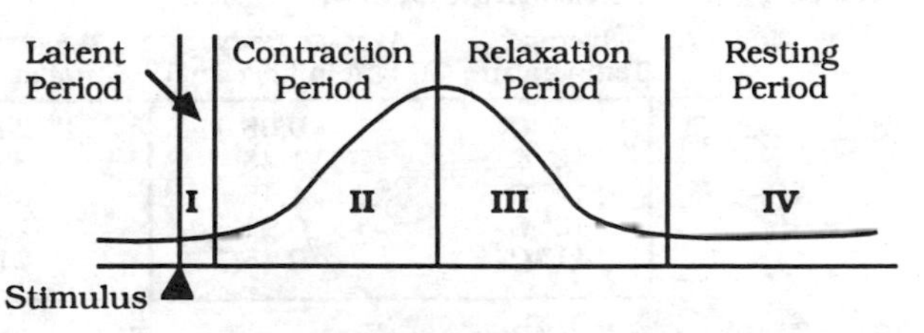

28. The energy for the movement during the contraction period (II) is provided by
    (A) ADP
    (B) ATP
    (C) FAD
    (D) NAD
    (E) vitamins

29. There is very little action during the latent period for which of the following reasons?
    (A) The stimulus is changing the permeability of the membrane.
    (B) ADP is being decomposed.
    (C) The resynthesis of glycogen is taking place.
    (D) The oxygen debt is being replaced.
    (E) Actin and myosin are being synthesized.

30. The conversion of ADP to ATP in a muscle takes place primarily during
    (A) the instant that contact is made with the stimulus
    (B) I and II
    (C) I and IV
    (D) II and III
    (E) III and IV

Questions 31-32 refer to the diagram of the food web at the right.

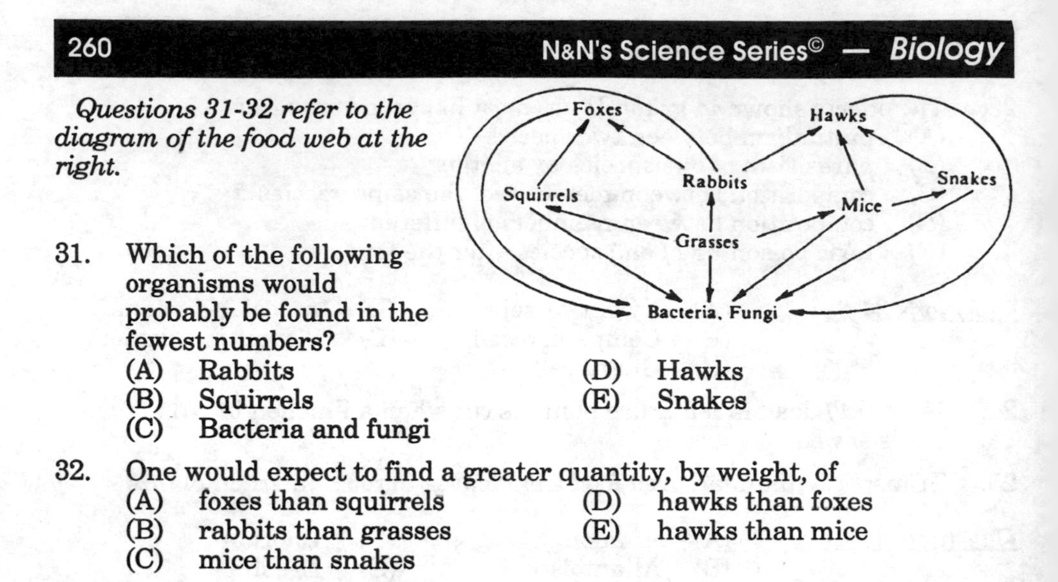

31. Which of the following organisms would probably be found in the fewest numbers?
    (A) Rabbits
    (B) Squirrels
    (C) Bacteria and fungi
    (D) Hawks
    (E) Snakes

32. One would expect to find a greater quantity, by weight, of
    (A) foxes than squirrels
    (B) rabbits than grasses
    (C) mice than snakes
    (D) hawks than foxes
    (E) hawks than mice

Question 33

To test the effects of temperature on the metamorphosis and reproduction of a certain species of invertebrate, some investigators collected larvae from a creek and raised them to adulthood in the laboratory. They determined the average adult body weight, the average number of eggs laid by each female, and the time required for the adults to emerge from the larvae under five different growth temperatures. The average temperature of the creek was 18°C. The data are given in the table below.

Average Temperature	Average Body Weight (milligrams)	Average Number of Eggs	Days to Adult
18° C	0.66	248	11
16° C	1.28	350	16
14° C	1.46	380	40
13° C	0.91	289	56
11° C	0.48	212	82

33. Theoretically, the greatest number of generations of offspring per year could be produced at which temperature?
    (A) 18°C (B) 16°C (C) 14°C (D) 13°C (E) 11°C

Questions 34-35

Four test tubes are set up as shown below. All of the tubes contain water to which a few drops of indicator have been added. The indicator used is yellow when the pH of the solution is less than 6.0, green when the solution pH is between 6.0 and 7.5, and blue when the solution pH is above 7.5.

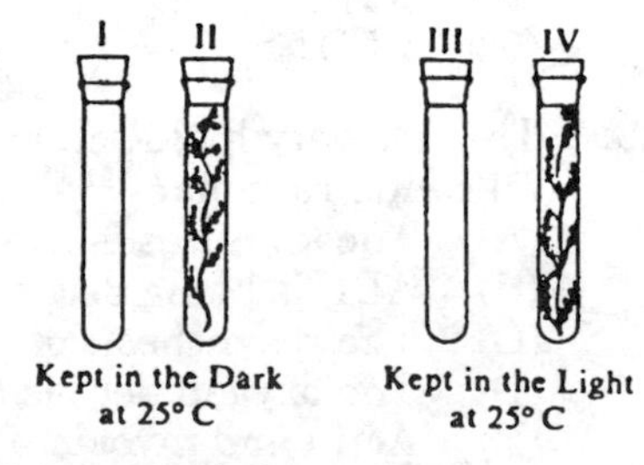

All tubes are greenish-blue at the beginning of the experiment. Tubes II and IV contain comparable sprigs of a green plant. Tubes I and II are kept in the dark; tubes III and IV are kept in the light. All are examined 24 hours later. At this time, the indicator in tube II is yellow; the indicator in tubes I, III, and IV is still greenish blue.

34. Which of the statements about the experiment (previous page) is correct?
    (A) Tubes I and III represent experimental sets.
    (B) Tube II is the control for tube III.
    (C) Temperature is the independent variable.
    (D) Light is the independent variable.
    (E) pH is the independent variable.

35. After 24 hours, the indicator in tube II is yellow because the
    (A) plant photosynthesized and gave off O_2, which lowered the pH
    (B) plant respired and gave off CO_2, which formed carbonic acid and lowered the pH
    (C) plant could not photosynthesize and died, causing the H_2O to become alkaline
    (D) plant excreted nitrogenous wastes which caused the medium to become more alkaline
    (E) indicator decomposed in the dark causing the medium to become more acidic

---

Achievement Test Answers and Explanations

Check your answers. For more information concerning the question, refer to the area of the text indicated or review the additional material provided in the supplementary section.

1.	**Answer C:**	Information included in Unit I	page 45 of text.
2.	**Answer D:**	Information included in Unit V	page 164 of text.
3.	**Answer E:**	Information included in Unit I	page 25 of text.
4.	**Answer B:**	Information included in Unit V	page 120 of text.
5.	**Answer D:**	Information included in Unit IV	page 130 of text.
6.	**Answer B:**	Information included in Unit I	page 8 of text.
7.	**Answer C:**	Information included in Unit I	page 8 of text.
8.	**Answer C:**	Information included in Unit IV	page 157 of text.
9.	**Answer A:**	Information included in Unit III	page 105 of text.
10.	**Answer C:**	Information included in Unit VII	page 205 of text.
11.	**Answer C:**	Information included in Unit V	page 153 of text.
12.	**Answer D:**	Information included in Unit II	page 57of text.
13.	**Answer C:**	Information included in Unit II	page 56 of text.
14.	**Answer E:**	Information included in Unit II	page 50 of text.
15.	**Answer C:**	Information included in Unit VII	page 213 of text.
16.	**Answer C:**	Information included in Unit IV	page 50 of text.
17.	**Answer B:**	Information included in Unit VII	page 181 of text.
18.	**Answer B:**	Information included in Unit IV	page 119, 126 of text.
19.	**Answer C:**	Information included in Unit I	page 22 of text.
20.	**Answer D:**	Information included in Unit VI	page 190 of text.
21.	**Answer D:**	Information included in Unit IV	page 130 of text.
22.	**Answer B:**	Information included in Unit Supplementary	page 255.
23.	**Answer D:**	Information included in Unit Supplementary	page 255.
24.	**Answer A:**	Information included in Unit II	page 51 of text.
25.	**Answer E:**	Information included in Unit II	page 51 of text.
26.	**Answer B:**	Information included in Unit IV	page 131 of text.
27.	**Answer A:**	Information included in Unit IV	page 131 of text.
28.	**Answer B:**	Information included in Unit II, III	page 55, 109 of text.
29.	**Answer A:**	Information included in Unit II, III	page 55, 109 of text.
30.	**Answer E:**	Information included in Unit II, III	page 55, 109 of text.
31.	**Answer D:**	Information included in Unit VII	page 205 of text.
32.	**Answer C:**	Information included in Unit VII	page 26, 206 of text.
33.	**Answer A:**	Information included in Unit LS	page 26, 245 of text.
34.	**Answer D:**	Information included in Unit LS	page 38, 56, 240 of text.
35.	**Answer B:**	Information included in Unit LS	page 38, 56, 240 of text.

SELF–HELP Question Answers and Page References

page 262

Col. 1 - **Question Number**; Col. 2 - **Answer**; Col. 3 - **Page Reference**
Skill Page Reference: see Unit pages or Skills - Pages 233 - 248

Unit 1

Core pg. 29

Question	Answer	Page
1.	2	8
2.	4	6
3.	4	5
4.	4	5
5.	3	5
6.	4	8
7.	2	7
8.	4	7
9.	4	7
10.	1	7
11.	2	22
12.	1	15
13.	3	16
14.	1	15
15.	4	16
16.	4	19
17.	4	19
18.	1	25
19.	2	20
20.	3	21
21.	4	22
22.	3	22
23.	1	22
24.	4	20
25.	2	25
26.	3	25
27.	2	25
28.	4	25
29.	3	27
30.	3	25

Extended pg. 31

Question	Answer	Page
1.	2	22
2.	4	22
3.	4	22
4.	3	27
5.	3	22
6.	4	22
7.	2	22
8.	3	21
9.	1	20
10.	4	18
11.	1	22
12.	2	15
13.	2	21
14.	5	20
15.	4	25
16.	3	27
17.	2	26
18.	4	26
19.	2	25
20.	1	19

Skill pg. 34

1. 3
2. 1
3. 1
4. 1
5. 1
6. 2
7. 2
8 #1 - The coarse adjustment is used to focus under low power.
#2 - The fine adjustment is used for final adjustment.
#3 - The low power objective is used first to focus on specimen.
#4 - The diaphragm is used to change the amount of light reaching the specimen.
9. | | **d** | |
10. 1
11. 2
12. 2
13. 1
14. 1
15. 1

Unit 2

Core pg. 72

Question	Answer	Page
1.	3	43
2.	2	39
3.	3	37
4.	2	38
5.	3	37
6.	4	43
7.	2	43
8.	1	44
9.	2	43
10.	1	43
11.	1	49
12.	4	49
13.	2	48
14.	1	48
15.	2	51
16.	4	48
17.	1	49
18.	2	50
19.	3	50
20.	2	51
21.	4	56
22.	1	60
23.	1	60
24.	2	66
25.	3	55
26.	3	55
27.	1	62
28.	2	62
29.	3	62
30.	3	63
31.	4	68
32.	2	66
33.	2	68
34.	2	69
35.	1	65
36.	2	66
37.	1	66
38.	2	70
39.	3	70
40.	2	70

Extended pg. 75

Question	Answer	Page
1.	2	20
2.	4	43
3.	3	20
4.	2	38
5.	1	38
6.	3	39
7.	1	38
8.	4	56
9.	3	63
10.	2	38
11.	3	38
12.	4	56
13.	2	38
14.	3	57
15.	1	56

Skill pg. 77

1. graph
2. graph
3. graph
4. 4
5. 2
6. 4
7. 1
8. 3
9 nutrients: glucose, maltose, or lactose
10 The amount of gas collected will increase because the maltose will be hydrolized to glucose.
11 oxygen
12 photosynthesis
13 A decrease in light would produce a decrease in photosynthesis and a decrease in the amount of oxygen produced.
14. 3
15. 3

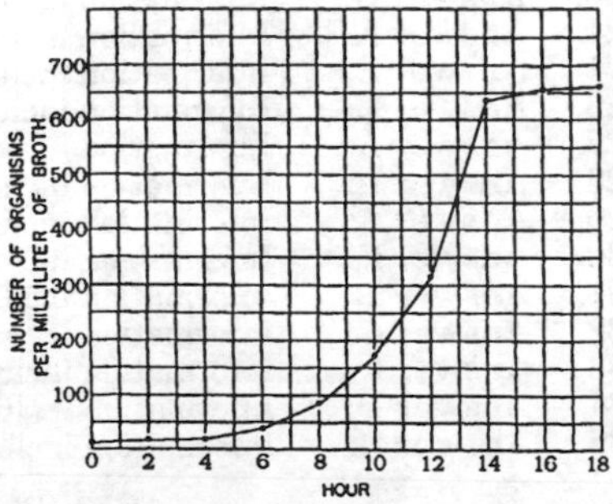

NUMBER OF ORGANISMS PER MILLILITER OF BROTH
700
600
500
400
300
200
100
0 2 4 6 8 10 12 14 16 18
HOUR

Unit 3

Core pg. 111

Question	Answer	Page
1.	3	84
2.	2	82
3.	4	82
4.	2	82
5.	3	82
6.	4	82
7.	4	91
8.	4	81
9.	2	89
10.	3	89
11.	4	89
12.	2	91
13.	1	87
14.	4	89
15.	2	90
16.	1	95
17.	4	95
18.	2	96
19.	2	95
20.	2	94
21.	1	98
22.	4	99
23.	2	98
24.	1	98
25.	3	98
26.	3	99
27.	4	102
28.	1	103
29.	4	102
30.	4	101
31.	1	101
32.	1	101
33.	3	102
34.	4	105
35.	3	106
36.	2	106
37.	4	102
38.	4	109
39.	3	109
40.	3	109

Extended pg. 113

Question	Answer	Page
1.	3	107
2.	1	103
3.	3	96
4.	5	86
5.	6	86
6.	1	91
7.	1	90
8.	4	90
9.	1	90
10.	4	90
11.	3	90
12.	2	102
13.	1	103
14.	1	103
15.	2	95
16.	1	95
17.	3	90
18.	4	94
19.	2	106
20.	4	88

Skill pg. 116

Question	Answer
1.	2
2.	3
3.	1
4.	2
5.	3
6.	4
7.	1
8.	1
9.	2
10.	4
11.	graph
12.	graph
13.	graph
14.	2
15.	4

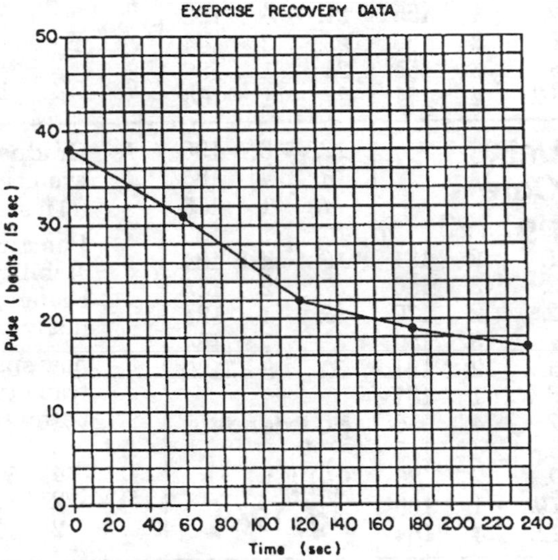

EXERCISE RECOVERY DATA

Pulse (beats/ 15 sec)

Time (sec)

Unit 4

Core pg. 141

Question	Answer	Page
1.	3	122
2.	4	121
3.	4	128
4.	4	122
5.	4	124
6.	2	120
7.	2	123
8.	3	121
9.	3	122
10.	4	126
11.	4	126
12.	4	138
13.	3	128
14.	1	121
15.	4	128
16.	1	128
17.	2	134
18.	1	135
19.	3	135
20.	3	134
21.	3	139
22.	1	138
23.	4	138
24.	3	138
25.	3	138

Extended pg. 143

Question	Answer	Page
1.	2	134
2.	2	134
3.	1	135
4.	1	135
5.	2	130
6.	1	129
7.	3	130
8.	4	136
9.	1	132
10.	2	134
11.	3	133
12.	4	133
13.	5	134
14.	2	134
15.	3	130
16.	2	130
17.	1	130
18.	4	130
19.	4	131
20.	3	131
21.	1	131
22.	2	131
23.	4	131
24.	3	131
25.	3	134

Skill pg. 145

Question	Answer
1.	graph
2.	graph
3.	graph
4.	1

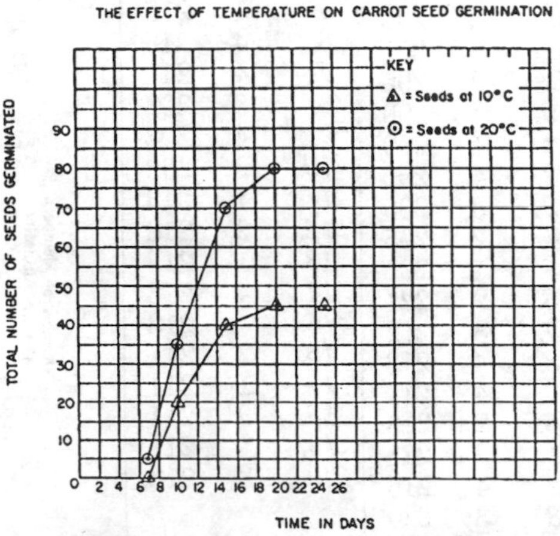

THE EFFECT OF TEMPERATURE ON CARROT SEED GERMINATION

KEY

△ = Seeds at 10°C

⊙ = Seeds at 20°C

TOTAL NUMBER OF SEEDS GERMINATED

TIME IN DAYS

Unit 5

Core pg. 172

1.	1	148
2.	4	148
3.	3	154
4.	1	154
5.	1	153
6.	3	154
7.	1	155
8.	2	152
9.	3	155
10.	4	154
11.	3	148
12.	3	148
13.	1	158
14.	3	158
15.	3	153
16.	3	153
17.	2	159
18.	3	159
19.	4	158
20.	3	162
21.	4	163
22.	2	164
23.	1	166
24.	1	164
25.	1	164

Extended pg. 174

1.	1	165
2.	3	164
3.	2	166
4.	4	98
5.	3	165
6.	1	159
7.	4	162
8.	2	162
9.	4	158
10.	3	164
11.	1	165
12.	2	166
13.	2	162
14.	1	165
15.	2	165
16.	3	166
17.	2	165
18.	1	166
19.	4	167
20.	2	169

Skill pg. 177

1.	4
2.	3
3.	1
4.	1
5.	2
6.	3
7.	1
8.	3
9.	1
10.	4

Unit 6

Core pg. 193

1.	2	181
2.	3	180
3.	2	179
4.	3	180
5.	2	181
6.	1	181
7.	4	186
8.	4	180
9.	2	186
10.	1	169
11.	4	184
12.	3	190
13.	4	185
14.	1	185
15.	1	180
16.	4	184
17.	2	185
18.	4	184
19.	1	184
20.	3	190

No Extended

Skill pg. 196

1 Gradualism - (a) continuous and slow evolution; (b) accumulation of small changes; (c) fossil record of transitional forms.

Punctuated Equilibrium - (a) long periods of stability; (b) brief periods of significant change; (c) fossil record of not having very many transitional periods.

2 new species evolution

3 transitional forms in some evolutionary pathways.

4.	4
5.	2
6.	2
7.	2
8.	4
9.	3
10.	2

Unit 7

Core pg. 224

1.	1	201
2.	2	212
3.	2	201
4.	2	201
5.	3	201
6.	2	209
7.	2	201
8.	2	201
9.	3	205
10.	1	200
11.	4	200
12.	4	205
13.	3	205
14.	2	205
15.	1	205
16.	2	221
17.	3	212
18.	4	213
19.	4	211
20.	2	213
21.	2	221
22.	4	215
23.	3	220
24.	4	219
25.	4	212

Extended pg. 227

1.	1	214
2.	3	213
3.	3	215
4.	4	215
5.	1	215
6.	4	215
7.	4	211
8.	3	202
9.	4	202
10.	3	202
11.	1	213
12.	2	213
13.	1	213
14.	4	213
15.	4	213
16.	2	211
17.	1	211
18.	3	211
19.	4	211
20.	3	213

Skill pg. 230

1. Example: Radioactive contamination could cause mutations, birth defects, and/or diseases, such as cancer in the human population.

2. 4

3. 1

4. 2

5. - 8. graph

9. 1

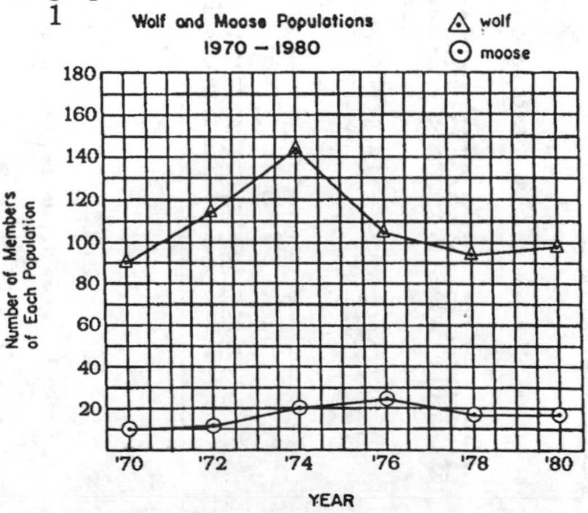

Wolf and Moose Populations
1970 – 1980
wolf
moose
180
160
140
120
100
80
60
40
20
Number of Members of Each Population
'70
'72
'74
'76
'78
'80
YEAR

Word Parts Prefixes and Suffixes

Many words used in biology get their roots (first meanings) from Greek and Latin. The following is a list of word parts that may help you to better understand some of the more difficult biological words.

For example, **biology** is defined as "the *study* of *living* things." It comes from two Greek words: **bio**, meaning life, and **logy**, meaning study. **Photosynthesis** is defined as "the process by which *light* energy is used by a green plant to *make* organic nutrients; the *conversion* of *light* energy to chemical bond energy." It also comes from two Greek words: **photo**, meaning light, and **synthesis**, meaning making.

When studying biology, separate the more complex words into their simpler *word parts*. This should help make the language of biology less difficult and your studying a little easier.

Word Part	Meaning	Sample Word
a-, an-	not, without	**a**biotic, **an**aerobic
ab-	from	**ab**ductor muscle
acro-	tip, end	**acro**petal
ad-	toward, to	**ad**ductor muscle
aero-	air	**aero**bic
ana-	up	**ana**phase
angio-	vessel	**angio**sperm
anti-	against	**anti**body
aqua-	water	**aqua**tic biome
arthro-	joint	**arthro**pod
-ase	"enzyme"	malt**ase**
astro-	star	**astro**cyte
auto-	self	**auto**nomic
auxi-	increase	**auxi**ns
bi-	two	**bi**cellular
bio-	living	**bio**logy
-blast	bud	osteo**blast**
brach-	arm	**brach**ial artery
cardi-	heart	**cardi**ogram
centi-	one hundredth	**centi**meter
centro-	center	**centro**some
chloro-	green	**chloro**phyll
chroma-	color	**chroma**tid
-cide	killer, death	bio**cide**
coni-	cone	**coni**ferous tree
corpus	body	**corpus** luteum
cutin	skin	**cutin**
cyto-	hollow, cell	**cyto**blast **cyto**plasm

-derm	skin	**epidermis**
di-	twice	**diarthritic**
dorsal	back	**dorsal** side
duct-	opening	**ductless** gland
ecto-	outside	**ectoderm**
endo-	within	**endoderm**
epi-	top, upon	**epicotyl, epiglottis**
erythro-	red	**erythrocyte**
exo-	without	**exocrine** gland
ex-	out	**ex**tensor muscle
extra-	beyond, outside	**extracellular fluid**
fauna	animal	**fauna**
fibro-	band	**fibrogen**
fila-	thread	**filament**
flora	flower	**flora**
gameto-	spouse	**gametophyte**
gastric-	stomach	**gastric** acid
gene-	descent, begining	**genetics**
-genesis	decent, beginning	**biogenesis**
geo-	earth	**geotropism**
germin-	bud start	**germination**
glyco-	sweet	**glycogen**
gran-	grain	**granular**
hap-	simple	**hap**loid number
hemo-	blood	**hemoglobin**
herb-	plant	**herbivore**
hetero-	different, other	**heterozygous**
histo-	tissue	**histology**
homeo-	alike	**homeostasis**
homo-	same	**homologous**
-hybrid	offspring	**dihybrid**
hydro-	water	**hydrolysis**
hyper-	above	**hyper**sensitivity
hypo-	under	**hypocotyl**
in-	not, into	**inhabitable** **inhalation**
inter-	between	**interneuron**
intra-	within	**intracellular**
iso-	equal	**isogenes**
juven-	youthful	**juven**ile hormone
karyo-	nucleus	**karyotype**
kilo-	thousand	**kilometer**
kineto-	to move	**kinetochore**

lac-	milk	**lac**togenic hormone
lateral	side	**lateral** meristem
leuco-	white	**leuco**cyte (WBC)
lipi-	fat	phospho**lipi**d
-logy	study, discourse	zoo**logy**
-lysis	to loose	hydro**lysis**
macro-	large	**macro**evolution
median	middle	**median** line
mela-	black	**mela**nin pigment
meristem	divided	apical **meristem**
meso-	middle	**meso**phyll layer
meta-	many, middle	**meta**morphic
		metaphase
micro-	small	**micro**scope
milli-	one thousandth	**milli**meter
mito-	thread	**mito**chondria
mono-	one	**mono**ploid
-morphic	form	meta**morphic**
multi-	many	**multi**cellular
myo-	muscle	**myo**cardium
-natal	birth	pre**natal**
neo-	new	**neo**virus
neuro-	nerve	**neuro**chord
-nomial	name	bi**nomial**
non-	not, without	**non**living
oo-	egg	**oo**genesis
osteo-	bone	**osteo**cyte
ovi-	egg	**ovi**duct
para-	beside	**para**thyroid glands
patho-	suffering	**patho**gen
peri-	around	**peri**stalsis
-phase	to appear	meta**phase**
pheno-	appearance	**pheno**type
photo-	light	**photo**synthesis
-phyll	leaf	chloro**phyll**
-plasm	formed	cyto**plasm**
plate	flat, broad	**plate**let
-ploid	form, structure	mono**ploid**
-pod	foot	arthro**pod**
poly-	many	**poly**peptide
-porus	passage	semi**porus**
post-	after	**post**natal
pre-	before	**pre**natal
prim-	first	**prim**itive earth
pro-	before	**pro**gesterone
pseudo-	false	**pseudo**pod
pube-	adult	**pube**rty

quad-	squared (foured)	**quadriceps**
radial	ray	**radial** symmetry
re-	again	**reproduce**
sapro-	rotten	**saprophyte**
scien-	knowledge	**science**
-scopic	view	**microscopic**
semi-	half, partial	**semipermeable**
soma-	body	**somatic cells**
-spore	seed	**angiospore**
sperm-	seed	**spermatocyte**
-sphere	global, round	**biosphere**
-stalsis	contraction	**peristalsis**
-stasis	standing	**homeostasis**
strat-	layered	**stratiform**
sub-	under	**subclavian**
super-	above	**superior**
sym-	with	**symbiosis**
-synthesis	composition, making	**photosynthesis**
tele-	far, away	**telescopic**
telo-	end	**telophase**
tetra-	four	**tetrad**
toxi-	poison	**toxic**
trans-	across	**transplant**
tri-	three	**triglyceride**
-trophic	nutrition (food)	**autotrophic**
-tropism	turn	**phototropism**
-type	pattern	**karyotype**
uni-	one	**unicellular**
vacu-	empty	**vacuole**
vas	vessel	**vas** deferens **vascular tissue**
vegetative	bring alive	**vegetative** propagation
vita-	life	**vital** force
-vitre	glass	**invitreous** fertilization
xantho-	yellow	**xanthophyll**
zoo-	animal	**zoology**
zygo-	yolk	**zygote**

Glossary and Index of Biology Terms

Listing (Index) of Scientists

Abiotic factor: (200) physical factor found in the environment such as soil, water, temperature, light, substrata, and gases.

Absorption: (44, 48, 83) process of taking in dissolved materials through the cell membrane, into the blood stream or into the lymphatic system.

Acid: (19) any substance that gives off hydrogen ions or protons.

Acid rain: (220) rain with below normal pH levels formed as a result of sulfur or nitrous compound being released into the atmosphere.

Acquired characteristics: (184) characteristics developed during the lifetime of an organism which are not inherited by succeeding generations.

Active Immunity: (88) immunity that develops through stimulation by a disease organism or a vaccination.

Active site: (25) section of enzyme that combines with substrate(s) and performs the catalytic reaction.

Active transport: (48) energy requiring process by which molecules move against the concentration gradient.

Adaptation: (37, 39, 44, 50, 60, 66, 62, 70, 184) gene or genes which allows an organism to survive and reproduce in its environment.

Adenine: (164) a nitrogenous base found in ATP, DNA, and RNA.

Adenosine diphosphate (ADP): (55) a low energy molecule composed of adenine, ribose, and two phosphates; with the addition of energy and a third phosphate, it becomes ATP.

Adenosine triphosphate (ATP): (39, 55, 94) energy carrying molecule; see cellular respiration.

ADP: (55) see adenosine diphosphate.

Adrenal cortex: (106) outer part of adrenal gland, steroid hormones.

Adrenal gland: (106) endocrine gland, located on each kidney, which releases hormones which aid the body in dealing with stress.

Adrenal medulla: (106) middle of adrenal gland, secretes adrenalin.

Adrenalin: (106) adrenal gland hormone; regulates general metabolic rates.

Aerobe: (192) organism that uses oxygen during respiration.

Aerobic respiration: (55, 57) method of producing ATP from the complete oxidation of glucose, with the addition of oxygen and producing carbon dioxide and water as waste products.

Aging: (109, 136) normal process which results in a decrease in all physiological functions.

Aggregate: (191) molecules formed as a stage in the origin of life; see heterotroph hypothesis.

AIDS: (91) acquired immunodeficiency syndrome; fatal disease caused by HIV (human immunodeficiency virus).
Air sac: (95) see alveolus.
Albinism: (158) genetic condition in which there is no color pigments.
Algae: (63, 70) unicellular or multicellular protists that carry on photosynthesis and have cell walls.
Alignment of chromosomes: (126, 127) during meiosis I and II.
Alimentary canal: (45) gastrointestinal tract through which ingested material moves.
Alkaline: (19) base; substance which releases OH ion groups.
Allantois: (131) in fertilized egg, respiratory and excretion functions.
Allele: (148) one of two or more gene variations for a trait.
Allergy: (88, 96) result of the body's release of histamine in response to foreign antigens.
Alternation of Generations: (254) in plants, generations move between sporophyte and gametophyte produced organisms.
Altitude vs. Latitude: (214) comparison of world biomes.
Alveolus: (95) moist respiratory membranes, one cell layer thick, located at the end of the bronchioles and responsible for the exchange of gases with the blood capillaries.
Ameba: (44) see amoeba.
Amino acid: (22, 83) molecule composed of an amino group, an R group and a carboxyl acid group, serves as the building blocks of protein.
Ammonia: (62) nitrogenous waste of cells.
Amniocentesis: (162) removal of fetal cells within a sample of amniotic fluid from a pregnant woman; useful in determining the presence of genetic malformations prior to birth.
Amnion: (131, 136) membranous sac which surrounds the embryo and is filled with fluid, serves as a shock absorber.
Amniotic (amnionic) **fluid**: (131) fluid found between embryo and amnion.
Amoeba (ameba): (44) protist organism which uses pseudopods and carries on phagocytosis active transport - ingestion.
Amylase: (83) enzyme which hydrolizes maltose into two glucose molecules.
Anabolic metabolism: (6) building up process.
Anaerobe: (57, 192) any prokaryotic organism (bacteria) that cannot or harmed by oxygen molecules.
Anaerobic respiration: (55, 57) process by which ATP is produced from glucose without the use of oxygen molecules.
Anal pore: (44) the opening found in a paramecium through which undigested materials are eliminated from the vacuole.
Anaphase: (120, 127) the stage of mitotic cell division in which chromosomes are pulled in opposite directions.
Anemia: (91) a blood condition which results from reduced amounts of RBC or hemoglobin molecules.
Angina pectoris: (91) pain in heart or the left side and arm due to reduced blood flow into the heart caused by a narrowing of the coronary arteries.
Animal: (7, 8, 60, 63, 70) heterotrophic kingdom of multicellular, motile organisms.
Annelida: (8) phylum of segmented worms such as the earthworm.
Anterior: head or front end of a structure.
Anther: (138) part of the stamen of a flower that produces pollen
Antibody: (88) Y–shaped protein released from the B–lymphocytes that binds to specific antigens; makes them harmless.
Anticodon: (168) consists of a sequence of three bases found in t-RNA that matches the codon found in m-RNA and specifies the amino acid it carries
Antigen: (88) any foreign protein that invades the body and stimulates WBC or an immune response.
Anus: (84) opening at the posterior end of the alimentary canal.

Aorta: (90) largest artery of the body that carries oxygenated blood from the heart to the major arteries of the body.
Aortic arches: (52) five pairs of contracting structures found in the earthworm that pump blood through its vessels.
Apical meristem: (140) region of rapidly dividing cells found in the tips of stems and roots of plants.
Appendicitis: (86) inflammation of the appendix.
Appendix: (82) sac-like structure found below the junction of the small and large intestine.
Aquatic biome: (213) major world water biome; marine or freshwater.
Aquatic environment: (131) existing in water.
Arteriole: (89) blood vessel which carries oxygenated blood from an artery to several capillaries.
Artery: (89) thick walled blood vessel which carries blood under high pressure from the heart to the organs. All arteries except the pulmonary artery carry oxygenated blood.
Arthritis: (110) inflammation of the cartilage around the joint.
Arthropoda: (8) phylum of animals having "jointed legs" and an exoskeleton, such as grasshoppers and lobsters.
Artificial selection: (159) inbreeding, hybridization, vegetative propagation.
–ase: (25) commonly used ending for naming an enzyme for its substrate.
Asexual reproduction: (119) form of reproduction that does not involve fusion. The process is based upon mitotic cell division and produces organisms exactly like the parent.
Asthma: (96) disease of inflamed respiratory structures.
Atom: (18) simplest part of an element that has the characteristics of that element.
ATP: (39, 55, 94) adenosine triphosphate, energy transfer compound in cells.
Atrium: (89, 90) thin-walled upper collecting chamber of the heart.
Autonomic nervous system: (103) the motor nerves and ganglia coming from the brain and spinal cord that control involuntary functions such as peristalsis.
Autosome: (154) group of chromosomes found in both male and female body cells, not including the sex chromosomes (*XX* and *XY*).
Autotrophic: (37, 201) refers to organisms with the ability to manufacture their own food molecules from inorganic matter and obtain their energy either from sunlight or chemical reactions.
Autotrophic nutrition: (37) type of nutrition performed by organisms which carry on either photosynthesis or chemosynthesis.
Auxin: (68) group of plant hormones control cell elongation and growth.
Axon: (66) part of a neuron that carries impulses away from the cell body.

Bacteria: (8) monerans, mostly heterotrophs, often disease causing or decomposers, generally identified by shape, such as bacillus (rod), coccus (round), and spirilum (spiral).
Base: (19) see alkaline
Behavior: (102, 251) state of learning, display of learned and instinctive attitudes.
Benedict's (Fehling's) solution: (240) indicator for sugars.
Bicarbonate ion: (96) primary carrier of carbon dioxide in blood.
Bile: (83) fluid produced by liver, stored in gall bladder, emulsifies fat.
Bilipid membrane: (48) formed by phospholipids, arranged so that their hydrophobic tails are facing each other.
Binary fission: (122) form of asexual reproduction in which a single cell organism divides mitotically into two equal parts.
Binomial nomenclature: (9) two name system of classification using the *Genus* and *species*.
Biodegradable: (221) process by which materials are chemically broken down into simple molecules by the activity of bacteria or other decay organisms.

Bioluminescence: (252) energy of light in certain organisms, such as the fire fly, some bacteria, etc.
Biome: (212, 213) large geographic area determined by the abiotic factors and containing a particular type of climax vegetation.
Biosphere: (199, 218) that part of planet Earth in which living organisms are found.
Biotic factors: (200, 201) refers to the living organisms found in a particular area and their relationship with other living things.
Birth: (136) process of labor ending in the delivery of an offspring and in placental animals, the placenta.
Blastocoel: (130) central, hollow cavity found with the blastula stage of cleavage.
Blastopore: (130) indented area of the gastrula stage of cleavage where cellular differentiation begins.
Blastula: (130) the hollow ball stage of cleavage.
Blood: (87) fluid tissue composed of RBC's, WBC's, platelets, plasma, and suspended materials.
Blood clotting: (88) process by which bleeding stops.
Blood platelet: (87) cell fragment which plays an important role with formation of a blood clot.
Blood pressure: (90) see hypertension.
Blood typing: (88, 154) ABO groups, proteins in the blood.
Blending inheritance: (152) see incomplete dominance
Blue green algae: (8) prokaryotic organism (Monera) that performs photosynthesis.
Bones: (108) living tissues that provide structure for certain animals.
Bowman's capsule: (98) area of filtration in the nephron of a kidney.
Brain, human: (102) large group of specialized neurons, which receive sensory input, store information, coordinate motor responses, and control all involuntary activities.
Brain, primitive: (66) fused ganglia of the earthworm, grasshopper.
Breathing: (95) involuntary function which allows for the exchange of air between the lungs and the atmosphere.
Bromthymol blue: (240) indicator for gas; blue color indicates oxygen, yellow color indicates carbon dioxide.
Bronchi: (94) air exchange tubes, left and right divisions of trachea, lined with mucous membrane and ringed with cartilage.
Bronchioles: (95) a system of tubes containing cartilaginous rings and mucus membrane which connect with the trachea.
Bronchitis: (96) Inflammation of the bronchi of the respiratory system.
Bryophytes: (8, 50) plant phylum of mosses, lacking true roots, stems, and leaves.
Bud: (122) in plants, a small swelling that will develop into a flower, leaf or branch; in animals (hydra), a new organism which develops asexually from the parent organism.
Budding: (122) form of asexual reproduction in which a single cell organism divided its cytoplasm into two unequal parts.
Bulb: (124) results from vegetative propagation on plants; it is an underground stem, composed of modified leaves with stored nutrients; produces a plant genetically the same as original plant.

Calcitonin: (106) hormone secreted by the thyroid gland that decreases the calcium levels in the blood.
Calorie (Food): (21, 81) amount of heat necessary to raise 1000 grams of water 1° Celsius.
Cambium: (140) see lateral meristem; growth region of a plant.
Cancer: (121) abnormal and rapid mitotic cell division; disease.
Capillary: (89) a microscopic blood vessel, one cell layer thick, that allows for the exchange of materials with tissue of the body.
Capillary action: (208) liquid's upward movement within a small diameter tube.

Carbohydrate: (19, 20) specific group of nutrients composed of sugars and starch and containing a 2:1 hydrogen - oxygen ratio.
Carbon: (18) essential element with hydrogen making organic compounds.
Carbon cycle: (208) the flow of carbon atoms from the inorganic form (CO_2) through various organic molecules and back to CO_2.
Carbon Dioxide: (55-57, 62) metabolic waste of cellular respiration.
Carbon Fixations: (39) see dark reactions.
Cardiac muscle: (109) form of striated muscle under involuntary control found only in the heart.
Carnivore: (201) animal which feeds upon other animals.
Carrier: (155) in genetics, a female that can give a sex-linked trait to her offspring although she does not show the trait herself.
Cartilage: (108) special type of dense but flexible fibrous connective tissue.
Catabolic metabolism: (6) refers to chemical reactions in which materials are broken down.
Catalyst: (25) any compound which speeds up rate of a chemical reaction and is not destroyed by the reaction.
Cell: (10, 15) the fundamental unit of structure and function of living things and performs all the physiological activities.
Cell, animal (15) unit of structure and function.
Cell, plant (15) unit of structure and function.
Cell membrane: (15, 48) bilipid membrane, protein embedded , semipermeable and separates external environment from cellular compartment.
Cell organelles: (15) subcellular structures; function in life activities.
Cell plate: (121) formed during telophase stage of plant mitotic cell division from the endoplasmic reticulum and used for cytokinesis.
Cell theory: (10, 16) cell is the basic unit of structure and function for all living things; cells arise from preexisting cells.
Cell wall: (16) rigid, porous structure composed of cellulose and lignin, that surrounds plant cells and other microorganisms.
Cellular respiration: (55, 94) the enzyme controlled process by which energy is released from food molecules.
Cellulose: (16, 21) structural carbohydrate found in the cell walls of plants.
Central nervous system: (66, 102) vertebrate animals: consists of brain and dorsal spinal cord; invertebrate animals: a brain and ventral nerve cord.
Centrifuge: (12) a machine which spins readily and separates materials based upon differences in density.
Centriole: (16, 121) cylindrical structure, composed of microtubes, located near the nucleus of animal cells and plays a role in cell division.
Centromere: (120) the area which holds two chromatids together.
Centrosome: (16, 121) animal cell structure containing two centrioles found near the nucleus.
Cerebellum: (102) part of the vertebrate brain which coordinates voluntary muscular activity.
Cerebral palsy: (103) group of congenital diseases characterized by uncontrolled motor functions.
Cerebrum: (102) that part of the vertebrate brain that receives sensory input, produces motor responses and stores information.
Channel proteins: (48) proteins which regulate the passage of specific ions across the cell membrane.
Chemical bonds and structures: (18) attractive force, due to shared electrons, holding atoms together.
Chemosynthesis: (38) autotrophic nutrution, making with chemical energy.
Chitin: (71) a complete polysaccharide found in the setae of earthworms and in the exoskeleton of arthropods.
Chlorophyll: (16, 38) is a protein molecule which absorbs specific wavelengths of light energy and is essential to the process of photosynthesis.

Chloroplast: (16, 38) organelle within which photosynthesis (photolysis) and carbon fixation reactions take place.
Cholesterol: (22, 85, 86) lipid-like molecule found in animal fat and associated with cardiovascular disease.
Chordata: (8) animal phylum including vertebrates having notochords in the dorsal area of the body.
Chorion: (131) outer membrane which surrounds the amnion of the embryo of higher vertebrates such as reptiles, birds, and mammals.
Chromatid: (120, 148) one of two identical chromosomes formed as a result of chromosome replication.
Chromatography: (38) method used to separate and identify chemical substances based upon differences in their solubility.
Chromosomal alteration: (157) change in the number of or structure of chromosomes, resulting in mutations.
Chromosome: (15, 120, 148) DNA surrounded by protein; it is the structure which contains the genetic information and is found in the nucleus.
Cilium-cilia: (44, 70) tiny hairlike structure found on the surface of cells with the ability to move back and forth.
Circulation: (49) flow of soluble materials within a cell or between various tissues in complex animals.
Citric acid cycle: (57) part of cellular respiration; see Kreb's cycle.
Classification: (7) process of organizing living things.
Cleavage: (130, 135) the series of rapid cell divisions that follows fertilization; during this time, there is a decrease in the size of the cell due to the reduced length of time for interphase.
Climax Community: (211) final stage of ecological succession; stable community determined by climax vegetation and animal life it can support.
Clone: (168) genetically identical copy of a particular organism.
Closed Circulatory System: (52) type of circulatory system in which blood always remains within its blood vessels.
Condensation: (208) part of water cycle, water vapor clings to solid particles in the air, formation of clouds, rain, etc.
Codominance: (153) involves the activities of two dominant alleles (both alleles switched on) and the phenotype of the offspring are different from both parents.
Codon: (166) a sequence of bases found in messenger RNA which determines the amino acid to be used and its sequence in the synthesis of protein.
Coelenterata: (8) animal phylum including *Hydra* having a two way digestive tract, nerve net, and two cell layers.
Coenzyme: (25) essential to the normal structure and function of an enzyme.
Colony: (122) in reproduction, bacteria and other organisms may reproduce asexually and cling to one another.
Color Blindness: (155) sex linked recessive gene found on the *X* chromosome and appears more often in males than females.
Commensalism: (202) symbolic relationship between organisms in which one is benefited and the other is not affected (+,0)
Common ancestry: (8, 180) inference from evolutionary evidence that organisms showing common structures may have had a common genetic history.
Community: (199, 211) all the plant and animal populations interacting in a given area.
Comparative evidences for evolution: (180, 181) cytology, biochemistry, anatomy, embryology,
Competition curve: (255) population comparison of growth in two species within the same niche.
Compound: (18) molecule made up of two or more different elements combined chemically.
Compound light microscope: (11, 237) microscope made up of two or more convex lenses; for example, an ocular and objective lens.

Concentration Gradient: (48) difference in concentration responsible for diffusion between an area of high concentration and an area of low concentration; this difference determines the direction of flow for diffusion.
Conducting tissue: (50) (Vascular tissue) specialized tissue responsible for transporting fluids in plants (xylem and phloem).
Conservation: (221) positive aspect of man in response to maintaining a balance in nature between biotic and abiotic factors.
Constipation: (86) condition in which the large intestine is emptied with difficulty.
Consumer: (205) heterotrophic organism - feeds upon other living matter.
Contractile vacuole: (16, 63) specialized structure, located in the cytoplasm of fresh water protists, which collects excess water and forces the water out of the organism by active transport.
Control: (245) that part of an experiment that is used as the standard for comparison.
Controlled experiment: (245) experiment in which one variable is changed at any one time.
Corm: (124) fleshy, bulb-like, underground stem of the crocus or gladiolus and an example of vegetative propagation.
Coronary circulation: (91) circulation of blood from the aorta of the heart into the cardiac muscle tissue and back into the right atrium.
Coronary thrombosis: (91) blood clot in the coronary arteries.
Corpus luteum: (134) yellow endocrine structure which formed in the ovary after the graafian follicle releases its ovum that secretes the hormone progesterone.
Cotyledon: (139) seed leaf; in dicotyledon, two generally store food; in monocotyledon, one generally functions as a digestive organ.
Covalent bond: (18) force of attraction between two atoms, produced as a result of shared electrons.
Crop: (45) structure used to temporarily store food in invertebrate animal, such as earthworm and insects.
Cross: (148, 153) mating of two organisms of the same species to determine the genetic traits in the offspring.
Crossing over: (153) exchange of genetic material between homologous chromosomes during synapse of meiotic cell division.
Cross-pollination: (139) transfer of pollen from the anther of one flower to the stigma of a different flower.
Cuticle: (40) outer covering of cells, protects against invasion.
Cutting: (124) form of artificial vegetative propagation in which roots develop from the undifferentiated cells of the cut stems.
Cyclosis: (49) flow of cytoplasm within a cell.
Cytokinesis: (119, 126) division of cytoplasm during mitotic cell division.
Cyton: (66) the part of a neuron which contains the nucleus.
Cytoplasm: (15) the area of the cell between the cell membrane and nucleus which contains water, the cytoskeleton, and organelles.
Cytoplasmic division: (120, 127) separation of cytoplasm following mitosis by cell plate formation (plant) or "pinching–in" or "furrowing" (animal).
Cytosine: (164) a nitrogenous base found in both DNA and RNA.

Dark reactions: (39) sequence of enzyme controlled reactions take place in the stroma of the chloroplast. $CO_2 + ATP + NADPH_2 \rightarrow C_6H_{12}O_6$.
Daughter cells: (120, 121) resulting identical cells of mitotic cell division.
Deamination: (22) The removal of the amine from an amino acid.
Decomposer: (206) a group of heterotrophic organisms which breakdown organic matter into inorganic molecules.
Dehydration synthesis: (19) enzyme controlled reaction, substrates are joined together to form a single larger substrate by the removal of water.
Denaturation: (26) three dimension shape of an enzyme or protein molecule is lost due to high temperatures or excess pH.

Dendrite: (66) that part of a neuron which picks up an impulse and carries it to the cyton.
Denitrifying bacteria: (209) bacteria that convert nitrates back into free atmospheric nitrogen.
Deoxyribonucleic Acid (DNA): (164–168) macromolecule within which all hereditary information is stored in triplet code.
Deoxyribose: (164) pentose or 5 carbon sugar found in DNA.
Diabetes: (107) high blood glucose levels due to a lack of insulin protection
Diaphragm, muscle: (94) a curved sheet of muscle between the chest and abdomen and responsible for breathing.
Diaphragm, microscope: (11) microscope part, controls light to specimen.
Diarrhea: (86) a condition caused by excess water in the large intestine; results in frequent and loose bowel movements.
Diastole: (90) condition or period of time which results from the relaxation of the cardiac (heart) muscle.
Dicotyledon: (139) seed with two cotyledons; also used as a system classification of plants.
Differentiation: (130) process by which cells specialize and develop into a specific type of cell.
Diffusion: (48) movement of particles, by random collision from an area of high concentration to an area of low concentration until a state of equilibrium takes place.
Digestion: (43, 83) (chemical) enzyme controlled reaction in which nutrients are chemically broken down by hydrolysis into their basic building blocks.
Digestion: (43, 83) (mechanical) process by which food is physically reduced in size making it easier for chemical digestion.
Dihybrid cross: (148) genetic cross, one organism, containing two contrasting traits, is mated with another organism with the same genetic makeup.
Dipeptide: (22) molecule with two amino acids joined by a peptide bond.
Diploid (2*n*) number: (126, 157) the normal number of chromosomes found in the somatic (body) cell of each member of a species.
Disaccharide: (20) a complex sugar composed of two monosaccharides or simple sugars chemically bonded together.
Disease: (255) loss of an organism's homeostasis due to inflammation, a virus, bacteria, or foreign protein.
Disjunction: (126, 127, 157) separation of the tetrad structure during the metaphase stage of meiosis I.
Dissecting microscope: (12) low power microscope used during dissection of small organisms.
Dissection: (244) process of "taking apart" a plant, animal, or other organism in order to study the structural parts.
DNA: (164–168) see deoxyribonucleic acid.
Dominance: (147) in genetics, phenotype condition which results from a gene always being switched on; when a pure dominant is crossed with a pure recessive - all of the offspring show the dominant trait.
Dorsal: the back or top surface of an organism.
Dominant allele: (148) the expressed gene in an allelic pair, masking the recessive gene of the pair.
Double Helix: (165) double spiral shape formed by the parallel sides of the DNA molecule.
Down's Syndrome: (158) human condition, results from a chromosomal mutation; either an enlarged chromosome or an extra 21st autosome.
Ductless gland: (68, 105) see endocrine
Duct gland: (81) see exocrine

Earthworm: (45, 51, 60, 63, 66, 70) representative of Phylum Annelida.
Ecological succession: (211) process by which pioneer organisms change the soil, resulting in a series of communities ending with the climax community.
Ecology: (199) study of the interrelationships between plants, animals and the physical environment.
Ecosystem: (199) stable continuing environment composed of a living community and a nonliving environment.
Ectoderm: (130) outer layer of cells in the simple invertebrate animals, or one of primary layers of cells found in gastrula of cleavage.
Effector: (65) cell, tissue, or organ which produces a response to a stimulus.
Egestion: (43) elimination of undigested wastes from digestive system.
Egg cell (ovum): (131, 134) female gamete containing the ***n*** or haploid number of chromosomes
Electron: (18) negatively charged particle which moves about the nucleus of an atom.
Electron microscope: (12) type of microscope which uses electron beams instead of light to photograph and magnify an object.
Element: (18) basic form of matter; pure substance that cannot be changed to another.
Embryo: (109, 130, 139) early or developing stage of any organism.
Emphysema: (96) lung condition caused by smoking and results in the breakdown of the walls between the alveoli.
Emulsify: (83) process by which a mass of fat or oil is separated into individual droplets.
Endocrine gland: (68, 105) mass of tissue which secretes its product (hormones) directly into the blood capillaries.
Endoderm: (130) innermost layer of cells found in simple animals or in the gastrula stage of development.
Endoplasmic reticulum: (15) series of interconnecting tubes, and sacs, located in the cytoplasm of the cell; composed of bilipid membrane with ribosomes sometimes attached to it.
Endoskeleton: (71) in vertebrate animals, a network of support structures, cartilage or bone located inside the organism.
Energy flow: (205) in an ecosystem, the use of producers, consumers, and decomposers to maintain energy in the system.
Environment: (159, 199, 218) physical, chemical, biological surroundings.
Environmental laws: (222) Farm Act, Clean Air, Clean Water, etc.
Enzyme: (25) large three dimensional protein molecule with an active site that acts as an organic catalyst.
Enzyme denaturation: (26) three dimension shape of an enzyme or protein molecule is lost due to high temperatures or excess pH.
Enzyme substrate complex: (25) physical union of the enzyme and its substrate(s) at the active site.
Epicotyl: (139) upper leaves and stem found in a plant embryo.
Epidermis: (40) outer layer of cells found in any organism; its function is protection.
Epiglottis: (94) flap-like structure which blocks the entrance to the trachea (glottis) when swallowing.
Equatorial plane: (120, 127) middle area of a cell where chromatids line up at metaphase.
Erythrocyte (red blood cell): (87) a non nucleated biconcave shaped cell filled with hemoglobin.
Esophagus: (82) a muscular tube that connects the pharynx with the next specialized section of the digestive tract. (Crop and gizzard or stomach).
Essential amino acid: (84) one of eight amino acids that can not be synthesized by the liver; must be included in the diet each day.
Estrogen: (106, 134, 135) hormone (steroid) responsible for secondary female sex characteristics.

Eukaryotic: (8) any cell with an organized nucleus.
Evaporation: (208) loss of water from a surface, results in cooling.
Evolution: (179) theory which explains the origin of life on the planet Earth; the series of changes by which simple organisms develop into complex more advanced types.
Excretion: (6, 62, 98, 208) physiological process by which organisms remove metabolic wastes.
Exhalation: (95) the process by which air is removed from the lungs.
Exocrine: (81) mass of tissue which releases its secretion into a tube and is transported to another part of the organism.
Exoskeleton: (70) support structure found on the outside of the organism (Chitin on the insect).
Exploitation: (219) poor use of environmental organisms.
Extensor: (109) a muscle which moves a limb away from the body.
External fertilization: (129) union or fusion of an egg (ovum) and sperm outside of the parents body.
Extracellular digestion: (43, 44) release of enzymes outside the organism into the food or into a digestive tube.
Eye piece (ocular lens): (11, 237) single lens or lens system through which an object is seen.

Fallopian tube (oviduct): (134) funnel shaped tube which transports ova from the ovary to the uterus.
Fat: (21, 85) organic molecule composed of three fatty acids and one glycerol; stored energy.
Fatty acid: (21, 85) lipid molecule composed of a long chain of carbons bonded to hydrogens, with a carboxyl acid group at one end and a methyl group at the other end.
Fauna: (212) in an ecosystem, all of the animals.
Feces: (84) undigested waste material eliminated through the anus.
Feedback mechanism: (96, 106, 135) a control system by which one structure can regulate the activities of another structure.
Female: (134, 154) organism with *XX* chromosomes and/or producing ova (egg cells).
Female reproductive system: (134) glands, organs, and hormones necessary for ovum production, fertilization, gestation, and birth.
Fermentation: (55) process by which glucose is broken down without oxygen into pyruvic acid and then into alcohol and carbon dioxide.
Fertilization: (129, 135) union of two gametes that restores the $2n$ chromosome number.
Fetus: (135) embryo after it has completed it second month of development and external genitalia have formed.
Field of vision: (238) area as seen through a microscope.
Filament: (138) with the anther, makes up the male part of the flower.
Filtration: (98) process by which materials are forced out of the blood and enter the tissue or nephron.
Flagellum: (70) long whip like structure rotates to produce movement.
Flexor: (109) any muscle which brings the limb closer to the body.
Flora: (212) in an ecosystem, all of the plant organisms.
Flower: (138) sexual reproductive organ of the plant.
Fluid mosaic model: (48) commonly accepted explanation of the structure of the plasma membrane.
Follicle stage: (134) see menstrual cycle, ovum matures, estrogen secreted.
Follicle Stimulating Hormone (FSH): (106, 134, 135) hormone which affects the ovaries by stimulating the growth of the follicle, and testes by stimulating sperm production.
Food chain: (205) sequence of organisms, beginning with producers through which food energy is passed throughout an ecosystem.

Food vacuole: (44) membranous structure, located in the cytoplasm, within which intracellular digestion takes place.
Food web: (205) series of interconnecting food chains that end with the decomposes.
Fossil: (180) the preserved remains of plant or animal organisms.
Founder effect: (188) in evolution, an isolated population will develop a different gene frequency than its original population.
Fraternal twins: (135) two eggs fertilized by two sperm forming two zygotes resulting in two fetus being born at the same time.
Freshwater biome: (215) climax ecosystem, ponds, lakes, and rivers.
Fruit: (139) structure which develops from plant ovary and contains seed.
FSH: (106, 134, 135) see follicle stimulating hormone.
Fungi: (7, 44, 60) kingdom of heterotrophic plantlike organisms that lack chlorophyll.

Gall bladder: (82) pouch structure located near the liver which concentrates and stores bile.
Gallstones: (86) cholesterol deposits which sometimes form within the gall bladder.
Gamete: (126, 127) any sex cell containing the haploid or (*n*) number of chromosomes.
Gametogenesis: (126, 133) meiotic cell division which results in the formation of gametes.
Ganglion: (66) ganglia (pl.) a small mass of neurons which relay impulses and control the involuntary function of organs.
Gastric caeca: (45) in the grasshopper, secretes digestive enzymes.
Gastric gland: (82) digestive gland in stomach wall, secrete enzymes and hydrochloric acid.
Gastric juice: (82) water, hydrochloric acid, and digestive enzymes released by the glands located in the stomach lining.
Gastrula: (130, 135) stage of cleavage in which cellular differentiation takes place resulting in the formation of ectoderm, endoderm, and, later, the mesoderm.
Gene: (147, 153) section of DNA, having the information (code) to synthesize a polypeptide.
Gene chromosome theory: (148) see Gregor Mendel's work.
Gene counseling: (161) information to help individuals understand their options concerning reproduction and heredity.
Gene frequency: (169) number of times or percentage that a particular gene appears in member of a population.
Gene linkage: (153) association of genes on the same chromosome.
Gene mutation: (157, 158, 168) change in the arrangement of base pairs which spell out the code in a gene.
Gene pool: (169) all of the genes found in a population.
Genetic engineering: (168) process by which DNA or genes are made and transferred into different organisms.
Genetics: (147) branch of biological science which studies DNA and the principles of heredity and variation in organisms.
Genotype: (148) genetic make up of an organism; hereditary material passed form generation to generation.
Genus: (8) group of related organisms containing certain common structural characteristics and consisting of two or more species.
Geographic isolation: (187) physical separation of a population so that they breed only within the population; this eventually results in speciation.
Geotropism: (68) Plant growth response to the force of gravity.
Germination: (139) growth and development of the plant embryo (seed) under the proper physical conditions.
Gestation period: (135) amount of time between fertilization of the ovum and the birth of the organism.
Gizzard: (45) section of the invertebrate digestive tube with which mechanical digestion of food takes place.

Glomerulus: (98) network of blood capillaries located within Bowman's capsule in a nephron.
Glucagon: (106) hormone secreted by the Islets of Langerhans (pancreas) which increases the level of glucose in the blood.
Glucose: (19) six carbon sugar (monosaccharide) that is the primary fuel for the production of cellular energy (ATP).
Glycogen: (84) polymer chain of glucose molecules stored within liver and muscle cells.
Glycolysis: (57) series of reactions by which one glucose molecule is broken down, without oxygen, to form two molecules of pyruvic acid with a net gain of 2 ATP.
Goiter: (107) enlargement of thyroid gland due to a lack of iodine in the diet.
Golgi complex: (16) also called golgi body, an organelle composed of several flattened membranes bound together; within which protein and carbohydrates are combined and secreted from the cell.
Gonad: (106, 126) in animals, the organ which produces gametes and sex hormones.
Gout: (99) disease of the inflammation of joints.
Gradualism: (188) slow, steady series of changes used to explain how evolution took place.
Grafting: (124) form of vegetative propagation by which a twig from one plant is attached to the stock of another plant.
Granum: (38) grana (pl.) stack of lamellae found within the chloroplast.
Graphing: (245) method of comparisons in a visual form.
Grasshopper: (45, 52, 60, 63, 66) representative of Phylum Arthropoda.
Growth: (6, 130, 135, 140) increase either in the size of a cell or in the number of cells.
Growth curve: (254) graphing of the stages of growth in an organism.
Growth hormone: (105, 135) hormone released by the anterior pituitary gland which stimulates protein synthesis and the growth of cells.
Guanine: (164) nitrogenous molecule found in DNA and RNA.
Guard cell: (40) one of two cells, located in the leaf, that expand and contract and control the size of the stomate.

Habit: (102, 252) voluntary action performed often, becomes automatic.
Habitat: (212) particular area in the ecosystem where a specific plant or animal organism lives.
Half-life (252) length of time it takes for half the atoms in a radioactive element to decay.
Haploid: (126) see monoploid, gametogenetic result of meiosis, having only one chromosome from each homologous pair of chromosomes.
Hardy Weinberg Principle: (169) in sexually reproducing organisms, the frequency of genes does not change when using a specific set of conditions.
Heart: (90) muscular pump which forces blood in one direction.
Heart attack: (91) any malfunction of the heart.
Hemoglobin: (52, 87) protein molecule composed of globins and heme found in RBC and transports oxygen.
Hemophilia: (155) sex linked genetic trait that prevents normal blood clotting, due to the absence of one or more essential factors.
Herbivore: (201) heterotroph that feeds only on plants.
Hermaphrodite: (126) organism that contains both the male and the female gonads (testes and ovary); examples: earthworm (cross-fertilization) and hydra (both self- and cross-fertilization).
Heterotroph: (43, 201) organism which can not synthesize its own organic molecules.
Heterotroph hypothesis: (190) explains how the first cells could result from the natural physical forces at work on the primitive planet earth.
Heterotrophic nutrition: (37, 43) type of nutrition carried on by those organism which must take in preformed organic molecules (nutrients).

Heterozygous: (149) the presence of a dominant and recessive gene or two different alleles for a trait.
High blood pressure: (91) see hypertension
Histamine: (88) chemical secretion of cells causing an allergic response, such as the irritation and swelling of mucous membranes.
Homeostasis: (6, 48, 87, 98) balanced and stable internal environment.
Homologous chromosomes: (126) pair of chromosomes, being the same size and shape and containing genes for the same set of traits.
Homozygous: (148) same two genes or alleles for a trait.
Hormone, animal: (68, 105, 134) chemical messenger (protein or steroid) released by endocrine glands directly into the circulatory system.
Hormone, plant: (68) plant's chemical messenger; see auxins.
Host: (202) any organism which provides nutrients to or an environment for a parasite.
Human: (45, 52, 81, 60, 63, 67, 71) structures and functions of systems.
Human systems and **diagrams**: (90 - circulatory, 82 - digestive, 105 endocrine, 95 - respiratory, 109 - skeletal/muscles, 99 - urinary))
Hybrid: (149) see heterozygous.
Hydra: (44, 51, 60, 63, 66, 70) representative of Phylum Coelenterata.
Hydrogen (H): (18) element needed with carbon to make organic compounds.
Hydrolysis: (19, 85) enzyme controlled reaction in which a substrate, with the addition of water, is separated into two substrates.
Hydrotropism: (68) movement of a plant to the stimulus of water.
Hypertension: (91) systolic and/or diastolic blood pressure higher than normal.
Hypocotyl: (139) portion of a seed that develops into the root.
Hypothalamus: (105) section of the brain that controls involuntary activities; attached to the pituitary gland.

Identical twins: (135, 159) two individuals formed from a single zygote; having exactly the same DNA or genetic makeup.
Immunity: (88) ability of leukocytes to produce antibodies and other factors resistant to disease (foreign antigens).
Implantation: (135) attachment of the gastrula (early embryo) to the endometrium (lining) of the uterus.
Imprinting: (252) upon birth learning.
Impulse: (65, 101) an electrochemical change which results from the flow of ions across the membrane of a neuron.
Incomplete dominance (partial dominance): (152) involves the activities of two alleles neither of which is completely turned on; the offspring show a phenotype midway between either parent.
Independent assortment: (147, 153) principle of Gregor Mendel, random segregation of chromosomes with different traits.
Indicator: (240) a substance when added in small amounts is used to indicate chemical conditions or changes, such as bromthymol blue.
Infectious disease: (254) diseases caused by viruses, bacteria, yeast, molds, protozoa, and certain worms, having harmful effects on life.
Ingestion: (43) process of taking food into the digestive system so that it may be hydrolized or digested.
Inference: (245) conclusion based on collected, organized data.
Inhalation: (95) activity by which air is forced into the lungs.
Inate behavior: (241) behavior passed through heredity; inc. reflex, instinct.
Inorganic compound: (19) compound NOT produced by plant or animal activities; usually is *not* composed of carbon and hydrogen.
Instinct: (251) unlearned, inborn behavior.
Insulin: (106) hormone, produced by Isles of Langerhans in the pancreas, that stimulates liver and muscle cells to remove glucose from the blood.

Intercellular fluid (ICF): (89) fluid obtained from plasma and located between the tissue and the blood capillary.
Intermediate inheritance: (152) when the offspring appearance is different than either parent, a combination or blending of parental genotype.
Internal fertilization: (129) fusion of an ovum and a sperm within the oviduct or fallopian tube of an animal organism.
Interneuron: (101) type of neuron found only in the brain and spinal cord.
Interspecies competition: (212) competition between different species within the same habitat.
Intestine: (45 - earthworm and grasshopper, 83-84 - human)
Intracellular digestion: (39, 43) within the cell digestion.
Involuntary muscle: (109) smooth muscle or cardiac muscle that is under the control of medulla oblongata (autonomic nervous system).
Ion: (18) atom that has gained or lost electron becomes either positively or negatively charged.
Ionic bond: (18) chemical bond in which electron moves closer to one atom.
Islets of Langerhans: (106) group of endocrine cells located in the pancreas which secretes the hormones glucagon and insulin.
Isotope: (18) group with extra neutrons in the nucleus.
Isotope, radioactive: (252) used in the dating of fossils.

Joint: (109) part of the skeleton where two or more bones meet and allow for movement.

Karyotyping: (162) process - chromosomes of an individual are studied.
Kidney: (98) principle organ of the urinary system.
Kidney disease: (99) diseases of the urinary system.
Kinetochore: (120) disc shaped structure which holds the chromatids together at the centromere.
Kingdom: (7, 8) largest category in the science of taxonomy. and having one common characteristic shared by all members.
Koch's Postulates: (254) German bacteriologist discovered requirements for causes of specific infectious diseases.
Kreb's cycle: (57) also called citric acid cycle, part of aerobic respiration.

Lab skills: (233) includes safety, required processes, etc.
Lacteal: (83) a lymphatic capillary located within the villus of the intestine.
Lactic acid: (55, 109) organic acid formed during anaerobic respiration from pyruvic acid.
Lactogenic hormone: (135) secreted by the anterior pituitary gland and stimulates the production of milk.
Lamella: (38) flat membrane sac located in the chloroplast, that contains chlorophyll and the antenna molecules.
Large intestine: (84) specialized compartment of the digestive tube designed to collect undigested materials and reabsorb water.
Larynx: (94) structure composed of cartilage, located above the trachea and containing the vocal chords.
Lateral meristem: (140) in plants, a region of rapidly dividing cells responsible for growth in the diameter of a woody stem.
Leaf: (40, 51, 60) photosynthetic organ of the plant.
Lenticel: (60) openings in the stems of woody plants; gas exchange.
Leukocyte: (88) (white blood cell) any one of five different types of nucleated cells that provide protection and immunity for the organism.
Leukemia: (91) form of cancer which results in the production of large amounts of abnormal leukocytes.

LH: (135) see Luteinizing Hormone.
Lichen: (211) pioneer organism(s) which is a combination of an alga and a fungus which exists in a relationship of mutualism.
Ligament: (110) tough, fibrous connecting tissue which connects and holds the bones together at the joints.
Light reaction: (38) in photosynthesis, a series of reactions that occur in the lamellae that involve chlorophyll, sunlight, water, ATP and $NADPH_2$.
Limiting factor: (200) physical factor that limits the life forms that can exist within a particular ecosystem.
Lipase: (85) hydrolytic enzyme that breaks down triglycerides into glycerol and 3 fatty acid molecules.
Lipid: (21, 22) class of molecules which include waxes, fats and oils.
Litmus paper: (240) indicator for pH.
Liver: (98) organ involved with excretion and the recycling of useable materials and the production of urea and bile.
Locomotion: (70, 108 ability to move from place to place.
Lock and key model: (26) explanation of the working of enzymes.
Lugol's solution: (12, 240) iodine indicator for starch; stain used to better observe cell parts.
Lung: (95, 98) in humans, two organs containing bronchioles and millions of alveoli - oxygen and carbon dioxide are exchanged with the blood.
Luteinizing Hormone: (135) secreted by the anterior pituitary and causes the graafian follicle to release the ovum (ovulation).
Lymph: (89) intercellular fluid (ICF) present in the lymphatic tubes.
Lymphatic circulation: (90) circulation of the lymph in the body.
Lymphocyte: (88) type of white blood cell, produce specific antibodies.
Lysosome: (16) organelle; a membrane sac filled with hydrolytic enzymes, found in cytoplasm of cells.

Magnify: (11, 237) ability to enlarge or to make larger.
Male: (133) organism that produces sperm cells.
Male reproductive system: (133) functions to secrete hormones, produce sperm, and deposit spermatozoa within the female reproductive tract.
Malpighian tubules: (63) organs of excretion found in insects and other arthropods.
Maltase: (85) hydrolytic enzyme which digests maltose into two monosaccharides.
Malthus, Thomas: (252) English economist best known for his essay on the Principle of Population.
Mammary glands: (132) found in mammals only, used to produce milk necessary for infants.
Marine biome: (215) largest and most stable of all biomes, it is composed of the oceans on this planet.
Marsupial: (132) pouch containing mammal, represents the change in form between external development and placental development.
Measurement: (12, 235) system of determining size, etc.
Medulla oblongata: (102) part of the brain that is connected to the spinal cord and controls all involuntary activities such as heart beat.
Meiosis: (126, 127, 157) see meiotic cell division.
Meiotic cell division: (126, 127, 157) type of cell division which occurs only in gonads and produces gametes.
Memory: (251) part of the learning process; see cerebrum.
Meningitis: (103) bacterial or viral infection of the meninges (membrane which surround the brain and the spinal cord).
Menopause: (134) time in a woman's life when she no longer has her menstrual cycle.
Menstrual cycle: (134) hormone controlled reproductive cycle which produces a new ovum and endometrial membrane every 28 days.
Menstruation: (135) breakdown of highly vascular endometrial membrane.

Meristem: (140) asexual reproduction and growth regions of plants.
Mesoderm: (130) layer of cells located between the ectoderm and endoderm.
Messenger RNA: (166) single helix, transcribed from DNA having codons.
Metabolism: (6, 62) total chemical reactions that occur in an organism.
Metamorphosis: (69) hormone controlled changes in certain animals.
Metaphase: (120, 127) stage of mitotic cell division when chromatids line up along equatorial plane and are attached to the spindle fibers.
Methylene blue: (12, 240) stain used to prepare microscopic specimens.
Microdissection: (12) any removal, addition and/or transfer of individual cell organelles such as the nucleus.
Micrometer (micron or **µm)**: (12, 235) one millionth of a meter; one thousand µ equal one millimeter.
Microscope: (11, 237) tool of biologists to observe small things.
Migration of chromatids: (127) see meiosis; movement to ends of cells.
Mineral salts: (19, 62) inorganic compounds, both useful in maintaining homeostasis and excreted as metabolic waste.
Mitochondrion: (15, 56) found in aerobic organisms in which aerobic respiration is completed.
Mitosis: (119, 127) see mitotic cell division.
Mitotic cell division: (119, 127) type of cell division which produces two cells having a complete set of homologous chromosomes (*2n*).
Modern theory of evolution: (186) based on Darwinian evolution, but incorporates variations and genetics.
Monera: (7, 60) Kingdom of prokaryotic organisms, such as bacteria and blue green algae.
Monocot: (139) group of plants whose seed contains a single cotyledon.
Monoploid: (126, 127) also called haploid; single set of chromosome resulting from meiosis in gametes.
Monosaccharide: (19) any simple sugar with empirical formula - $C_nH_{2n}O_n$.
Morula: (130) mass of cells stage of cleavage.
Motile: (70) ability to move.
Motor neuron: (101) special nerve which carries impulses from the central nervous system to a muscle or gland (effector).
Mucous or **mucus**: (60) substance composed of protein-monosaccharides and water secreted by cells found in respiratory and digestive membranes.
Multiple alleles: (154) three or more different genes for a trait such as blood typing A, B, O genes.
Muscle: (70, 109) tissue containing actin and myosin and, therefore, the ability to contract when stimulated.
Mutagenic agent: (158) any substance that can change DNA structure by either supplying energy or chemically reacting with it.
Mutation: (157, 168) any change either in the structure of DNA or chromosomes or in the number of chromosomes.
Mutualism: (202) symbiosis in which both organisms benefit from each other's activities.

Nasal cavity: (94) first opening for the respiratory system.
Natural selection: (184, 187) Darwin's theory that only those organism containing the best variations can survive and reproduce in a given environment.
Negative feedback system: (95, 106, 135) in the endocrine system, a decrease in function in response to a stimulus; a self-regulating system for hormone level control.
Nephridium: (63) nephridia (pl.) earthworm organs of excretion.
Nephrons: (63, 98) microscopic structures found in the kidney that are responsible for removing urea and maintaining water balance.
Nerve: (65, 101) group of axons, surrounded by protective tissue, which carry impulses to and from the central nervous system.

Nerve cell: (65) see neuron.
Nerve net: (66) arrangement of simple neurons found in a hydra.
Neuron: (65, 101) specialized cell for the production of impulses which result from the flow of ions through its membrane.
Neurotransmitter: (66) chemical messenger secreted by neurons which stimulates the production of an impulse in the next neuron.
Niche: (212) role played by an organism in an ecosystem.
Nitrifying bacteria: (209) group of bacteria which convert ammonia into nitrates.
Nitrogen cycle: (209) flow of nitrogen from the environment through living organisms and back to the environment.
Nitrogen fixing bacteria: (209) type of bacteria, found in the nodules of legumes, which convert free nitrogen into nitrates.
Nitrogenous base: (164) parts of nucleotides; in DNA: adenine, thymine, guanine, and cytosine; in RNA replace thymine with uracil.
Nitrogenous waste: (62) result of metabolism of amino acids (proteins).
Nomenclature: (9) see binomial nomenclature; system of classification.
Nondisjunction: (157) the failure of tetrads to separate properly during meiotic cell division.
Nonplacental mammal: (132) see marsupial.
Nucleic acid: (19, 164) large organic polymer composed of nucleotides such as DNA and RNA.
Nucleolus: (15) structure found in nucleus where ribosomes are produced.
Nucleotides: (164) molecule composed of phosphate, a pentose, and a nitrogen base and used to make nucleic acids.
Nucleus: (15) found in eukaryotic cells; a large membrane bound structure that contains the chromosomes (heredity); in the atom, it is composed of protons and neutrons.
Nutrient: (5, 81, 201) inorganic or organic compound used by the cell for its metabolic activities.
Nutrition: (37, 81, 191) life activity; see nutrients.

Objective lens: (11, 237) a single convex lens or lens system attached to the revolving nose piece of a microscope.
Ocular lens: (11, 237) single convex lens or lens optic system.
Omnivore: (201) a heterotrophic organism which feeds upon plant and animal matter.
One gene — one polypeptide hypothesis: (168) see chromosomes.
Oogenesis: (128) series of cell divisions (meiosis) and other changes that converts a diploid ($2n$) sex cell into a monoploid (n) ovum and three polar bodies.
Open circulatory system: (52) one in which blood leaves the heart and blood vessels and flows over the tissue.
Oral cavity: (82) first opening of the digestive system.
Oral groove: (44) in paramecium, scoop like structure which guides food particles into the mouth and gullet.
Organ transplants: (89) see applications of immunity and allergies.
Organelle: (15) structure located within the cell that is essential to the life activities of the cell.
Organic compound: (18) any molecule composed of carbon and hydrogen or is found in plant and animal organisms.
Organic evolution: (179) change in living things through time.
Organism: (8) a single living thing.
–ose: (21) common ending when naming carbohydrates, such as glucose.
Osmosis: (48, 62) movement of water molecules from an area of high to an area of low concentration through a semipermeable membrane.
Ovary: (106, 128, 134, 138) the gonad within which the ovum (egg) and the sex hormones are produced.

Overhunting: (219) destruction of wildlife.
Oviduct: (134, 135) see fallopian tube.
Ovulation: (134) that part of the menstrual cycle, when the graafian follicle releases the ovum in response to the luteinizing hormone.
Ovule: (139) in plants, the structure within the ovary which contains the egg nucleus and upon fertilization, develops into the seed.
Ovum: (127, 128, 134) an egg or gamete containing the haploid or (n) number of chromosomes.
Oxyhemoglobin: (96) chemical combination of oxygen and hemoglobin in the blood; transport of oxygen.
Oxygen cycle: (208) process of recycling oxygen in an ecosystem.

Palisade cells: (40) primary photosynthetic cells of upper leaf layer.
Pancreas: (83) an organ which secretes both digestive enzymes (exocrine) and hormones (endocrine).
Pancreatic fluid: (82) a mixture of water, and digestive enzymes having an alkaline (basic) pH.
Paramecium: (44, 62) protozoan having cilia and contractile vacuole.
Parasite: (202) any organism which lives on or in another organism and is harmful to it.
Parathormone: (106) a hormone secreted by the parathyroid gland and increases the calcium levels in the blood.
Parathyroid glands: (106) four seed sized structures located on thyroid gland; secrete parathormone which regulates calcium metabolism.
Parthenogenesis: normal development of an unfertilized ovum into an adult organism (drone or male bee).
Passive immunity: (88) short term protection produced when donor antibodies are injected into an individual.
Passive transport: (48) movement of materials across the cell membrane without the use of ATP.
Pedigree chart: (155) chart of relationships between generations.
Penis: (133) the male external genital organ containing the urethra through which both urine and sperm leave the body.
Peptide bond: (22) electron bond that holds two amino acids together.
Peripheral nervous system (PNS): (66, 103) all of the nerves and ganglia outside of the central nervous system (brain and spinal cord).
Peristalsis: (81) series of involuntary wave like muscular contractions which move food along the digestive tube.
Perspiration: (98) mixture of water, ions and urea released by the sweat glands to cool the body when it is heated.
Petal: (138) brightly colored, sweet scented structures designed to attract pollinating insects.
pH: (19, 27, 240) scale to measure the amount of acid or base (alkaline).
PGAL - phosphoglyceraldehyde: (39) stable molecule found in both photosynthesis and cellular respiration.
Phagocytosis: (44, 49, 88) process by which ameba and some WBC's engulf organisms.
Pharynx: (45, 94) throat; part of the digestive and respiratory systems.
Phenotype: (149) outward appearance of an organism, the result of its genetic makeup.
Phenylketonuria (PKU): (162) genetic condition, resulting from a homozygous recessive gene in which the normal enzyme is absent - results in brain damage and mental retardation if not treated during infancy.
Phloem: (51) tissue used to transport sap up and down the plant.
Phosphate: (164) inorganic compound PO_4H_3 essential for all life functions.
Photochemical reactions: (39) see photosynthesis.

Photolysis: (39) breaking down with use of light; see photosynthesis.
Photosynthesis: (38, 208) process by which carbon dioxide and water are converted into organic molecules through sunlight and chloroplast.
Phototropism: (68) growth of a plant towards sunlight due to effects produced by the unequal distribution of auxin.
Phylum: (8) in taxonomy, a major subdivision of Kingdom.
Pigment: (38) protein molecule with the ability to absorb all wave lengths of light except the one it reflects (example, chlorophyll).
Pinocytosis: (49) process by which large molecules are taken into the cell.
Pioneer organism: (192, 211) first organism to live in an area.
Pistil: (138) female reproduction system found in a flower.
Pistilate flower: (138) type of imperfect flower because it contains only the female reproductive system.
Pituitary gland: (105) "master gland of the body" is attached to the hypothalamus of the brain and controls the activities of other endocrine glands.
Placenta: (131, 135) a mass of tissue, present in the uterus only during pregnancy, through which nutrients and wastes are exchanged by mother and fetus.
Plant: (7, 8, 50, 60, 63) autotrophic kingdom of multicellular organisms.
Plasma: (87) fluid part of blood, consisting of water, ions, proteins.
Plasma membrane: (15, 48) see cell membrane.
Platelet: (87) see blood platelet.
Polar bodies: (128) non-reproductive gametes produced by meiosis.
Polio (infantile paralysis): (103) virus which attacks motor pathways of the brain and spinal cord.
Pollen grain: (139) small, tough structure which contains the male gamete produced by the anther of the flower.
Pollen tube: (139) tube produced by the tube nucleus, through which sperm nuclei enter the ovule.
Pollination: (139) transfer of pollen from the anther to the stigma of a flower.
Pollution: (219) the release of harmful substances into the environment; including water, air, chemical, waste disposal, etc.
Polypeptide: (22) section of a protein molecule composed of amino acids.
Polyploid: (158) organism whose cells contain multiples of the normal number of chromosomes. (example, $4n$)
Polysaccharide: (20) carbohydrate composed of a long chain of monosaccharides (simple sugars).
Population: (169, 199, 218, 221) number of individuals of the same species, living in a given habitat.
Postnatal development: (136) see human development; after the birth.
Precipitation: (213) release of water from the atmosphere in the form of rain, snow, sleet, hail, dew, or fog.
Predator: (201) carnivore which hunts, kills and feeds on its prey.
Pregnancy: (135, 136) temporary condition resulting from the development of an embryo fetus within the uterus of the mother.
Prenatal development: (135) see human development; before the birth.
Primate: (253) animals including the monkeys, apes and humans.
Producer: (205) see autotroph.
Progesterone: (134, 135) hormone released by the corpus luteum of ovary and used to develop blood vessels within the endometrium of the uterus.
Prokaryotic: (7) organism without an organized nucleus.
Protease: (85) enzyme which hydrolyzes protein into amino acids.
Protein: (19, 22) molecule composed of polypeptides.
Protein synthesis: (167) at ribosomes, the production of proteins.
Protista: (7, 60, 62) Kingdom composed of simple eukaryotic organisms.
Protozoan: (44, 62, 70) see protista; single cell "first animals."
Pseudopod: (44, 70) temporary cytoplasmic projection produced by amoeba to move; used in phagocytosis.

Puberty: (134) the time in an individual's life when they begin to produce gametes and sex hormones.
Pulmonary circulation: (90) flow of blood from the heart to the lungs and back to the heart.
Pulse: (89) in arteries, pressure of blood surge at systolic heart contractions.
Punctuated equilibrium: (188) theory of evolution that proposes that species change quickly after millions of years of stability.
Punnett square: (148) checkerboard method for determining various results in a genetic cross, including genotypes, phenotypes, and probability.
Pyramid of biomass: (206) amount of food matter is greatest at the producer level and decreases at each consumer level.
Pyramid of energy: (206) total amount of energy found in food is greatest at the producer level and decreases with each consumer level.
Pyruvic acid: (57) three carbon molecule, containing energy that remains after glycolysis or anaerobic respiration is completed.

Radioactivity: (253) process by which an atom breaks down into another atom or isotope by giving off radiation or particles.
Random segregation: (149) see segregation.
Receptor sense: (65, 101) specialized neuron or organ which receives stimuli and sends impulses to the central nervous system.
Recessive gene: (147) gene which is "switched off" in the presence of a dominant gene and its trait does not appear in the phenotype.
Recombination: (149) see crossing over.
Recombinant DNA: (168) DNA which has been changed by the addition or deletion of other sections of DNA.
Recycling: (200) the process by which materials are broken down and then reused to produce a new substance.
Red blood cell: (87) see erythrocyte.
Reduction division: (126) chemical reaction – atom receives electrons.
Reflex: (102, 251) involuntary, automatic response to a stimulus.
Reflex arc: (102) automatic change in the pathway taken by a nerve impulse in response to a stimulus.
Regeneration: (123) replacement of lost body tissue in animals.
Regulation: (6, 65, 101) process by which an organism controls all functions in order to maintain homeostasis.
Replication: (120, 127, 165, 166) process by which DNA make a perfect copy of itself.
Reproduction: (6, 119, 133, 185, 192) process by which organisms produce copies of themselves.
Reproductive isolation: (188) inability of some members of a species to reproduce with others of same species because of geographical barriers.
Resolving power: (11) in optic microscopes a limit to magnification.
Respiration: (6, 55, 94, 208) energy conversion; chemical bond to ATP.
Respiratory surface: (55, 94) moist member through which diffusion of oxygen and carbon dioxide can take place.
Response: (65) reaction to stimulus performed by a muscle or gland.
Rhizoid: (44) individual hypae of molds and mushrooms which enter the food that it is growing on.
Ribonucleic Acid (RNA): (166) molecule composed of RNA nucleotides transcribed from DNA and used to synthesize polypeptides.
Ribose: (166) five carbon sugar found only in RNA.
Ribosomal RNA: (166) type of RNA found in the ribosomes.
Ribosome: (15, 167) structure in which translation of protein synthesis takes place.
Root: (50, 60) plant part that grows downward, stores nutrients, and absorbs water and minerals.
Root hair: (50, 60) modified epidermal cell which increases the surface area for absorption.

Roughage: (81) fiber or cellulose found in vegetables - cannot be digested.
Runner: (124) horizontally growing stem by which the strawberry plant reproduces asexually by vegetative propagation.

Saliva: (82) solution of water, ions, and salivary amylase secreted by the salivary glands.
Salivary gland: (45, 82) parotid, sublingual, and submandibular glands which secrete saliva into the mouth.
Salts: (19) inorganic compounds necessary for cell activities.
Saprophyte: (201) heterotrophic organism (mold, mushroom, fungi, bacteria); obtains its nutrients from dead plant and animal matter.
Saturated fatty acid: (22, 85) fully hydrogenated with no double bonds.
Scavenger: (201) carnivore, feeds upon dead animals that it did not kill.
Scientific method: (245) way to solve problems through a hypothesis, controlled experimentation, verification of data, and logical thought.
Screening: (162) see genetic counseling.
Scrotum: (133) sac of skin, below male genital organ, containing two testis.
Secondary sex characteristics: (133, 134) changes produced within the body as a result of the secretion of sex hormones.
Secretion: substance released by cells or organs and used to maintain homeostasis in itself or offspring (bile, hormones, milk).
Seed: (139) develops after fertilization from the ovule and contains the plant embryo and the reserve food supply.
Segregation: (147) in meiosis, the random separation of homologous chromosomes and the alleles; they are composed of into different gametes; each gamete contains one homologue or one allele for a trait.
Self-pollination: (139) transfer of pollen from the anther to the stigma of the same flower.
Semen: (134) in male, combination of sperm and fluids.
Sensory neuron: (101) nerve cell carries impulses to CNS.
Sepal: (138) specialized petal for protection of bud.
Sessile: (70) term describing inability of an organism to move through its own action.
Setae: (70) in earthworm, paired bristles used for locomotion.
Sex chromosomes: (154) *X* and *Y* chromosomes; *XX*-female, *X*-male.
Sex linkage: (155) genes located on the *X* chromosome.
Sexual reproduction: (126) reproduction by the fusion of gametes produces variety.
Sickle cell anemia: (162) disease of the blood having irregular RBC's.
Skeletal muscle: (109) see voluntary or striated muscles.
Skin: (98) covering of body.
Small intestine: (83) long, narrow convoluted compartment of the digestive tube within which all chemical digestion is completed and nutrients are absorbed.
Somatic nervous system (SNS): (103) subdivision of the peripheral nervous system and consists of nerves that control the voluntary muscles of the skeleton.
Somatic tissue: all types of body tissue except that which produces gametes.
Speciation: (185, 187) process by which geographic isolation results in the formation of a new species.
Species: (9) group of organisms with the ability to interbreed.
Sperm cell: (128) in animals, the male gametes with the *n* number of chromosomes.
Sperm nuclei: (139) in plants, male gametes which result from the pollen grain.
Spermatogenesis: (128) see male gametogenesis.
Spinal column: (102) in vertebrate animals, a series of structures composed of cartilage or bone, separated by a cartilage disk and used to protect the spinal cord.
Spinal cord: (102) continuous with the brain, contains interneurons, the center for reflex actions, and relays impulses between the brain and other nerves.
Spindle fibers: (120) composed of microtubules and connect the centromere of the chromosome with the centriole or polar region during mitosis.

Spiracle: (60, 63) the opening in the exoskeleton of insects through which air passes into the tracheal tubes.
Spongy cells: (40) photosynthetic cells of lower portion of leaf.
Spore: (123) very small, tough reproductive cell produced by molds and other organisms.
Sporulation: (123, 253) process of producing spores; asexual reproduction in yeasts and other organisms.
Stain: (12, 240) dye which adheres to the organelles of a cell and makes them visible; including Lugol's iodine, methylene blue.
Stamen: (138) male reproductive structure of the flower.
Stamenate flower: (138) flower containing the stamen but not the pistil.
Starch: (20) carbohydrate, a polymer composed of glucose molecules.
Stem: (51, 60) main support structure of plants; xylem and phloem.
Stigma: (138) in a flower, the top part of the pistil which captures the pollen.
Stimulus: (65) change in the external or internal environment that causes an organism to respond.
Stomach: (82) expandable compartment of the digestive tube within which protein digestion begins.
Stomate: (40, 60) in the epidermis of a leaf, the opening through which gases are exchanged with the environment.
Striated muscle: (109) any muscle tissue (skeletal or cardiac) which shows parallel bands or lines (striations).
Stroke: (103) brain tissue damage due to a blood vessel hemorrhage or embolism in the brain.
Stroma: (39) see chloroplast; spaces between grana and lamella.
Style: (138) that part of the pistil that connects the stigma with the ovary.
Substrate(s): (25, 27) molecule which temporarily binds to the active site of an enzyme and is changed as a result.
Succession: (211) see ecology; pioneer organisms to a climax community.
Survival of the fittest: (185) see evolution and natural selection.
Sweat gland: (98) in skin, a bulb-like structure connected to a long tube through which perspiration flows to the surface of the skin.
Symbiosis:(201, 202) association or living together of two unlike organisms where neither is harmed and one or both benefited.
Synapse: (66) the distance between the terminal branch of one neuron and the dendrite of another neuron, through which neurotransmitters diffuse.
Synapsis: (126) in meiosis, the coming together and twisting of homologous chromosomes (chromatids) to form a tetrad.
Synthesis: (6) process by which small molecules are joined together as a result of dehydration synthesis to form a larger, more complex molecule.
Systemic circulation: (90) flow of blood from the heart to the body tissues and back to the heart.
Systole: (90) contraction of the atrial and ventricular chambers of the heart.
Systolic blood pressure: (90) highest blood pressure produced in the blood vessels as a result of the contraction of the ventricle.

Taxonomy: (7) study of classification.
Tay-Sachs disease: (163) homozygous recessive genetic disorder characterized by deterioration of the brain cells.
Telophase: (120, 127) stage of mitotic cell division within which the chromosomes elongate, a nuclear membrane appears and cytokinesis is completed.
Temperature, effect on enzymes: (26) denaturation due to relatively high temperatures; reduction in rate of activity due to low temperatures.
Tendon: (110) tough connective tissue by which muscle is attached to bone.
Tendonitis: (110) injury or inflammation to the tendon, usually at the bone juncture.
Terminal branches: (65) end brushes of axons in neurons.

Terrestrial biome: (213) major plant and animal associations on land determined by major climate zones of the world.
Terrestrial environment: (131) existing on land.
Test cross: (149) in genetics, a process used to determine the genotype of an individual showing the dominant trait.
Testis: (106, 128, 133) male gonad within which produces sperm and testosterone.
Testosterone: (106, 134) hormone secreted by the testis at puberty and results in the production of the secondary male sex characteristics.
Tetrad: (127, 157) in meiosis, the structure formed when homologous chromosomes (chromatids) come together and twist.
Thymine: (164) nitrogenous base found in DNA only.
Thyroid gland: (105) endocrine gland, located on the trachea, which secretes thyroxine and calcitonin.
Thyroid Stimulating Hormone (TSH): (106) hormone secreted by the anterior pituitary gland and used to stimulate the thyroid gland.
Thyroxine: (106) hormone secreted by the thyroid gland and used to regulate body metabolism.
Tools of science: (242) see microscope and laboratory skills sections.
Trachea: (94) tube containing cartilage rings, through which air is exchanged between the pharynx and bronchi going to the lungs.
Tracheal tubes: (60, 63) in grasshopper, used to transport respiratory gases into and out of the body through spiracles.
Tracheophytes: (8, 50) phylum of plants having true roots, stems, and leaves; the higher plants.
Transcription: (167) process by which messenger RNA is made from DNA.
Transfer RNA: (166) molecule in the shape of a three leaf clover which contains a specific anticodon and binds to its specific amino acid.
Translation: (167) process by which the ribosome, m-RNA, and the t-RNA with its amino acid come together to synthesize a polypeptide.
Translocation: (158) transfer of a section of one chromosome to a non homologous chromosome.
Transpiration pull: (50, 208) process of transport in a plant.
Transport: (6, 48, 87) life activity of absorption and circulation.
Triglyceride: (20) see lipids; a glycerol and three fatty acids.
Tropism: (68) movement (response) by plants to stimuli.
Tuber: (124) asexual reproducing structure in some plants.

Ulcer: (86) irritation or opening of lining of digestive tract.
Ultracentrifuge: (12) tool used to separate materials by weight.
Umbilical cord: (132, 136) connection of a fetus to the placenta.
Unsaturated fatty acid: (22, 85) see lipids; a fatty acid with double bonds, not completely filled with C-H bonds.
Uracil: (166) nitrogenous base only found in RNA.
Urea: (62) nontoxic soluble nitrogenous waste present in urine and perspiration.
Uric acid: (62) nitrogenous waste that is relatively harmless.
Urinary bladder: (99) muscular organ which temporarily stores urine prior to its elimination from the body.
Urine: (98) fluid with an acid pH made of urea, uric acid, ions, water.
Ureters: (99) tubes that carry urine from the kidneys to the urinary bladder.
Urethra: (99, 133) tube carrying urine from urinary bladder to outside of the body.
Use and disuse: (184) see Lamarck and evolution.
Uterus: (134) muscular organ found in females within which the fetus develops.

Vacuole: (16, 44) ball like organelle composed of bilipid membrane containing nutrients or substances essential to the cell.

Vagina: (134) muscular tube which connects the uterus to the outside of the female body through which the fetus is born.
Variable: (245) in an experiment, the one factor that is different from the conditions found in the control.
Variation: (184, 186 trait found in one organism of a species that is not found in other organisms of that species.
Vascular bundle: (50) in monocotyledon plants, the structure composed of the xylem and phloem surrounded by the bundle sheath.
Vascular tissue: (34) any tissue through which a fluid moves.
Vegetative propagation: (123, 159) in plants, form of asexual reproduction by which undifferentiated cells produce new plants exactly like the original plant.
Vein, animal: (89) blood vessel found in animals which carries blood, under low pressure, back to the heart
Vein, plant: (40) transport structure of leaf composed of xylem and phloem.
Vena cava: (90) superior and inferior, major veins - enter r. atrium of heart.
Ventral: the bottom surface of an organism.
Ventral nerve cord: (66) in earthworms and grasshoppers, the position of the nerve cord located below the digestive tube.
Ventricle: (89) large muscular chamber which pumps blood out of the heart.
Venule: (89) blood vessel which carries blood from the capillary to the vein.
Villus: (83) narrow, finger-like projection found in the small intestine which increases the surface area for absorption.
Virus: (16) often disease causing "non-living microbe," an extremely small particle, composed of DNA or RNA surrounded by a protein jacket and capable of reproducing only inside of a cell.
Visceral muscle: (109) involuntary muscle that does not show striations or parallel discs.
Vitamin: (25) organic molecule, which serves either as a coenzyme to some enzymes or as an integral part of a complex molecule.
Voluntary muscle: (109) see skeletal muscle.

Water: (19, 62) end product of cellular respiration, photosynthesis, dehydration synthesis; necessary for cell activities and composition.
Water cycle: (208) movement of water molecules from the surface of the earth into the atmosphere and back again as precipitation.
White blood cells: (87) see leukocyte.

X - chromosome: (154) in diploid cells of the human, the sex chromosome, found as a pair in females and a single chromosome in males; associated with sex-linked genetic recessive characteristics.
Xylem: (50) hollow, narrow, dead plant cells which conduct water upward.

Y - chromosome: (154) in diploid cells of the human male.
Yolk: (128) stored food found in animal ova and used for embryological development.
Yolk sac: (131) the membrane which surrounds the yolk.

Zygote: (126, 129, 139, 157) fertilized egg containing the diploid number of chromosomes.

Biology Practice Exam 1

Part I

Answer all 59 question in this part. (65)

Directions (1-59): for *each* statement or question, select the word or expression that, of those given, best completes the statement or answers the question.

1 The organ system represented in the diagram contains specialized cells. The function of these cells is most closely associated with the process of
1 regulation 3 synthesis
2 growth 4 hydrolysis

2 Members of a population of grey squirrels, *Sciurus carolinensis*, are classified in the same species because they
1 obtain their food in the same manner
2 can mate and produce fertile offspring
3 produce enzymes by synthesis
4 live in the same area

3 A fine network of channels within cells aids in the movement of substances. This network is known as the
1 endoplasmic reticulum 3 cell wall
2 mitochondria 4 ribosomes

4 Which statements listed below are associated with the cell theory?
(*A*) Cells are the basic units of structure in living things.
(*B*) Cells are the basic units of function in living things.
(*C*) Cells come from preexisting cells.

1 statements *A* and *B*, only 3 statements *B* and *C*, only
2 statements *A* and *C*, only 4 statements *A*, *B*, and *C*

5 Which substance is classified as an inorganic compound?
1 glucose 2 fat 3 water 4 protein

6 A chain of chemically bonded amino acid molecules forms a compound known as a
1 lipid 3 nucleotide
2 monosaccharide 4 polypeptide

7 After it is produced by an autotroph, glucose can be
1 converted into storage products by hydrolysis
2 utilized to capture light energy from the Sun
3 combined with three fatty acids to form a lipid
4 used as an energy source in cellular respiration

8 Which structures are adaptations for nutrition in the paramecium?
1 pseudopodia and food vacuoles 3 flagellum and eyespot
2 oral groove and cilia 4 contractile vacuoles and oral groove

9 The diagram represents three steps in the hydrolysis of sucrose. In this diagram, structure *X* is most likely

1 a molecule of oxygen
2 an organic catalyst
3 the end product
4 the substrate

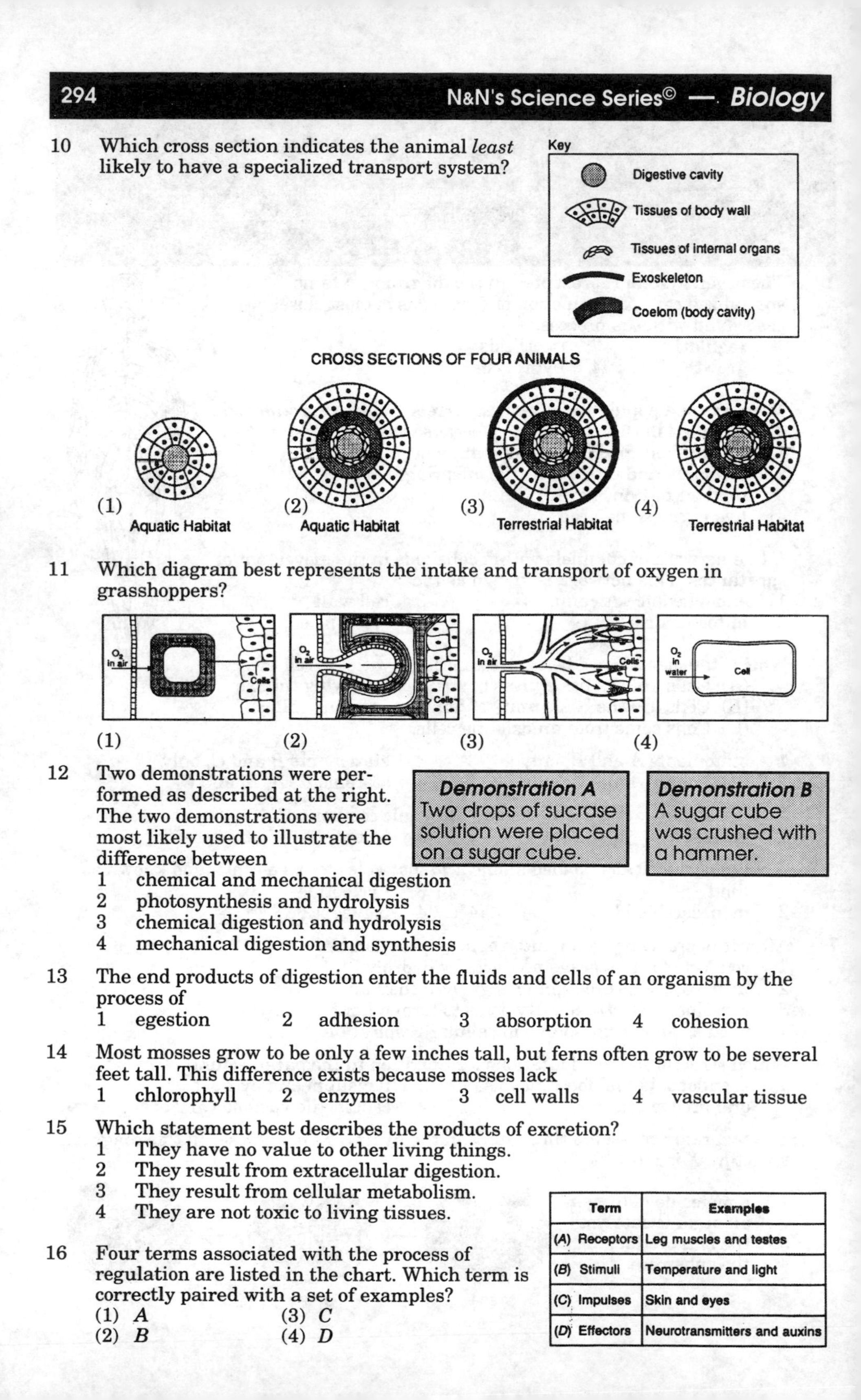

10 Which cross section indicates the animal *least* likely to have a specialized transport system?

11 Which diagram best represents the intake and transport of oxygen in grasshoppers?

12 Two demonstrations were performed as described at the right. The two demonstrations were most likely used to illustrate the difference between

1 chemical and mechanical digestion
2 photosynthesis and hydrolysis
3 chemical digestion and hydrolysis
4 mechanical digestion and synthesis

13 The end products of digestion enter the fluids and cells of an organism by the process of
1 egestion 2 adhesion 3 absorption 4 cohesion

14 Most mosses grow to be only a few inches tall, but ferns often grow to be several feet tall. This difference exists because mosses lack
1 chlorophyll 2 enzymes 3 cell walls 4 vascular tissue

15 Which statement best describes the products of excretion?
1 They have no value to other living things.
2 They result from extracellular digestion.
3 They result from cellular metabolism.
4 They are not toxic to living tissues.

16 Four terms associated with the process of regulation are listed in the chart. Which term is correctly paired with a set of examples?
(1) *A* (3) *C*
(2) *B* (4) *D*

Term	Examples
(*A*) Receptors	Leg muscles and testes
(*B*) Stimuli	Temperature and light
(*C*) Impulses	Skin and eyes
(*D*) Effectors	Neurotransmitters and auxins

17 The diagram represents a cyclic event in the process of cellular respiration. This cycle is important to living organisms because it
1 is a nonreversible process that produces energy for the cell
2 converts the potential energy of organic molecules to a form that can readily be used or stored
3 transforms adenosine triphosphate molecules into protein molecules that cells can use for energy
4 is the only process in cells that does not require enzymes

18 In the human nervous system, synapses are located directly between
1 axons and terminal branches
2 impulses and receptors
3 dendrites and cytons
4 terminal branches and dendrites

19 The diagram represents a cell in water. Formulas of molecules that can move freely across the membrane are shown. Some molecules are located inside the cell and others are in the water outside the cell. Based on the distribution of molecules, what would most likely happen to these molecules after a few hours?

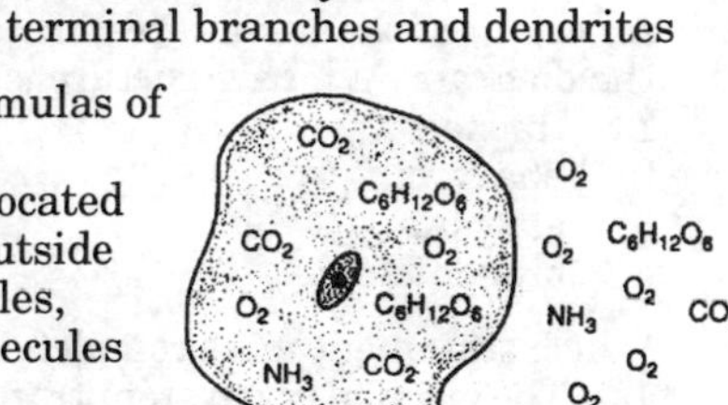

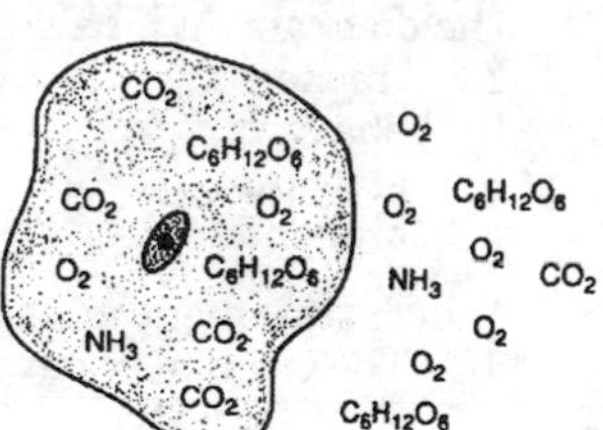

1 The concentrations of $C_6H_{12}O_6$ will increase inside the cell.
2 The concentration of CO_2 will increase outside the cell.
3 The concentrations of NH_3 will increase inside the cell.
4 The concentrations of O_2 will increase outside the cell.

20 The diagram represents a demonstration involving two plant seedlings, *A* and *B*.

The demonstration was most likely set up to show the effect of
1 sunlight on stomate size
2 darkness on lenticels
3 sunlight on auxins
4 darkness on water transport

21 A portion of a plant stem is represented in the diagram. The structures indicated by letter *A* function most directly in the process of

1 gas exchange
2 intracellular digestion
3 meiotic division
4 hormone synthesis

22 Locomotion in the earthworm is accomplished by the combined action of
1 cilia and aetae
2 muscles and setae
3 muscles and nephrons
4 Malpighian tubules and muscles

23 Which statement best describes the nervous system of the organism shown in the diagram?

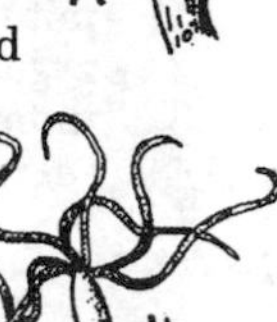

1 It has a highly developed brain and a dorsal nerve cord.
2 It has a primitive brain and few neurons.
3 It has fused ganglia and a ventral nerve cord.
4 It has a nerve net and no brain cells.

24 Which substance is normally absorbed by the large intestine?
1 water 2 glycogen 3 protein 4 cellulose

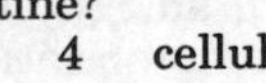

25 The diagram represents a portion of the human digestive tract. The muscular contractions that occur in the region labeled *X* are known as
1 cyclosis
2 hydrolysis
3 synthesis
4 peristalsis

26 Which statement accurately describes human capillaries?
1 They have walls one cell thick.
2 They have valves to prevent backflow of blood.
3 They filter bacteria out of the blood.
4 They contract to assist flood flow.

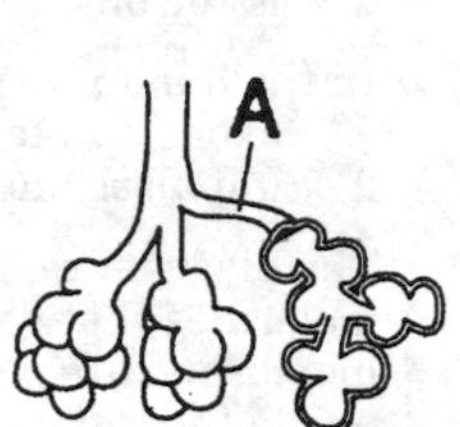

27 A portion of the human respiratory tract is represented in the diagram. Which structure is indicated by letter A?
1 trachea
2 pharynx
3 alveolus
4 bronchiole

28 Which statement most completely describes the function of regulation in humans?
1 The circulatory system transports hormones to various glands, stimulating their activity.
2 The nervous system sends impulses to coordinate all body systems.
3 The nervous and endocrine systems work together to maintain homeostasis.
4 The large intestine stores undigested materials before they are eliminated from the body.

29 Which letter in the diagram indicates the organ that is involved in the deamination of amino acids?
(1) *A*
(2) *B*
(3) *C*
(4) *D*

30 Which structures shown in the diagram contract when the arm moves?
(1) 1 and 5
(2) 1 and 6
(3) 2 and 4
(4) 3 and 4

31 Which process is represented in the photographs?
1 mitotic cell division
2 zygote formation
3 internal fertilization
4 segregation and recombination

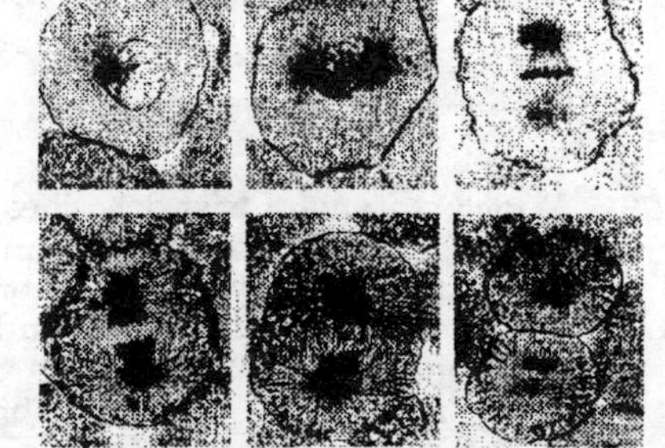

32 The diagrams represent two different organisms classified in the same kingdom. Both of these organisms reproduce asexually by means of
1 budding
2 binary fission
3 sporulation
4 bulb production

33 Which event normally occurs in meiosis but *not* in mitosis?
1 chromosome replication
2 synapsis of homologous chromosomes
3 nuclear membrane disintegration
4 movement of chromosomes to opposite poles

34 Which diagram best represents spermatogenesis in humans?

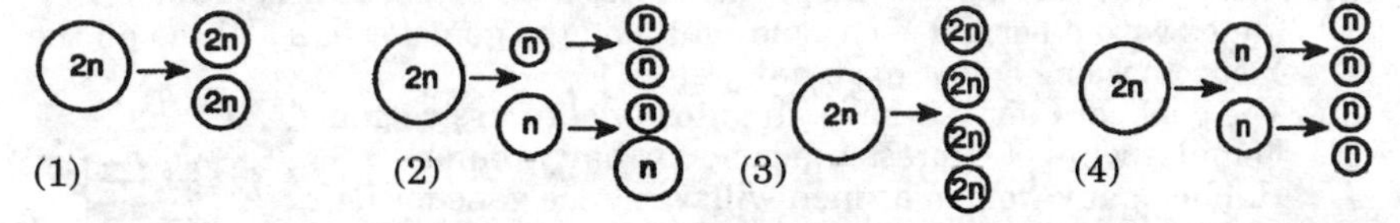

35 Reproduction in many vertebrates is characterized by external fertilization. This type of fertilization is usually associated with the production of large numbers of eggs. These statements best describe reproduction in
1 reptiles 2 birds 3 amphibians 4 mammals

36 The diagram represents a longitudinal section of an apple. In the diagram , the structure indicated by letter *X* normally contains a
1 diploid pollen grain
2 monoploid seed
3 diploid embryo
4 monoploid zygote

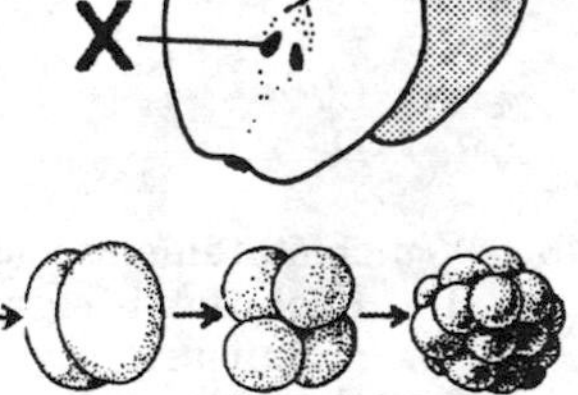

37 Which embryonic process is illustrated in the diagram?
1 cleavage 3 specialization
2 fertilization 4 meiosis

Base your answers to questions 38 and 39 on the information below and on your knowledge of biology.

The pedigree chart shows the inheritance of handedness in humans over three generations. The gene for right-handedness (*R*) is dominant over the gene for left-handedness (*r*).

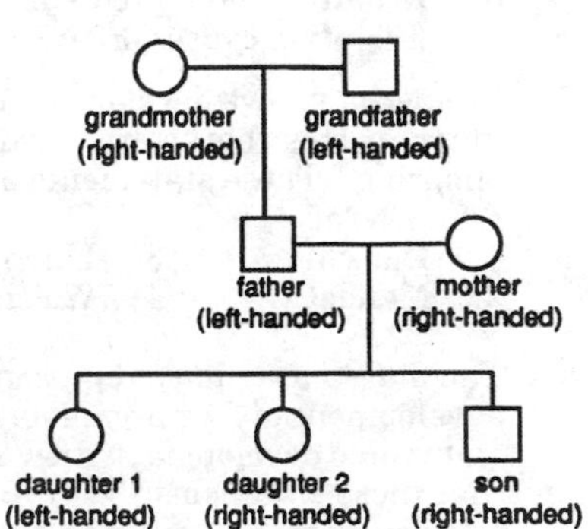

38 For which individual is *Rr* the most probable genotype?
1 grandfather 3 father
2 grandmother 4 daughter 1

39 Which two individuals have identical genotypes for handedness?
1 grandmother and grandfather 3 daughter 1 and daughter 2
2 mother and father 4 mother and son

40 In flowering plants, gametogenesis occurs within the
1 root tip and cambium layer
2 ovary and anther
3 anther and filament
4 xylem tissue and phloem tissue

41 Mendel's laws of heredity are best explained by the
1 fluid-mosaic model
2 gene-chromosome theory
3 lock-and-key model
4 heterotroph hypothesis

42 Which statement describes an effect of crossing-over during meiosis?
1 It increases the chance for variation in zygotes.
2 It interrupts the process of independent assortment.
3 It causes incomplete dominance within the gametes.
4 It inhibits segregation of homologous chromosomes.

43 Some human *Y*-chromosomes contain a gene for a trait called hairy pinna, which produces massive hair growth on the outer ear. Since the corresponding gene is not found on the *X*-chromosome, which statement is most likely true?
1 Males with a normal *X*-chromosome and the gene for hairy pinna on the *Y*-chromosome do not have hairy ears.
2 The gene for hairy pinna is an autosomal recessive gene.
3 Females will not express the gene for hairy pinna.
4 Half as many women as men will carry the gene for hairy pinna.

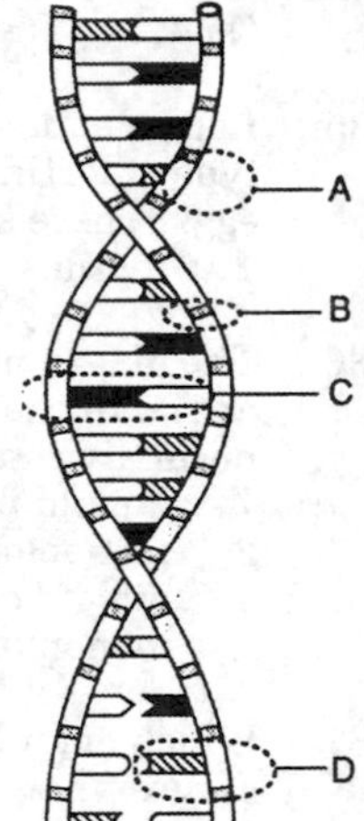

44 A gene mutation may be transmitted to offspring if the mutation occurs within
1 an egg cell
2 muscle cells
3 cells of the uterus
4 blood cells

45 In the diagram (right), which letter indicates a section of the molecule that includes all the components of a nucleotide?
(1) *A*
(2) *B*
(3) *C*
(4) *D*

46 Which statement best describes evolution?
1 Evolution is a predictable change from simple to complex organisms.
2 Evolution is a process of change through time.
3 Evolution often proceeds from complex to simpler organisms.
4 Evolution causes organisms to develop characteristics they need.

47 "Spider monkeys developed a long grasping tail as a result of their need to feed in trees. This characteristic will appear in their offspring." These statements are most in agreement with the ideas of
1 Darwin
2 Mendel
3 Lamarck
4 Weismann

48 The diagrams (right) represent stages in the embryonic development of four organisms. The similarities in embryonic development shown in the diagrams suggest that these organisms
1 are all members of the same species
2 all undergo external development
3 may have evolved from a common ancestor
4 have adaptations for the same environment as adults

49 In the diagram of a whale, the bones labeled "pelvis" and "femur" appear to be useless. The possibility that these bones were once useful gives support to the

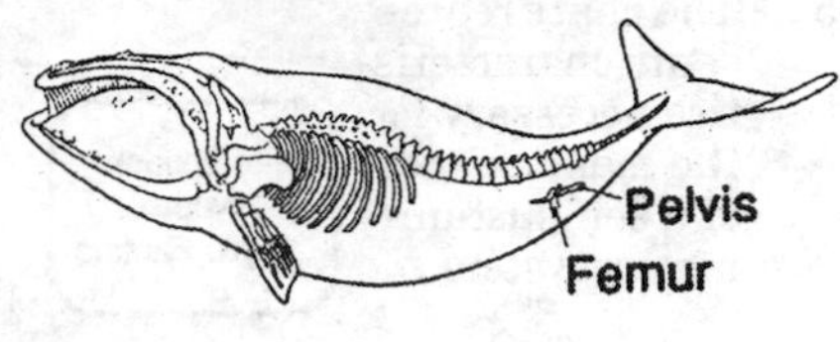

1 modern theory of evolution
2 concept of fossil formation
3 heterotroph hypothesis
4 concept of stable gene frequencies

50 One assumption about the environment of the primitive Earth, as stated in the heterotroph hypothesis, is illustrated in the diagram. Which inference can correctly be drawn from the diagram?

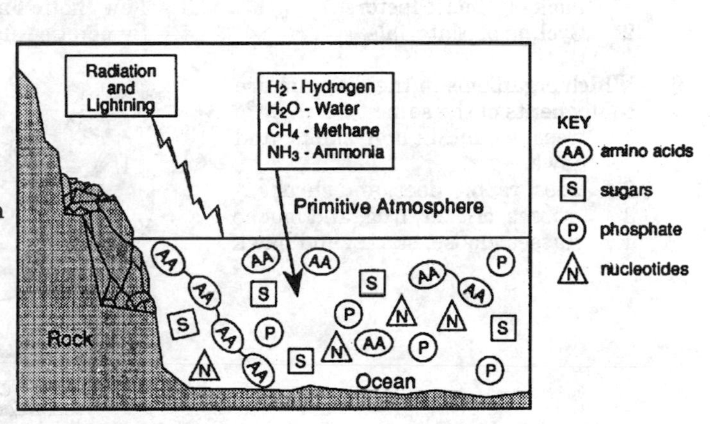

1 No chemical reactions occurred.
2 Energy from the environment most likely contributed to the formation of chemical bonds.
3 Atmospheric oxygen was present.
4 Organic substances could not have been produced under the conditions present.

51 Which graph best represents the rate of evolution described by the concept of punctuated equilibrium?

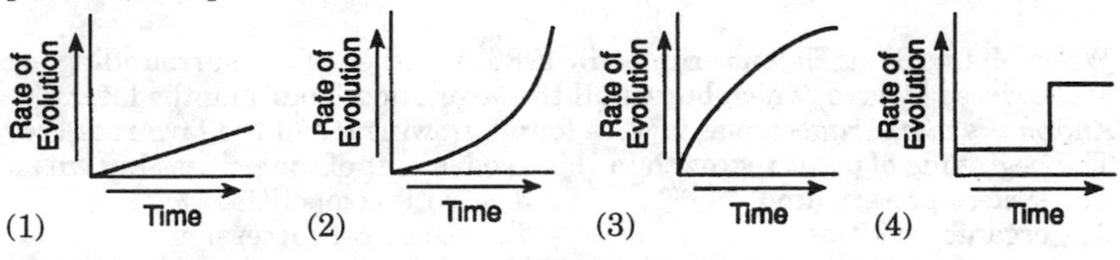

52 Which factor may have played a role in the development of the polar bear in Alaska and the brown bear in Russia into separate species?
1 geographic isolation
2 mitotic cell division
3 asexual reproduction
4 artificial selection

53 An example of a population is all the
1 field mice living in a barn
2 field mice and owls living in a barn
3 animals in a barn and their surroundings
4 animals in a barn and their food

54 Which statement describes an activity that is directly involved in the nitrogen cycle?
1 A crow uses carbohydrates for cell metabolism.
2 A trout excretes carbon dioxide into the water.
3 Soil bacteria convert ammonia into materials usable by autotrophs.
4 Plants release water into the atmosphere from the process of transpiration.

55 The chart represents characteristics necessary for the maintenance of a self-sustaining ecosystem.

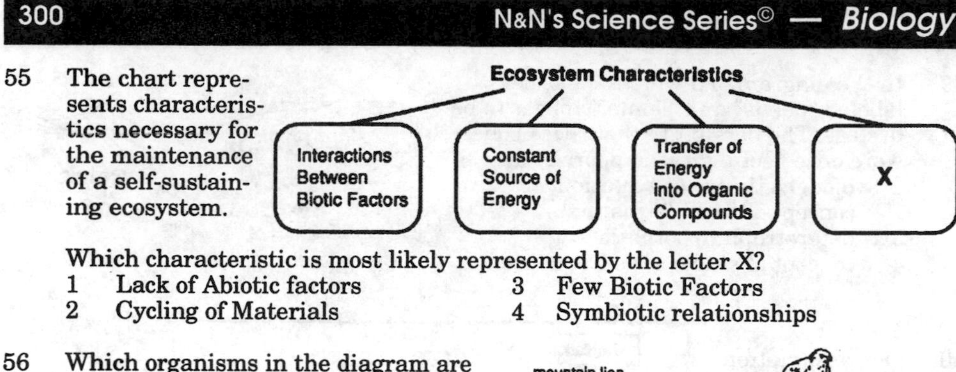

Which characteristic is most likely represented by the letter X?
1 Lack of Abiotic factors
2 Cycling of Materials
3 Few Biotic Factors
4 Symbiotic relationships

56 Which organisms in the diagram are components of the same food chain?
1 trees, mountain lion, snake, and hawk
2 trees, rabbit, deer, and shrubs
3 grasses, cricket, frog, and mouse
4 grasses mouse, snake, and hawk

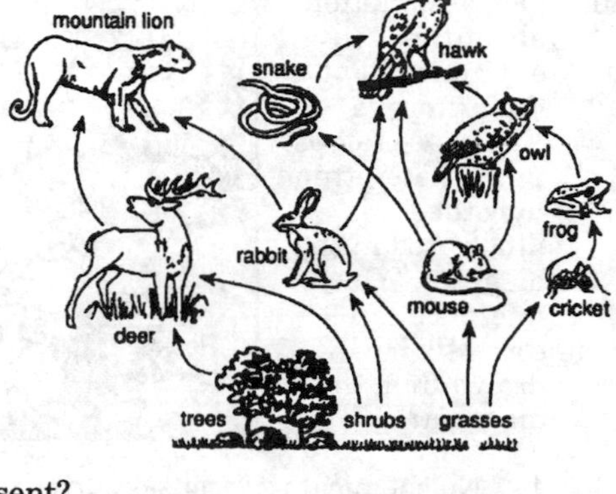

57 Which concept does the diagram represent?
1 the relative elevation of different organisms living at high altitudes
2 the direct relationship between abiotic and biotic factors in an ecosystem
3 the reduction in energy at each successive feeding level of a food chain
4 the imbalance in ecosystems caused by low numbers of carnivores

carnivores
herbivores
grass

58 When Mount Saint Helens erupted in 1980, a portion of the surrounding area was covered by lava, which buried all the vegetation. Four months later, *Anaphalis margaritacea* plants were found growing out of the lava rock crevices. The beginning of plant regrowth in this area is part of the process known as
1 species preservation
2 organic evolution
3 biotic competition
4 ecological succession

59 The graph shows data on the average life expectancy of humans. The change in life expectancy is most likely the result of

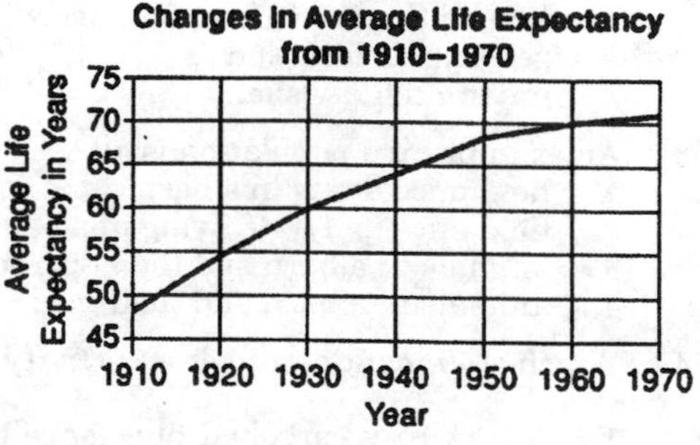

1 poor land-use management that has affected the quality of topsoil
2 technological oversights that have had an impact on air quality
3 a decrease in natural checks such as disease on the population
4 widespread use of biocides such as DDT in water supplies

Part II

This part consists of five groups, each containing ten questions. Choose two of these five groups. Be sure you answer all ten questions in each group chosen.

Group 1 – Biochemistry
If you choose this group, be sure to answer questions 60-69.

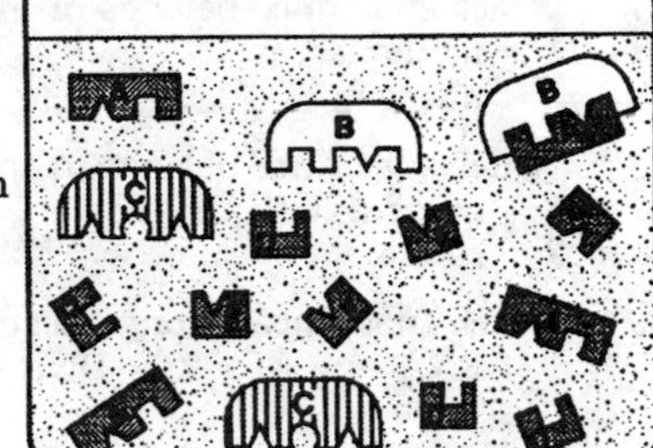

Base your answers to questions 60 and 61 on the diagram and on your knowledge of biology. The diagram represents a beaker containing a solution of various molecules, some of which are involved in the breakdown of molecule *A*.

60 Which structure represents an enzyme functioning in this reaction?
(1) *A* (3) *C*
(2) *B* (4) *D*

61 Which molecule is not associated with the reaction that is occurring in the solution?
(1) *A* (2) *B* (3) *C* (4) *D*

62 The fact that amylase in the human small intestine works best at normal body temperature suggests that
1 amylase is denatured at temperatures below 37°C
2 amylase can function only in the small intestine
3 the optimum temperature for amylase is 37°C
4 the lock-and-key model of enzyme action does not apply to amylase

Directions (63–64): For *each* phrase in questions 63 and 64 select the summary word equation, *chosen from the list below*, that is best described by that phrase. Then record its number.

Summary Word Equations

(1) water + carbon dioxide → glucose + oxygen + water
(2) glucose + glucose → maltose + water
(3) glucose → alcohol + carbon dioxide + ATP
(4) glucose + oxygen → carbon dioxide + water + ATP

63 _____ The process that most directly supplies energy for metabolic activities in humans.

64 _____ A process that directly requies light energy

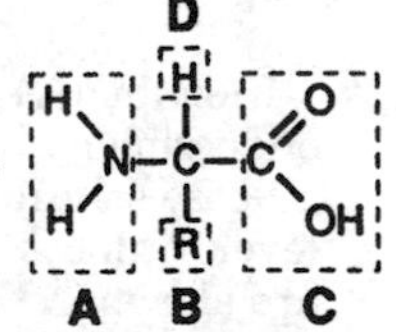

65 The diagram represents an organic molecule. Which portion of the molecule accounts for variations between amino acids?
(1) *A* (3) *C*
(2) *B* (4) *D*

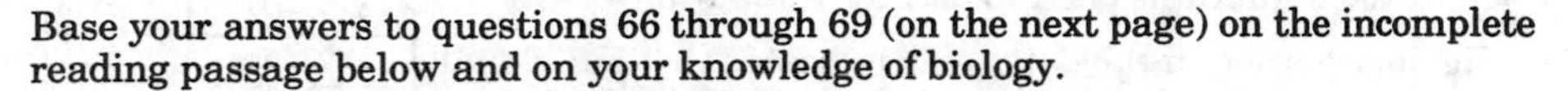

Base your answers to questions 66 through 69 (on the next page) on the incomplete reading passage below and on your knowledge of biology.

Biologists generally agree that the process of photosynthesis can be divided into two major reactions. These are often referred to as the light and dark reactions. During the light reaction or photochemical reaction, molecules of ___*A*___ are "split," producing hydrogen atoms and a gas, ___*B*___. During this reaction extra energy is stored in molecules of ___*C*___.

The dark reaction, also called the carbon-fixation reaction, requires no light because it is powered by the energy molecules made in the light reaction. A three-carbon compound known as ___*D*___ is formed during this reaction.

66 Which substance belongs in space *A*?
1 carbon dioxide
2 water
3 nitrogen
4 carbohydrate

67 Which substance belongs in space *B*?
1 methane
2 nitrogen
3 carbon dioxide
4 oxygen

68 Which substance belongs in space *C*?
1 adenosine triphosphate
2 ribose
3 deoxyribose
4 lactic acid

69 Which substance belongs in space *D*?
1 RNA 2 DNA 3 PGAL 4 ADP

Group 2 – Human Physiology
If you choose this group, be sure to answer questions 70-79.

Base your answers to questions 70 through 72 on the diagram and on your knowledge of biology.

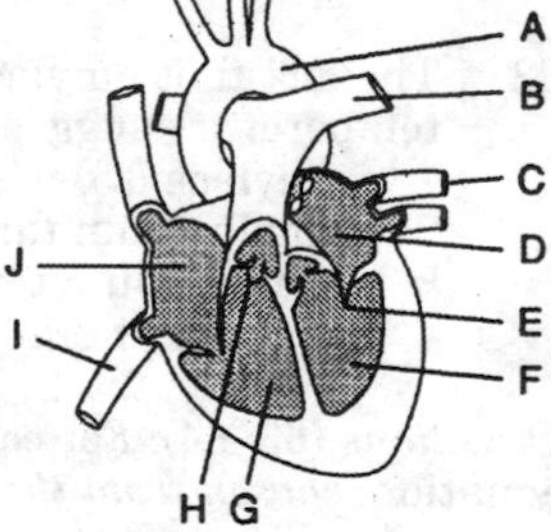

70 A valve that prevents the backflow of blood into an atrium is indicated by letter
(1) *I*
(2) *H*
(3) *C*
(4) *E*

71 The right ventricle is indicated by letter
(1) *J*
(2) *G*
(3) *F*
(4) *D*

72 Which structures contain oxygenated blood?

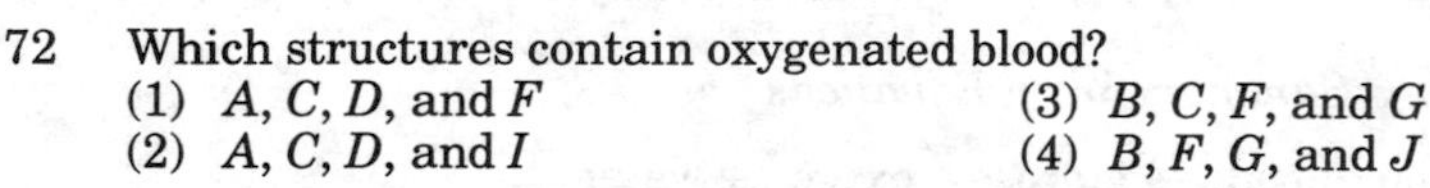

(1) *A, C, D,* and *F*
(2) *A, C, D,* and *I*
(3) *B, C, F,* and *G*
(4) *B, F, G,* and *J*

73 Which is a respiratory disease aggravated by cigarette smoking?
1 emphysema 2 meningitis 3 gout 4 leukemia

74 Antigen A is present on the red blood cells of which blood types?
(1) A and O (2) A and AB (3) B and AB (4) B and O

Base your answers to questions 75 and 76 on the information below and on your knowledge of biology.

Individual *A* and individual *B* were suspected of having endocrine malfunctions. Each drank an equal amount of glucose solution. Every half hour for the next 4 hours, the level of glucose in their blood was measured. The results are shown in the data table. [Normal blood glucose level is 80 to 100 milligrams per 100 deciliters of blood.]

Time (hours)	Blood Glucose Levels (mg/100 dL)	
	Individual *A*	Individual *B*
0.5	120	140
1.0	140	170
1.5	110	190
2.0	90	180
2.5	85	170
3.0	90	160
3.5	85	150
4.0	90	140

75 The information in the data table indicates that individual *B* has a condition that is most likely due to a malfunction of the
1 testes
2 parathyroids
3 Islets of Langerhans
4 ovaries

76 The information in the data table indicates that individual A produces enough
1 insulin
2 follicle-stimulating hormone
3 growth-stimulating hormone
4 parathormone

Base your answers to questions 77 through 79 on the diagram and on your knowledge of biology.

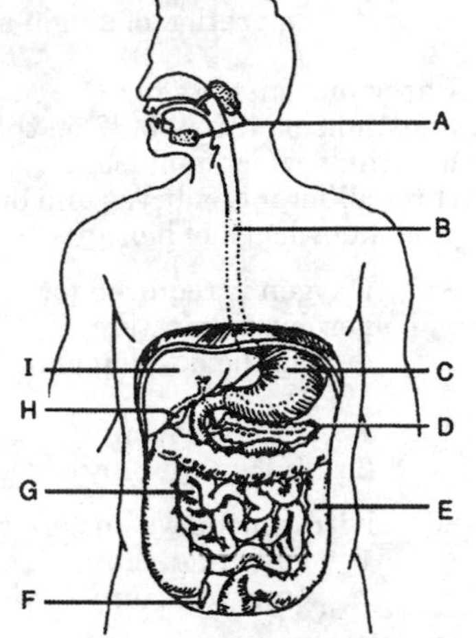

77 The end products of lipid digestion diffuse into the lacteals that are present in the lining of structure
(1) *H*
(2) *G*
(3) *C*
(4) *E*

78 Which organs are involved in the storage of glycogen?
(1) *A* and *E*
(2) *B* and *F*
(3) *C* and *H*
(4) *D* and *I*

79 Clinical studies have indicated that an increase in retention time of material in structure *E* is related to the occurrence of colon and rectal cancer. What could a person do to help decrease retention time in this structure?
1 limit the amount of physical activity before meals
2 increase protein consumption and decrease carbohydrate consumption
3 consume more fruits, vegetables, and grains
4 decrease intake of unsaturated fats and increase intake of saturated fats

Group 3 – Reproduction and Development
If you choose this group, be sure to answer questions 80-89.

Base your answers to questions 80 through 83 on the diagram sand on your knowledge of biology.

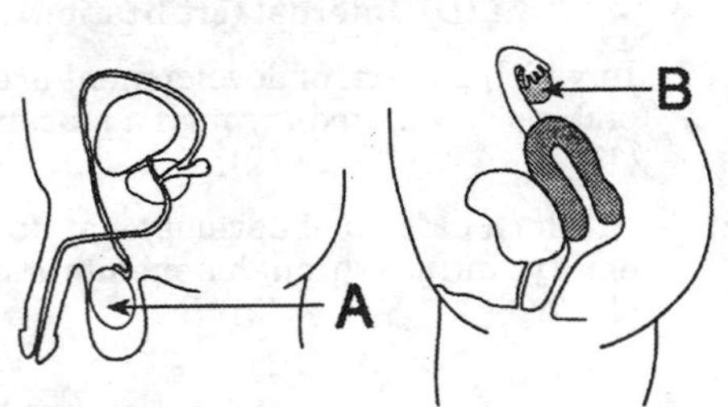

Directions (80–83): For *each* statement in questions 80 through 83 select the sentence, *chosen from the list below*, that best applies to that statement. Then record its *number*. [A number may be used more than once or not at all.]

Sentences

(1) The statement is correct for both structures *A* and B.
(2) The statement is incorrect for both structures *A* and *B*.
(3) The statement is correct for structure *A*, only.
(4) The statement is correct for structure *B*, only.

80 _____ Motile gametes are produced.

81 _____ Estrogen and progesterone are produced.

82 _____ A substance is produced that influences the development of secondary sex characteristics.

83 _____ This structure may contain a corpus luteum.

84 The restoration of the species number of homologous chromosomes occurs during the
1 formation of male sex cells
2 fertilization of an egg by a sperm
3 replication of centromeres
4 migration of single-stranded chromosomes

Base your answers to questions 85 through 87 on the diagrams which represent three different embryos and on your knowledge of biology.

85 Oxygen is required for aerobic respiration during the development of
1 embryo *A*, only
2 embryo *B*, only
3 embryos *A* and *B*, only
4 embryos *A*, *B*, and *C*

86 Nitrogenous wastes are removed by the mother during the development of
1 embryo *A*, only
2 embryo *B*, only
3 embryos *C*, only
4 embryos A and B

87 Which process essential to the survival of the embryo takes place in region 1 of diagram *B*?
1 exchange of blood cells
2 exchange of genes
3 diffusion of nutrients
4 digestion of food

Base your answers to questions 88 and 89 on the patterns of development listed below and on your knowledge of biology.

Patterns of Development
(*A*) External fertilization and external development
(*B*) External fertilization and internal development
(*C*) Internal fertilization and external development
(*D*) Internal fertilization and internal development

88 In which pattern of development are the wastes that are formed by a vertebrate embryo eliminated through a placenta?
(1) *A* (2) *B* (3) *C* (4) *D*

89 In which pattern of development do the wastes that are formed by a vertebrate embryo diffuse through a membrane into the external aquatic habitat?
(1) *A* (2) *B* (3) *C* (4) *D*

Group 4 – Modern Genetics

If you choose this group, be sure to answer questions 90-99.

Base your answers to questions 90 through 92 (next page) on the diagram and on your knowledge of biology. The diagram represents some steps in a metabolic process.

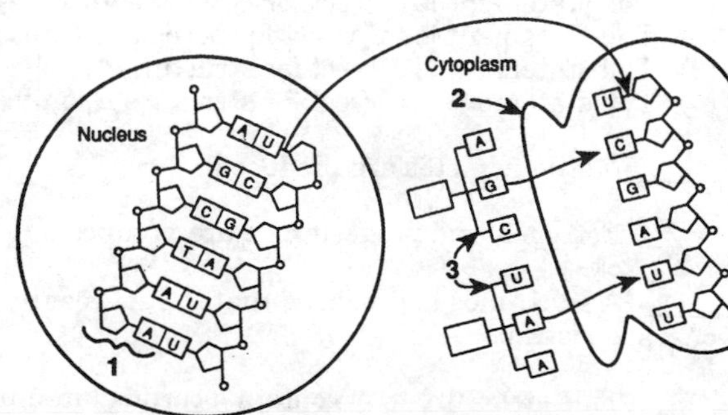

90 The structure indicated by number 1 most likely represents
1 an amino acid molecule
2 a messenger-RNA molecule
3 part of a transfer-RNA molecule
4 part of a DNA molecule

91 The structure indicated by number 2 is known as a
1 mitochondrion
2 nucleolus
3 ribosome
4 vacuole

92 The process represented in the diagram is most closely associated with
1 protein synthesis
2 starch hydrolysis
3 gene replication
4 artificial selection

93 The molecules indicated by number 3 are most likely
1 enzymes
2 transfer-RNA molecules
3 messenger-RNA molecules
4 proteins

94 In a stable population in which the gene frequencies have been constant for a long period of time, the rate of evolution would
1 increase
2 decrease
3 remain the same
4 increase, then decrease

95 The results of the process of cloning are most similar to the results of the process of
1 pollination 2 budding 3 fertilization 4 gametogenesis

Base your answers to questions 96 through 98 on the photographs of chromosomes from a human male and on your knowledge of biology.

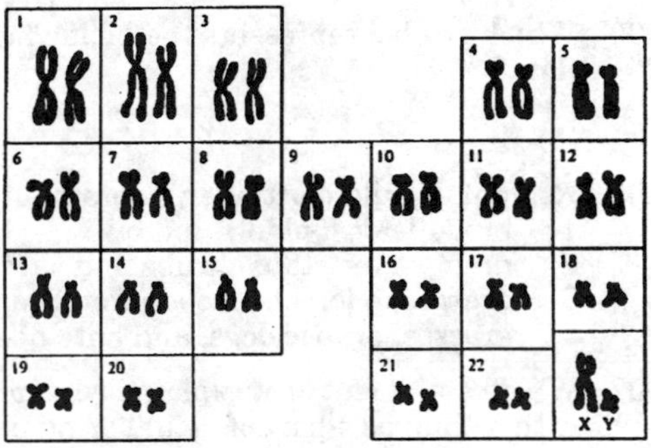

96 The arrangement of chromosomes in the diagram is known as a
1 karyotype
2 centromere
3 mutation rate
4 genotype

97 The arrangement of chromosomes in the diagram is a method used in the identification of
1 phenylketonuria
2 blood type
3 Down syndrome
4 hemophilia

98 If the chromosomes of a female were arranged in chart form like the chromosomes of this male, the chart would
1 be identical to that of the male
2 appear different in one chromosome pair
3 contain more chromosomes
4 have one-half the number of chromosomes

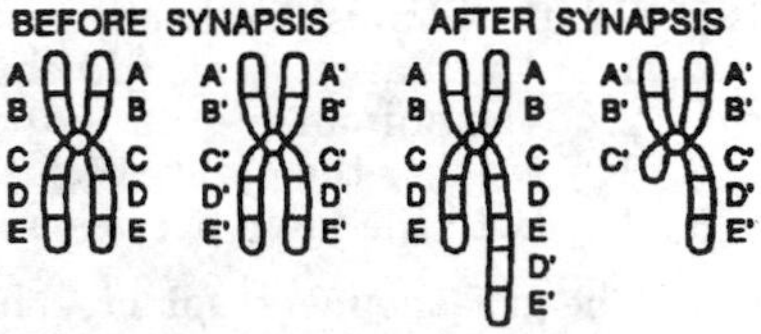

99 The diagrams represent a pair of homologous chromosomes before and after synapsis. The letters represent alleles on the pair of chromosomes. Which types of mutations are represented by the homologous chromosomes after synapsis?
1 translocation and crossing-over
2 polyploidy and translocation
3 base substitution and deletion
4 addition and deletion

Group 5 – Ecology

If you choose this group, be sure to answer questions 100-109.

Base your answers to questions 100 through 101 on the map and on your knowledge of biology. The illustrates various terrestrial biomes in slected areas of North and South America.

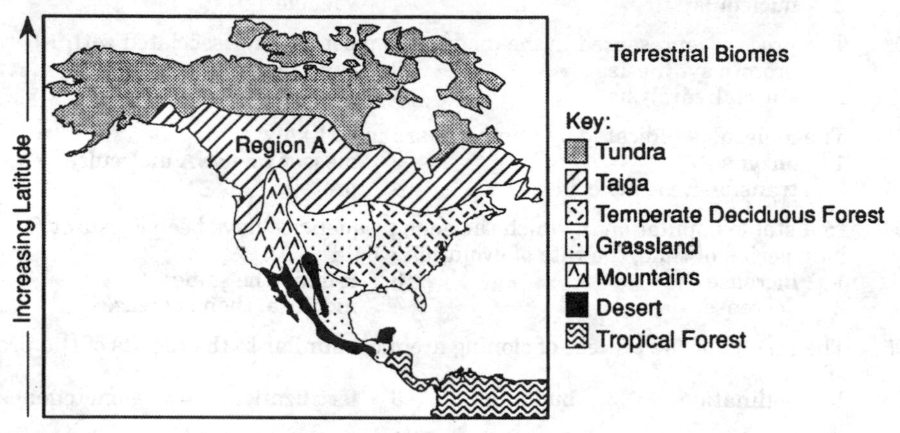

100 Which symbol represents the biome having mosses and lichens as the climax flora?

(1) (2)  (3) (4)

101 Region *A* would most likely contain an abundance of
1 broad-leaved plants and bison
2 drought-resistant shrubs and succulent plants
3 evergreen forests, moose, and bears
4 grasses, prairie dogs, and antelope

102 Which statement best explains why mountain ranges are shown separately rather than as a part of neighboring biomes?
1 Mountains are totally covered by snow.
2 Species of plants and animals vary with altitude in mountainous regions.
3 Few animals are found in mountainous regions.
4 Mountainous regions constantly undergo ecological succession.

Base your answers to questions 103 through 105 on the information below and on your knowledge of biology.

For 25 years, hay was cut from the same 10 acres on a farm. During these years, shrews, grasshoppers, spiders, rabbits, and mice were seen in this hayfield. After the farmer retired, he no longer cut the hay and the field was left unattended.

103 What will most likely occur in the former hayfield over the next few decades?
1 The plant species will change, but the animal species will remain the same.
2 The animal species will change, but the plant species will remain the same.
3 Neither the plant species nor the animal species will change.
4 Both the plant species and the animal species will change.

104 The grasshoppers, spiders, shrews, and other organisms, along with the soil, minerals, and amount of rainfall, constitute
1 a community
2 a population
3 an ecosystem
4 a food web

105 Just before he retired, the farmer determined the population size of several of the field species during the months of May, July, and August. The results are recorded in the table. Which graph best represents the relative population size of the field species for May?

Field Species	Number of Organisms		
	May	July	August
grasshoppers	1,000	5,000	1,500
birds	250	100	100
grasses	7,000	20,000	6,000
spiders	75	200	500

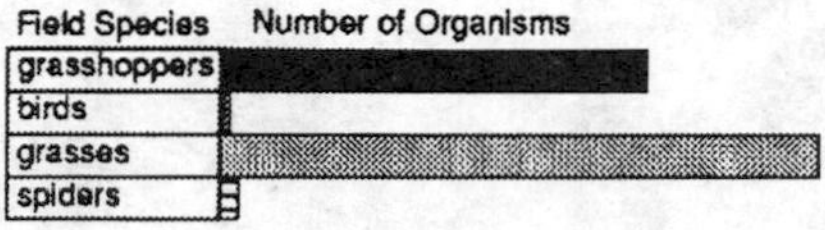

Field Species Number of Organisms
grasshoppers
birds
grasses
spiders

(1)

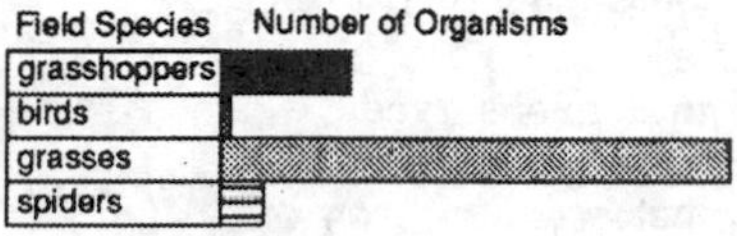

Field Species Number of Organisms
grasshoppers
birds
grasses
spiders

(3)

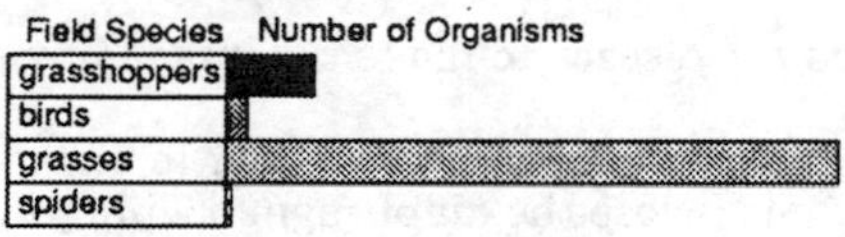

Field Species Number of Organisms
grasshoppers
birds
grasses
spiders

(2)

Field Species Number of Organisms
grasshoppers
birds
grasses
spiders

(4)

106 Although three different bird species all inhabit the same type of tree in an area, competition between the birds rarely occurs. The most likely explanation for this lack of competition is that these birds
1 have different ecological niches
2 share food with each other
3 have a limited supply of food
4 are unable to interbreed

Base your answers to questions 107 through 109 on the list of symbiotic relationships below and on your knowledge of biology.

Symbiotic Relationships
(*A*) Barnacles on whales (+,0)
(*B*) Nitrogen-fixing bacteria in the roots of legumes (+,+)
(*C*) Athlete's-foot fungus on humans (+,–)
(*D*) Protozoa in termite digestive tracts (+,+)
(*E*) Orchids on tropical trees (+,0)
(*F*) Tapeworms in dogs (+,–)

107 Which relationships are examples of mutualism?
(1) *A* and *E*
(2) *B* and *D*
(3) *C* and *F*
(4) *E* and *F*

108 Which organism is a parasite?
1 barnacle
2 legume
3 orchid
4 fungus

109 Lamprey eels attach to the skin of lake trout and absorb nutrients from the body of the trout. Which symbols best represent this relationship?
(1) (+,–)
(2) (+,0)
(3) (0,0)
(4) (+,+)

Part III

This part consists of five groups. Choose three of these five groups. Be sure to record your answers in accordance with the directions given in the question.

Group 1

If you choose this group, be sure to answer questions 110-114.

Base your answers to questions 110 and 111 on the photographs and on your knowledge of biology. The photographs show two views of a microorganism observed with a compound light microscope.

View 1 View 2

110 Which adjustment must be made to the microscope to make the image in view 1 as clear as the image in view 2?
1 remove the eyepiece
2 turn the fine adjustment
3 change from low power to high power
4 close the diaphragm opening

111 If the organism indicated by the pointer is 100 micrometers long before fission, the length of each new individual immediately after fission would be closest to
(1) 1.0 millimeter
(2) 100 millimeters
(3) 50 millimeters
(4) 0.05 millimeter

112 When compared to the image of a specimen observed with a compound light microscope under low power, the image of the specimen observed under high power will appear
1 larger and lighter
2 smaller and lighter
3 smaller and darker
4 larger and darker

113 Photographs of cells from two different organisms are show below.

A B

Which cell feature is a part of the cells shown in photograph *B*, but is not a part of the cells shown in photograph *A*?
1 plasma membrane
2 cytoplasm
3 cell wall
4 nucleus

114 The diagram represents a compound light microscope. The observation of a specimen under both high and low power could be interfered with by a spot on the surface of which structure?
(1) 1
(2) 2
(3) 3
(4) 4

Group 2

If you choose this group, be sure to answer questions 115-119.

Base your answers to questions 115 through 118 on the information and data table and on your knowledge of biology.

Certain chemicals cause mutations in cells by breaking chromosomes into pieces. Cells containing such broken chromosomes are known as mutated cells. Certain nutrients, such as beta carotene (a form of vitamin A), have the ability to prevent chromosome breakage by such mutagenic chemicals.

Data Table

Amount of Beta Carotene Added to Diet per kg of Hamster's Body Weight	Percentage of Mutated Cells
0 mg	11.5
20 mg	11.0
30 mg	8.0
40 mg	7.0
50 mg	4.5
75 mg	3.5
100 mg	2.0
150 mg	1.5

The results of an investigation of the effect of beta carotene in preventing chromosome damage are represented in the data table. In this investigation, varying amounts of beta carotene per kilogram of body weight were added to the diets of hamsters. A mutagenic chemical was also added to the diets of the hamsters at a constant dose rate.

Directions (115-116): Using the information in the data table, construct a line graph, following the directions below.

115 Mark an appropriate scale on each of the labeled axes.

116 Plot the data from the data table. Surround each point with a small circle and connect the points. Example:

Percentage of Mutated Cells

Amount of Beta Carotene Added to Diet (mg/kg of body weight)

117 The greatest effect of added beta carotene on the percentage of mutated cells occurred as the dose rate increased from
(1) 0 to 10 mg (2) 20 to 30 mg (3) 50 to 75 mg (4) 100 to 150 mg

118 Vitamin A was used in this experiment. Which conclusion can best be made concerning the effect of vitamin E on the production of mutated cells in hamsters exposed to this mutagenic chemical?
1 Vitamin E will increase the percentage of mutated cells produced.
2 Vitamin E will decrease the percentage of mutated cells produced.
3 There will be no measurable effect of vitamin E on the percentage of mutated cells produced.
4 No valid conclusion can be made concerning the effect of vitamin E on the percentage of mutated cells produced.

119 Which experimental procedure would best determine the effectiveness of a vaccine for preventing a certain disease in pigeons?
1 Expose 100 pigeons to the disease and then inoculate all 100 pigeons with the vaccine.
2 Expose 100 pigeons to the disease and then inoculate 50 of these pigeons with the vaccine.
3 Inoculate 10 pigeons with the vaccine and 90 pigeons with a harmless solution and then expose all 100 pigeons to the disease.
4 Inoculate 50 pigeons with the vaccine and 50 pigeons with a harmless solution and then expose all 100 pigeons to the disease.

Group 3

If you choose this group, be sure to answer questions 120-124.

Base your answers to questions 120 through 123 on the reading passabe below and on your knowledge of biology.

Get the Lead Out

Researchers have recently determined that children scored better on intelligence tests after the amount of lead in their blood was reduced. This study offers hope that the effects of lead poisoning can be reversed.

Lead poisoning can cause mental retardation, learning disabilities, stunted growth, hearing loss, and behavior problems. Scientists estimate that at least 3 million children in the United States have lead concentrations above the danger level of 10 micrograms per deciliter of blood. Researchers found an average increase of one point on an index scale for intelligence for every decrease of 3 micrograms per deciliter in blood concenteration.

A common source of lead poisoning is peeling or chipping paint in buildings constructed before 1960. Also, soil near heavily traveled roads may have been contaminated by the exhaust from older cars burning leaded gasoline.

In a recent related study, another group of researchers concluded that removing lead-contaminated soil does not reduce blood lead levels enough to justify its cost. The children in the study began with blood levels of 7 to 24 micrograms per deciliter. Replacing the lead-contaminated soil resulted in a reduction in blood lead levels of 0.8 to 1.6 micrograms per deciliter in 152 children under the age of 4.

These studies are not conclusive. Results indicate a need for further studies to determine if reducing environmental lead levels will significantly reduce lead levels in the blood.

120 One effect of lead poisoning is
1 an increase in growth
2 a decrease in platelet numbers
3 a decrease in learning problems
4 an increase in behavior problems

121 A decrease of 9 micrograms per deciliter in blood lead level would most likely lead to an average
1 increase in one point on an index scale for intelligence
2 increase of three points on an index scale for intelligence
3 decrease of three points on an index scale for intelligence
4 decrease in six points on an index scale for intelligence

122 The part of the nervous system most affected by high levels of lead in the blood is the
1 cerebrum 2 cerebellum 3 spinal cord 4 medulla

123 Using one or more complete sentences, state one practice that could be used to reduce lead in the home environment.

124 A laboratory demonstration was set up and maintained at 2°C for 2 hours. The results are shown in the diagram.

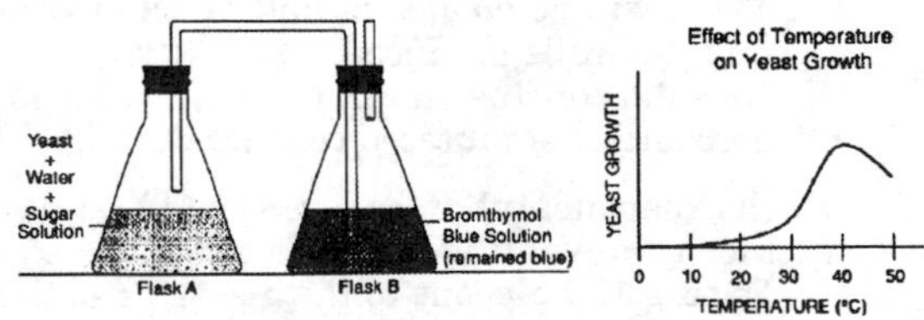

According to the graph, if the demonstration had been maintained at 40°C for 2 hours, the contents of flask *B* would most likely have
1 turned yellow, and no gas bubbles would have been produced
2 remained blue, and gas bubbles would have been produced
3 turned yellow, and gas bubbles would have been produced
4 remained blue, and no gas bubbles would have been produced

Group 4
If you choose this group, be sure to answer questions 125-129.

125 The diagram shows four setups used in an attempt to investigate the release of a gas during photosynthesis. Each setup was maintained at 25°C for a period of 10 hours.

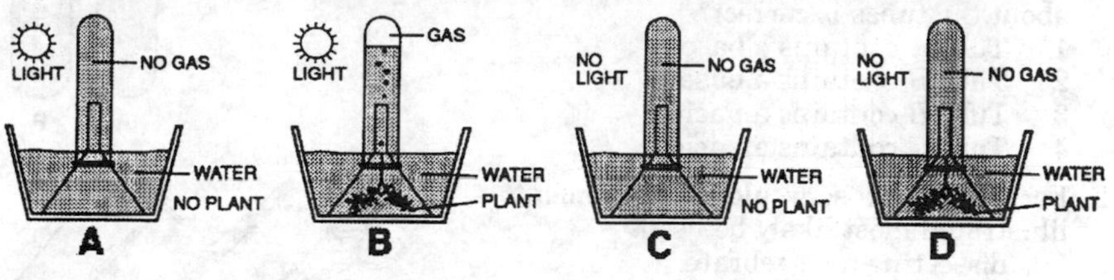

What was a variable in this experiment?
1 water
2 temperature
3 time
4 light

126 Which statement correctly describes the location of structures *A* and *B* in the diagram?
1 Structure *A* is dorsal to structure *B*.
2 Structure *B* is posterior to structure *A*.
3 Structure *A* is ventral to structure *B*.
4 Structure *B* is anterior to structure *A*.

127 The best estimate of the number of ovules found in the ovary of a gladiolus flower would be determined by averaging the data from
(1) 10 students each counting the ovules in 10 different ovaries
(2) 2 students each counting the ovules in the same ovary 50 times
(3) 1 student counting the ovules in the same ovary 100 times
(4) 10 students each counting the ovules in the same ovary 10 times

128 To investigate the effect of a substance on plant growth, two bean plants of the same species were grown under identical conditions with the substance added to the soil of one of the plants. At the end of 2 weeks, the plant grown with the substance was 12.5 centimeters tall and the plant grown without the substance was 12.2 centimeters tall. The researcher concluded that the presence of the substance causes plants to grow taller. Using one or more complete sentences, state one reason that this conclusion may not be valid.

129 Data from measurements of transpiration and water uptake in an ash tree are plotted in the graph. Which conclusion can correctly be drawn from the information in the graph?
1 The rate of transpiration is not related to the amount of water uptake.
2 Transpiration and water uptake are similar processes.
3 Transpiration reaches a maximum rate faster than water uptake.
4 As transpiration increases, the amount of water uptake decreases.

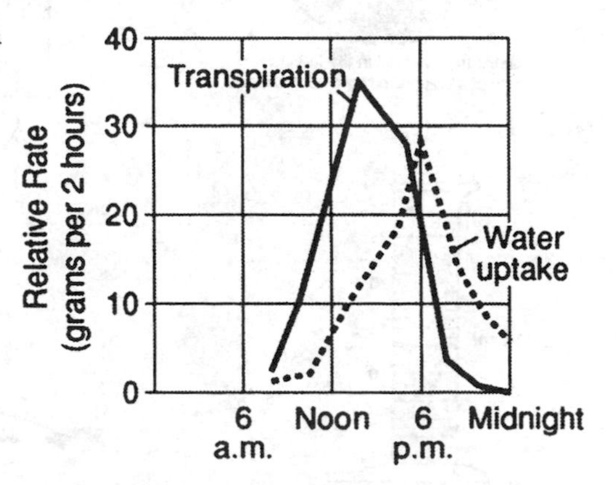

Group 5

If you choose this group, be sure to answer questions 130-134.

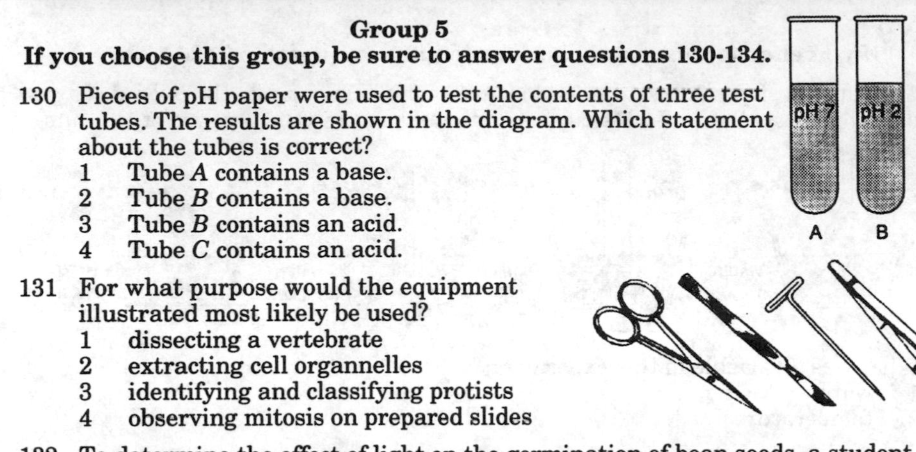

130 Pieces of pH paper were used to test the contents of three test tubes. The results are shown in the diagram. Which statement about the tubes is correct?

1 Tube *A* contains a base.
2 Tube *B* contains a base.
3 Tube *B* contains an acid.
4 Tube *C* contains an acid.

131 For what purpose would the equipment illustrated most likely be used?

1 dissecting a vertebrate
2 extracting cell organnelles
3 identifying and classifying protists
4 observing mitosis on prepared slides

132 To determine the effect of light on the germination of bean seeds, a student selected 12 healthy bean seeds and placed 6 of them in a petri dish in a dark closet and 6 of them in a petri dish on a windowsill. The experiment was conducted for 10 days at room temperature and no germination took place. Using one or more complete sentences, describe one error in the experimental setup that resulted in the failure of the seeds to germinate.

133 A scientist performed an experiment using the following steps:

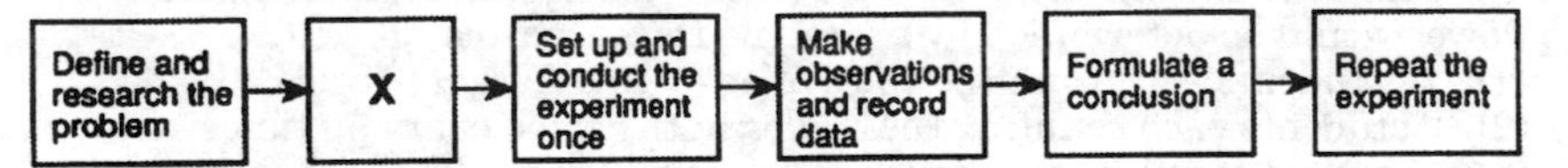

Using one or more complete sentences, identify the step that belongs in box *X*.

134 Which diagram shows a correct measurement?

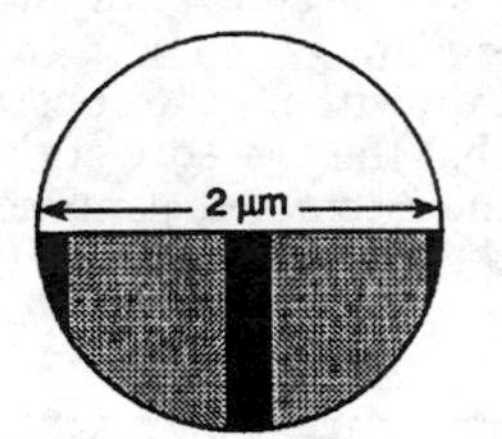

Metric ruler viewed in the low-power field of a compound light microscope

(1)

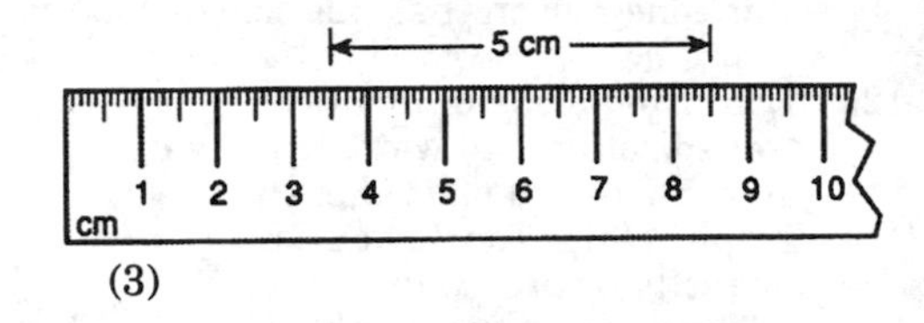

(3)

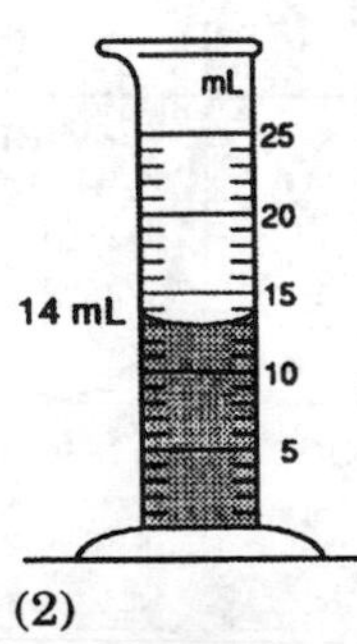

(2)

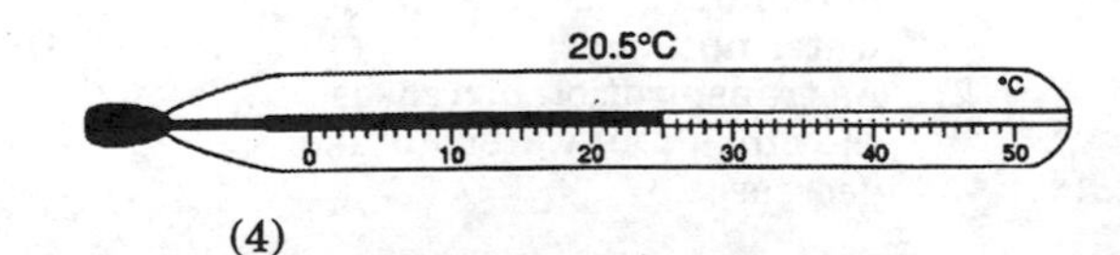

(4)

Biology Practice Exam 2

Part I

Answer all 59 question in this part. (65)

Directions (1-59): for *each* statement or question, select the word or expression that, of those given, best completes the statement or answers the question.

1 When an individual goes without eating for a day, his or her blood sugar level remains about the same throughout the day. This relatively constant condition is maintained by
1 homeostatic control
2 egestion
3 reproduction
4 growth of cells

2 The photograph at the right shows the skeleton of an organism.

This organism is classified as a member of which phylum?
1 annelids
2 coelenterates
3 chordates
4 bryophytes

3 Which factor contributed most to the development of the cell theory?
1 the discovery of many new species during the last century
2 the development of advanced techniques to determine the chemical composition of substances
3 the increase in knowledge concerning factors influencing the rate of evolution
4 the improvement in microscopes and microscopic techniques during the last two centuries

4 A white blood cell ingests, then digests, a number of bacteria. Which cell organelles were directly responsible for the digestion of the bacteria?
1 centrioles
2 lysosomes
3 ribosomes
4 mitochondria

5 The results of an experiment to determine the chemical composition of the cytoplasm of organism *X* are summarized in the data table at the right.
What percentage of the cytoplasm is composed of organic material?
(1) 15
(2) 20
(3) 22
(4) 92

Data Table

Substance	Percent by Mass in Cytoplasm
Water	77
Proteins	15
Fats	5
Carbohydrates	2
Mineral salts	1

6 Which two chemical processes are represented by the equation below?

$$\text{starch + water} \underset{}{\overset{\text{enzymes}}{\rightleftharpoons}} \text{sugars}$$

1 hydrolysis and dehydration synthesis
2 photosynthesis and dehydration synthesis
3 photosynthesis and respiration
4 hydrolysis and respiration

7 Which factor is *least* likely to influence the rate of enzyme activity in cells?
1 the pH at the reaction site
2 the number of Golgi complexes near the reaction site
3 the concentation of the substrate at the reaction site
4 the concentration of the enzyme at the reaction site

8 Most of the food and oxygen in the environment is produced by the action of
1 saprophytic bacteria
2 heterotrophic organisms
3 aerobic protozoans
4 autotrophic organisms

9 The graph at the right represents the absorption spectrum of chlorophyll. The graph indicates that the energy used in photosynthesis is most likely obtained from which regions of the spectrum?
1 yellow and orange red
2 violet blue and green
3 orange red and violet blue
4 green and yellow

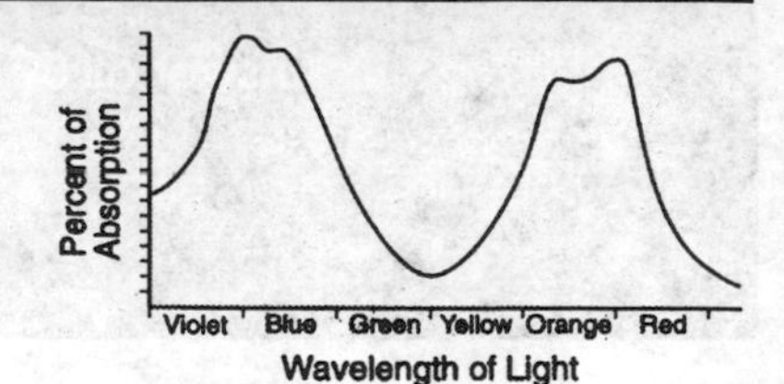

10 In the earthworm and the grasshopper, the gizzard increases the surface area of food for faster chemical digestion. In humans, this function is accomplished by the action of
1 the large intestine 2 enzymes 3 the esophagus 4 teeth

11 A student has a hamburger, french fries, and soda for lunch. Which sequence represents the correct order of events in the nutritional processing of food?
1 ingestion → digestion → absorption → egestion
2 digestion → absorption → ingestion → egestion
3 digestion → egestion → ingestion → absorption
4 ingestion → absorption → digestion → egestion

12 Which cell structure is represented by the three-dimensional diagram at the right?
1 chloroplast
2 mitochondrion
3 plasma membrane
4 replicated chromosome

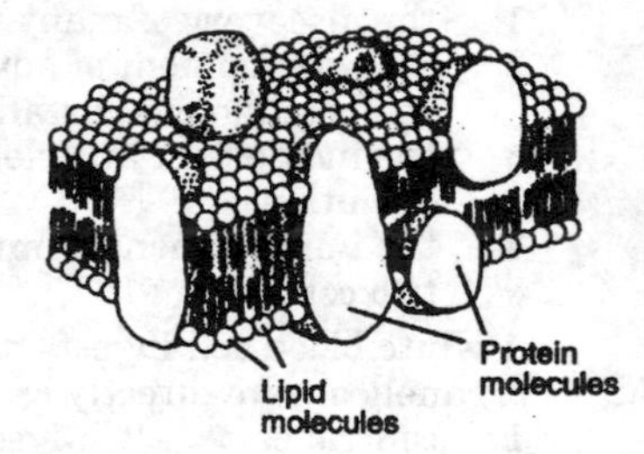

13 The roots, stems, and leaves of a bean plant all contain
1 lenticels and a cuticle
2 xylem and phloem tissues
3 palisade cells and xylem tissues
4 guard cells and lenticels

14 The transfer of energy from nutrients to ATP is accomplished most directly by the process of
1 cyclosis
2 diffusion
3 cellular respiration
4 glucose synthesis

15 The movement of a gas into moist intercellular spaces through stomates is an adaptation for respiration in
1 protists
2 plants
3 grasshoppers
4 earthworms

16 Which structure in a grasshopper has a function similar to that of the alveoli of a human?
1 chitinous exoskeleton
2 tracheal tube
3 gastric caecum
4 salivary gland

17 Excretion is best described as the removal of
1 metabolic wastes from a cell
2 toxic wastes by the process of cyclosis
3 water molecules from dipeptite hydrolysis
4 undigested material from the digestive tract

18 The principal excretory organs of the earthworm are the
1 skin and nephridia
2 Malpighian tubules and spiracles
3 setae and nephrons
4 anus and ganglia

19 The diagram at the right represents the functional unit of a nervous system.

Which letter indicates a structure that secretes a neurotransmitter?

(1) *A* (3) *C*
(2) *B* (4) *D*

20 The diagram at the right represents an arthropod.

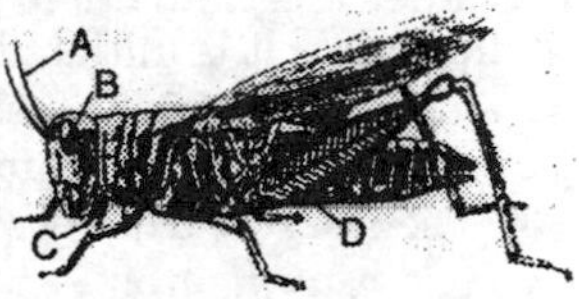

All of the lettered structures shown in the diagram are most closely associated with the life function of

1 egestion 3 regulation
2 locomotion 4 respiration

21 The growth of a geranium plant toward a light source depends most directly on the influence of

1 hormones 2 nerve fibers 3 endocrine glands 4 roots

22 The diagram below represents an ameba attempting to engulf an escaping paramecium.
Which structures are involved in these activities?

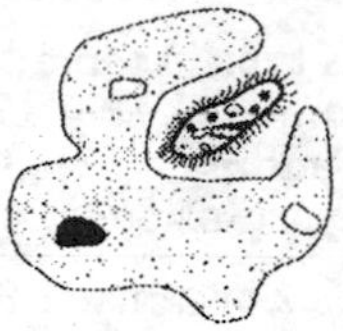

1 tentacles and cilia
2 setae and pseudopods
3 pseudopods and cilia
4 tentacles and flagella

23 In humans, excess fluid and other substances durrounding the cells are returned to the blood by

1 lymphocytes 2 arteries 3 platelets 4 lymph vessels

24 Diagrams *A* and *B* at the right represent structures found in the human body.

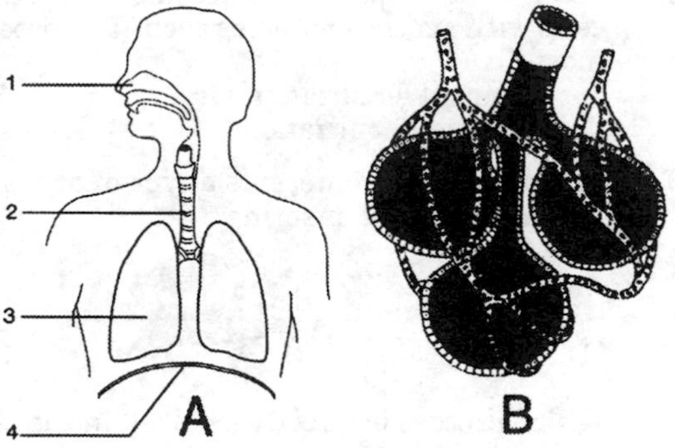

Diagram *B* represents the functional unit of which structure represented in diagram *A*?

(1) 1
(2) 2
(3) 3
(4) 4

25 The chart below lists two groups of nutrients.

Group A	Group B
Vitamins	Complex carbohydrates
Minerals	Protein
Water	Lipids

Which statement correctly describes what happens to the nutrients in these groups as they move through the human digestive system?

1 Group *A* is hydrolyzed, but group *B* is not.
2 Group *B* is hydrolyzed, but group *A* is not.
3 Both group *A* and group *B* are hydrolyzed.
4 Neither group *A* nor group *B* is hydrolyzed.

26 Which substance causes fatigue when it accumulates in human muscles?

1 excess oxygen 3 lactic acid
2 carbon dioxide 4 adenosine triphosphate

27 Which structure is correctly paired with its function?
1 urethra – eliminates urine from the bladder
2 neuron – filters the boood
3 ventricle – pumps blood directly into atria
4 liver – produces intestinal amylase

28 In order to stimulate an effector in a toe, which pathway does a nerve impulse follow after it is initiated at a receptor?
1 interneuron → sensory neuron → motor neuron
2 interneuron → motor neuron → sensory neuron
3 sensory neuron → motor neuron → interneuron
4 sensory neuron → interneuron → motor neuron

29 In humans, red bone marrow provides
1 structural support for the body
2 a source of new blood cells
3 an attachment site for muscle tissue
4 a stie to trap bacteria

30 Which type of reporduction is illustrated in the diagram at the right?
1 budding 3 gametogenesis
2 sporulation 4 regeneration

31 Complex organisms produce sex cells that unite during fertilzation, forming a single cell known as
1 an embryo 3 a gonad
2 a gamete 4 a zygote

32 A cell with a diploid chromosome number of 12 divided two times, producing four cells with six chromosomes each. The process that produced these four cells was most likely
1 internal fertilization 3 mitotic cell division
2 external fertilization 4 meiotic cell division

33 Which diagram represents a type of organism that uses external fertilization for the production of offspring?

(1)

(2)

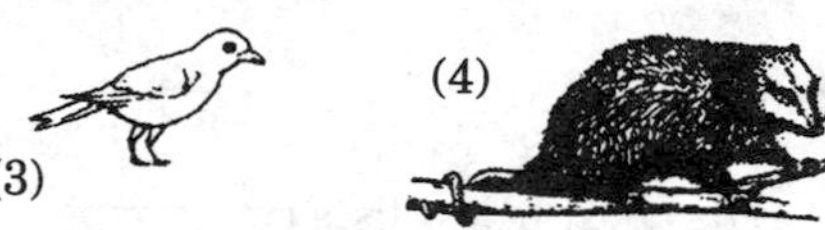
(3) (4)

34 Which process is directly involved in the change of a fertilized egg into an embryo?
1 spermatogenesis 3 cleavage
2 oogenesis 4 vegetative propagation

35 Which phrase best describes the process represented in the diagram at the right?

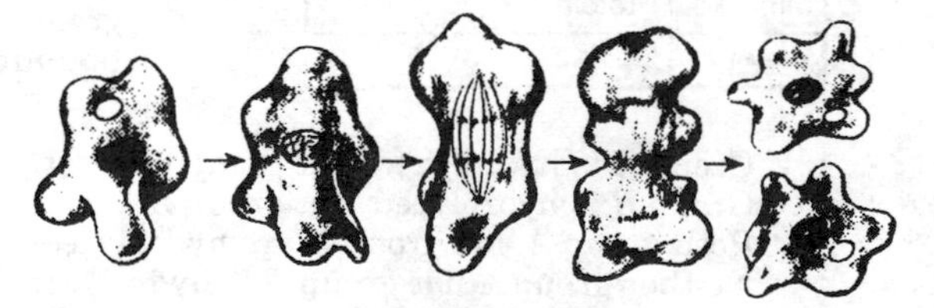

1 germination of a pollen grain in a flower
2 indentical gametes being formed by mitotic cell division
3 development of seeds in an ovule
4 daughter cells being formed by mitotic cell division

36 A difference between marsupials and placental mammals is that in marsupials
1 the young do not receive nourishment from mammary glands
2 fertilization occures outside the body of the female
3 external development occurs without direct nourishment from the mother
4 the embryo is born at a relatively immature stage of development

37 The diagram at the right represents some parts of a flower.
Both meiosis and fertilization occur within
(1) *A* and *F*
(2) *B*
(3) *C*
(4) *D* and *E*

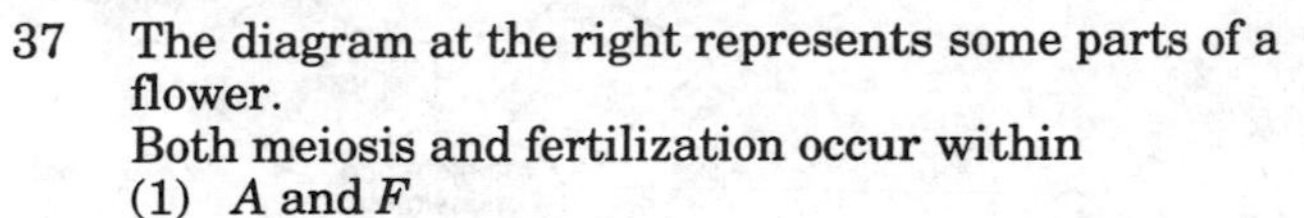

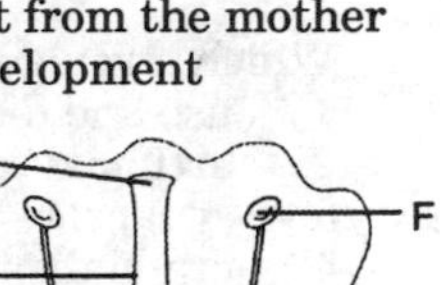

38 Which diagram represents a pair of homologous chromosomes?

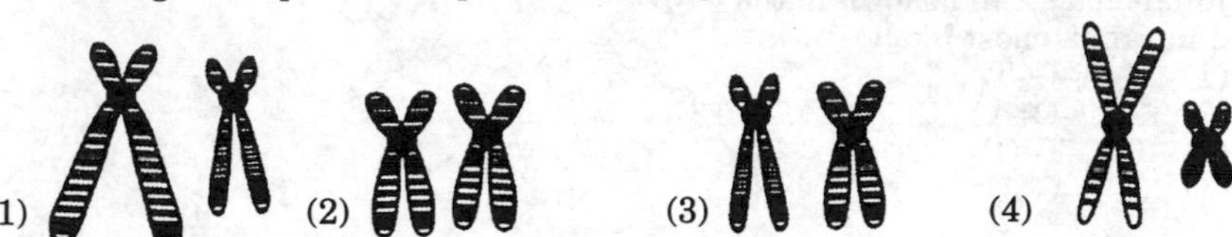

39 In minks, the gene for brown fur *(B)* is dominant over the gene for silver fur *(b)*. Which set of genotypes represents a cross that could produce offspring with silver fur from parents that both have brown fur?
(1) *Bb* x *Bb* (2) *BB* x *bb* (3) *BB* x *Bb* (4) *Bb* x *bb*

40 In cats, a pair of X-chromosomes will produce a female, while an *X*- and a *Y*-chromosome will produce a male. Fur color is controlled by a single pair of alleles. The alleles for orange fur and for black fur are located on the X-chromosome, only. Tortoiseshell cats have one allele for orange fur and one for black fur. Which statement concerning fur color in cats is true?
1 Only female cats can be tortoiseshell.
2 Only male cats can be tortoiseshell.
3 Males need two alleles to have orange fur.
4 Males need two alleles to have black fur.

41 The diagram at the right represents four beakers, each containing an equal number of two colors of beads.

Red and green beads
Yellow and blue beads
Black and white beads
Purple and orange beads

One bead was removed at random from each of the four beakers, and the colors were recorded. The beads were then returned to the original beakers. When the procedure was repeated several times, different combinations of colored beads were obtained. This activity could best be used to illustrate
1 mitotic cell division
2 sex linkage
3 crossing-over
4 independent assortment

42 In humans, Down syndrome is often a result of the
1 disjunction of homologous chromosomes during meiotic cell division
2 nondisjunction of chromosome number 21 in one of the parents
3 combination of an egg and sperm, each carrying a recessive allele for this disorder
4 fusion of two $2n$ gametes during fertilization

43 The diagram at the right represents some methods used by plant growers to produce and maintain desirable varieties of plants.

Which term belongs in area *X*?
1 use and disuse
2 artificial selection
3 synapsis
4 gradualism

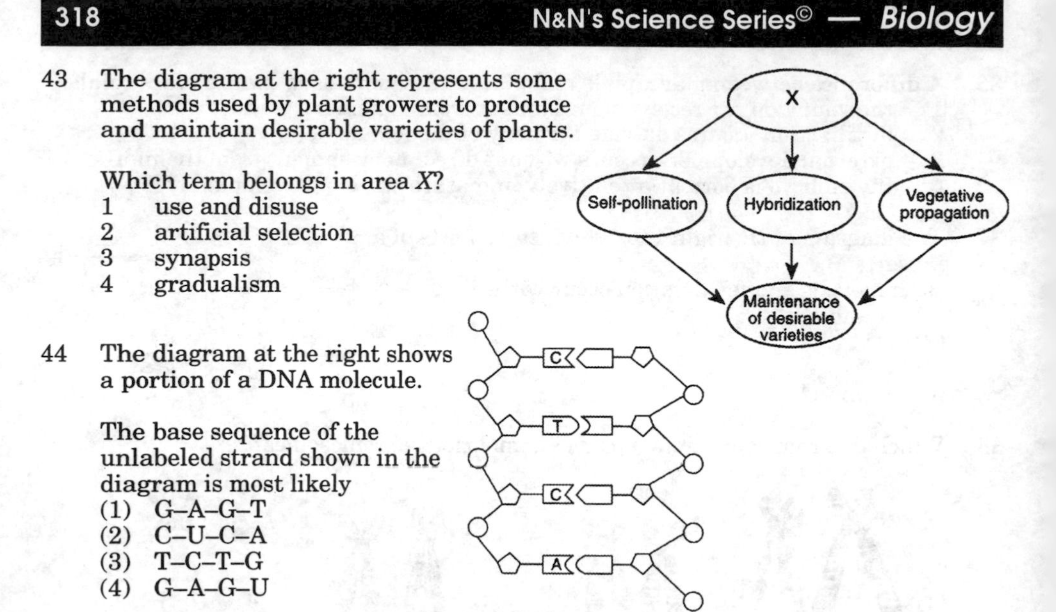

44 The diagram at the right shows a portion of a DNA molecule.

The base sequence of the unlabeled strand shown in the diagram is most likely
(1) G–A–G–T
(2) C–U–C–A
(3) T–C–T–G
(4) G–A–G–U

45 When a certain pure strain of fruit fly is cultured at a temperature of 16°C all of the flies will develop straight wings. The offspring of these flies will develop curly wings when they are raised at 25°C. This pattern of development indicates that
1 acquired characteristics can be passed from generation to generation
2 curly-winged flies will have straight-winged offspring regardless of the environment
3 the environment can affect the expression of certain genes
4 straight-winged flies will have curly-winged offspring regardless of the environment

46 Which organisms are considered most closely related, based on similarities in their embryological development?
1 hydra, earthworm, and frog
2 human, zebra, and grasshopper
3 human, dog, and mouse
4 chicken, bat, and butterfly

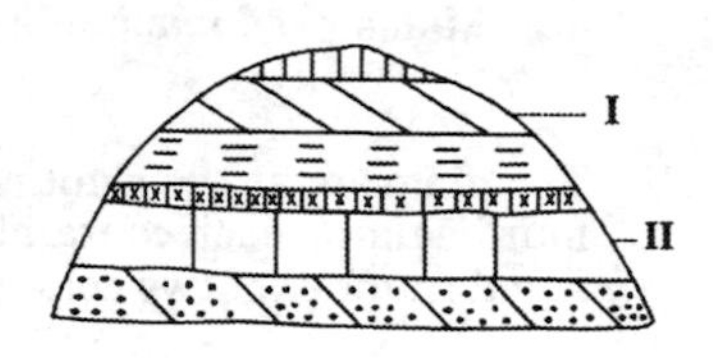

47 The diagram at the right shows the side of a hill exposed by an excavation. The rock layers have different thicknesses, colors, and textures. These geologic layers have been undisturbed since their formation.

The fossils in layer I resemble the fossils in layer II, although they are more complex. This observation suggests that
1 all fossils belong to the same kingdom and phylum
2 simple forms of life may have evolved from more complex forms
3 many different organisms have similar proteins and enzymes
4 modern forms of life may have evolved from older forms

48 The changes in foot structure in a bird population over many generations are shown in the diagram at the right.
These changes can best be explained by the concept of
1 evolution 3 stable gene frequencies
2 extinction 4 use and disuse

Time
Time
Present day

49 In 1889, August Weismann, a German biologist, conducted an experiment attempting to produce mice without tails. He cut the tails off adult mice and then permitted them to mate. All offspring had long tails. He repeated the experiment many times, always with the same results. This experiment helped to disprove the concept of
1 overproduction in a species
2 inheritance of acquired characteristics
3 survival of the fittest
4 struggle for existence

50 Genetic variations are the raw material for evolution. These variations cannot be acted upon by natural selection factors unless they
1 produce only unfavorable characteristics
2 produce only favorable characteristics
3 are found in fossil records of the population
4 are in the phenotype of the organism

51 Which statement is part of the heterotroph hypothesis?
1 Heterotrophs evolved before autogtrophs.
2 Aerobes evolved before anaerobes.
3 Atmospheric oxygen was present before carbon dioxide.
4 Proteins were present before amino acids.

52 Each point in the graphs below represents a new species. Which graph best represents the concept of gradualism?

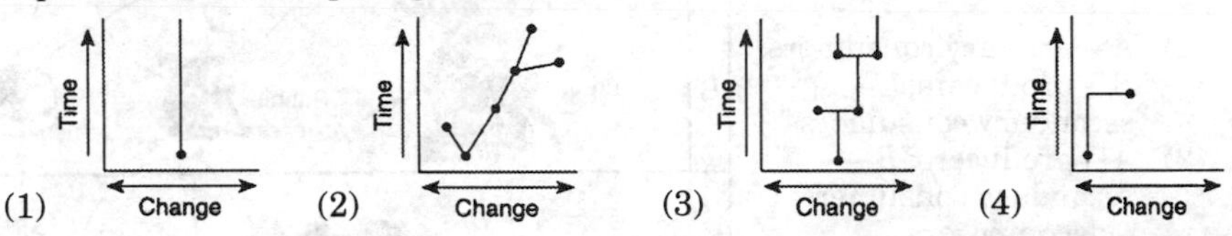

53 Which statement is a valid inference that can be made from the cartoon shown at the right?
1 Wildlife agents regulate reproduction rates of animal species in wildlife refuges.
2 Wildlife agents prevent the importation of organisms to areas where they have no natural enemies.
3 Some human activities have led to the endangerment of numerous animal species.
4 Biological control of pest species is prevented by laws.

Suddenly, Fish and Wildlife agents burst in on Mark Trail's poaching operation.

54 The diagram below illustrates some ecological interactions.
The diagram best represents the

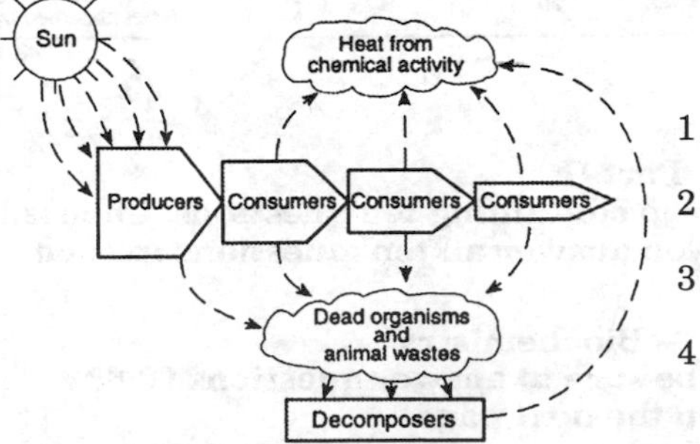

1 number of populations in the biosphere
2 ecological succession in a biotic community
3 pathway of energy as it is transferred through an ecosystem
4 physical interactions between the organisms in a population

55 The structural changes that occurred in certain plants over time, enabling them to thrive in dry habitats, are examples of
1 nutritional relationships
2 adaptations
3 succession
4 energy-flow relationships

56 Some organisms living in a vacant lot include rye grass, dandelions, grasshoppers, slugs, shrews, and ladybugs. Collectively, these organisms represent
1 an ecosystem
2 a biome
3 a population
4 a community

57 Competition between the members of a woodchuck population in a large field could be expected to increase as a result of an increase in the
1 woodchuck reproduction rate
2 spread of disease among the woodchucks
3 number of woodchucks killed by cars
4 number of secondary consumers

58 A food web is shown at the right.
Which components of the food web are indicated by letters *A, B,* and *C?*

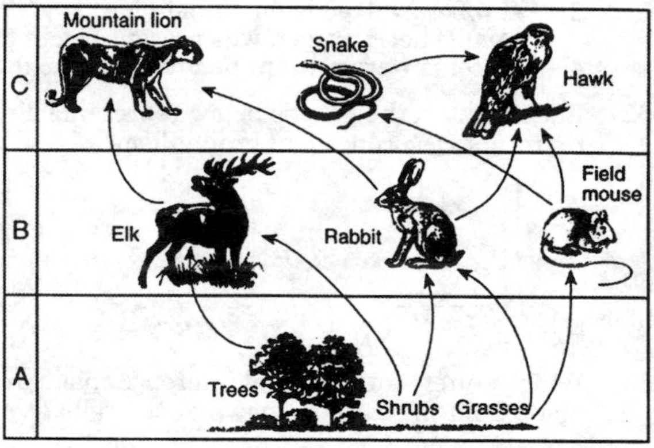

(1) *A* – primary consumers; *B* – producers; *C* – secondary consumers
(2) *A* – producers; *B* – secondary consumers; *C* – decomposers
(3) *A* – producers; *B* – primary consumers; *C* – secondary consumers
(4) *A* – primary consumers; *B* – secondary consumers; *C* – producers

59 The letters in the diagram at the right represent the processes involved in the water cycle.

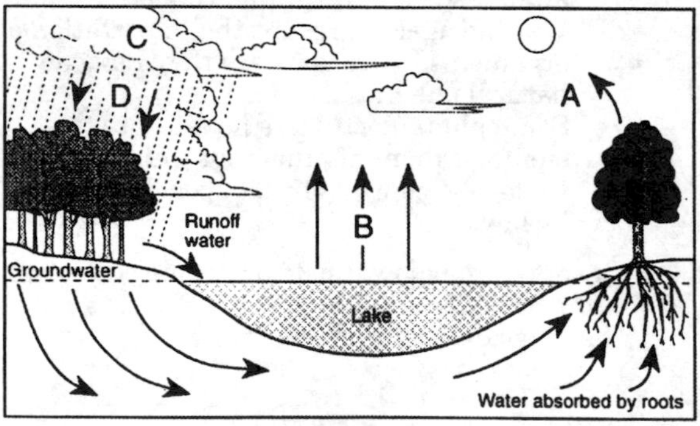

Which letter represents the process of transpiration?

(1) *A*
(2) *B*
(3) *C*
(4) *D*

Part II

This part consists of five groups, each containing ten questions. Choose two of these five groups. Be sure that you answer all ten questions in each group chosen. [20]

Group 1 — Biochemistry
If you choose this group, be sure to answer questions 60-69 found on the next page.

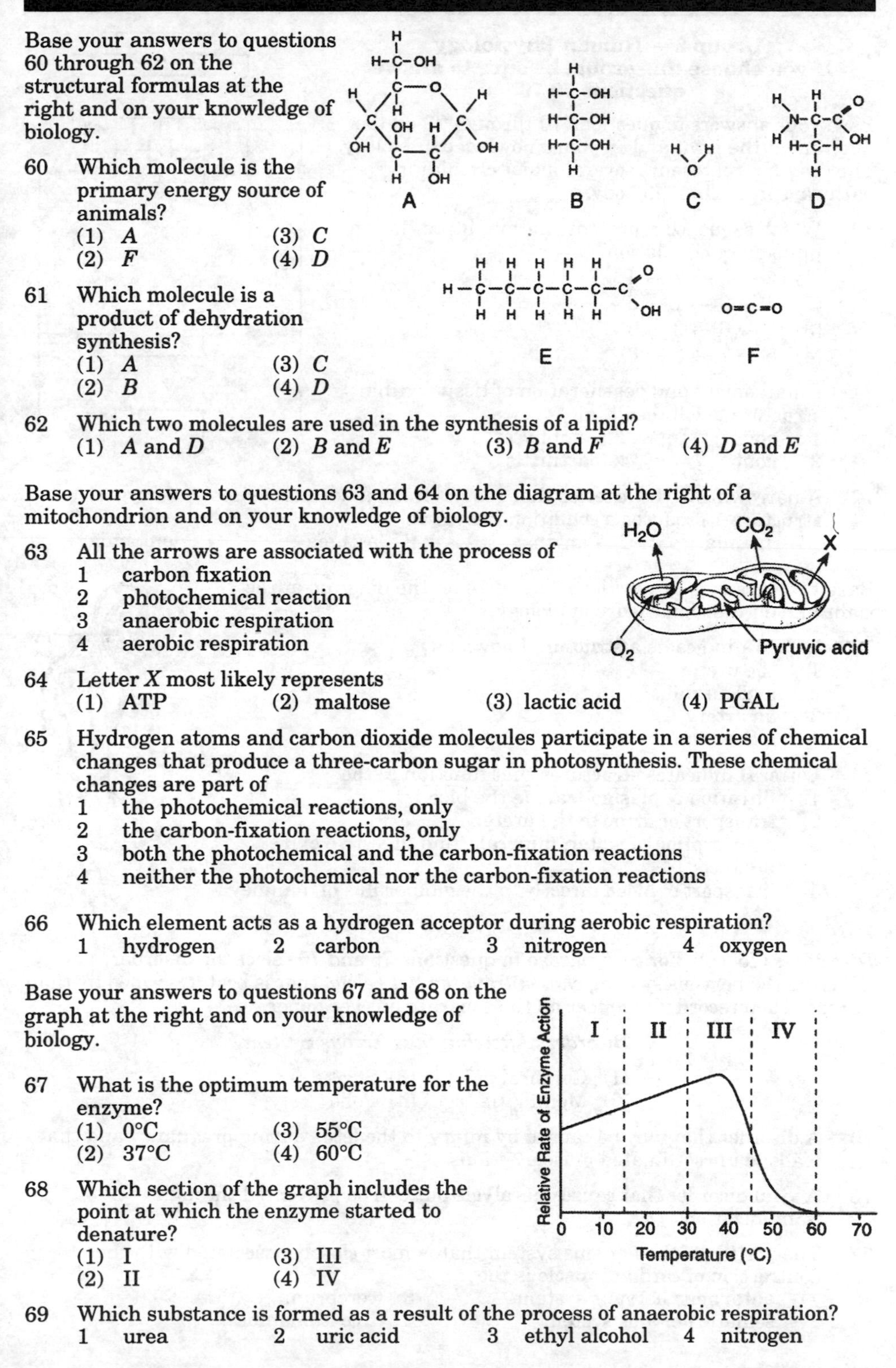

Base your answers to questions 60 through 62 on the structural formulas at the right and on your knowledge of biology.

60 Which molecule is the primary energy source of animals?
(1) *A* (3) *C*
(2) *F* (4) *D*

61 Which molecule is a product of dehydration synthesis?
(1) *A* (3) *C*
(2) *B* (4) *D*

62 Which two molecules are used in the synthesis of a lipid?
(1) *A* and *D* (2) *B* and *E* (3) *B* and *F* (4) *D* and *E*

Base your answers to questions 63 and 64 on the diagram at the right of a mitochondrion and on your knowledge of biology.

63 All the arrows are associated with the process of
1 carbon fixation
2 photochemical reaction
3 anaerobic respiration
4 aerobic respiration

64 Letter *X* most likely represents
(1) ATP (2) maltose (3) lactic acid (4) PGAL

65 Hydrogen atoms and carbon dioxide molecules participate in a series of chemical changes that produce a three-carbon sugar in photosynthesis. These chemical changes are part of
1 the photochemical reactions, only
2 the carbon-fixation reactions, only
3 both the photochemical and the carbon-fixation reactions
4 neither the photochemical nor the carbon-fixation reactions

66 Which element acts as a hydrogen acceptor during aerobic respiration?
1 hydrogen 2 carbon 3 nitrogen 4 oxygen

Base your answers to questions 67 and 68 on the graph at the right and on your knowledge of biology.

67 What is the optimum temperature for the enzyme?
(1) 0°C (3) 55°C
(2) 37°C (4) 60°C

68 Which section of the graph includes the point at which the enzyme started to denature?
(1) I (3) III
(2) II (4) IV

69 Which substance is formed as a result of the process of anaerobic respiration?
1 urea 2 uric acid 3 ethyl alcohol 4 nitrogen

Group 2 — Human Physiology
If you choose this group, be sure to answer questions 70-79.

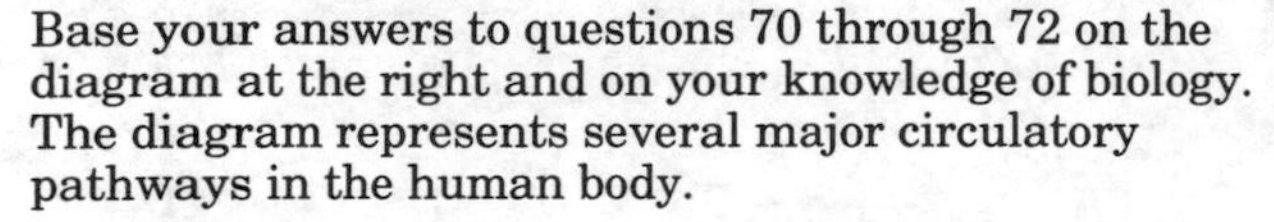
Base your answers to questions 70 through 72 on the diagram at the right and on your knowledge of biology. The diagram represents several major circulatory pathways in the human body.

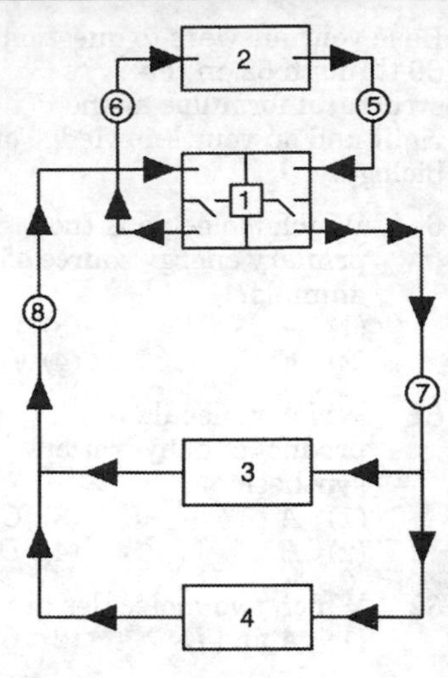

70 Which sequence represents normal blood flow in pulmonary circulation?
(1) 2 → 5 → 7 → 4
(2) 4 → 8 → 1 → 6
(3) 1 → 6 → 2 → 5
(4) 5 → 7 → 3 → 8

71 Enlargement and degeneration of tissue within structure 2 will result in
1 goiter
2 gout
3 emphysema
4 arthritis

72 A narrowing of the arteries that results in an inadequate supply of oxygen to structure 1 produces a condition known as
1 anemia
2 angina
3 asthma
4 leukemia

Base your answers to questions 73 and 74 on the diagram at the right and on your knowledge of biology.

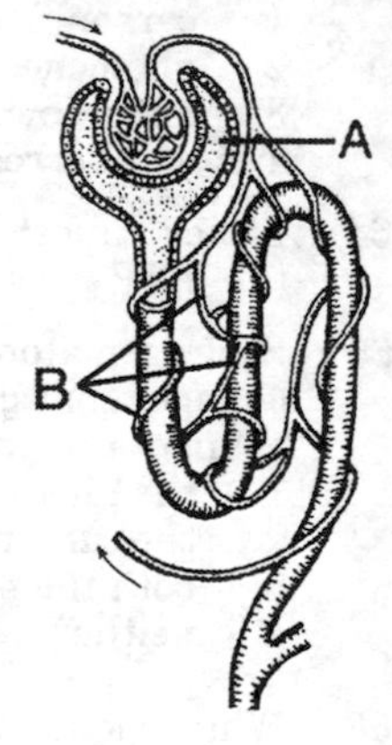

73 Letter *A* indicates a structure known as
1 the ureter
2 a glomerulus
3 an artery
4 a Bowman's capsule

74 Letter *B* indicates structures that function in the
1 filtration of plasma leaving the blood
2 transport of urine to the ureter
3 reabsorption of water, minerals, and digestive end products
4 transport of blood directly to the glomerulus of a kidney

Directions (75-76): For *each* phrase in questions 75 and 76 select the disorder affecting the nervous system, *chosen from the list below,* that is best described by that phrase. Then record its *number* on the separate answer paper.

Disorders Affecting the Nervous System

(1) Cerebral palsy
(2) Meningitis
(3) Stroke
(4) Polio

75 A disorder that may be caused by injury to the fetus during pregnancy and that leads to uncoordinated muscle actions

76 A viral disorder that causes paralysis but can be prevented through immunization

77 The portion of the nervous system that is most closely associated with the contraction of cardiac muscle is the
1 autonomic nervous system
2 somatic nervous system
3 cerebrum
4 hypothalamus

78 The graph at the right represents the level of glucose in a student's blood from 11:00 a.m. until 10:00 p.m. At 3:30 p.m. the student ran in a cross-country meet, and at 5:30 p.m. the student ate dinner.

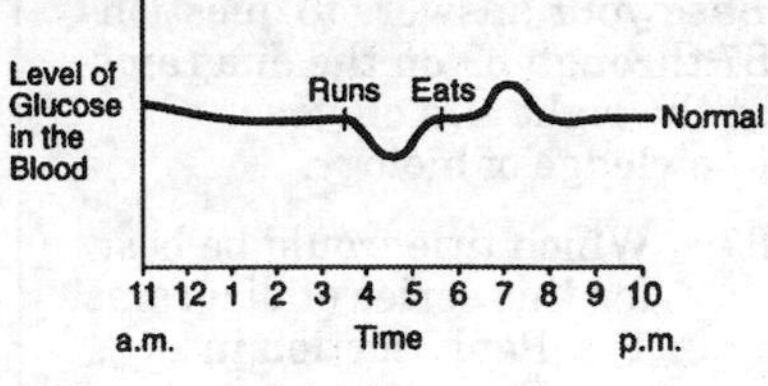

Which hormones were primarily responsible for producing some of the changes in blood sugar level between 4:30 p.m. and 8:00 p.m.?

1 insulin and glucagon
2 estrogen and progesterone
3 parathormone and TSH
4 adrenalin and FSH

79 The breaking apart of platelets in the blood helps in the
1 synthesis of hemoglobin
2 formation of a clot
3 release of antibodies
4 deamination of amino acids

Group 3 — Reproduction and Development
If you choose this group, be sure to answer questions 80-89.

Base your answers to questions 80 through 83 on the diagrams at the right , which represent the male and female productive systems in humans, and on your knowledge of biology.

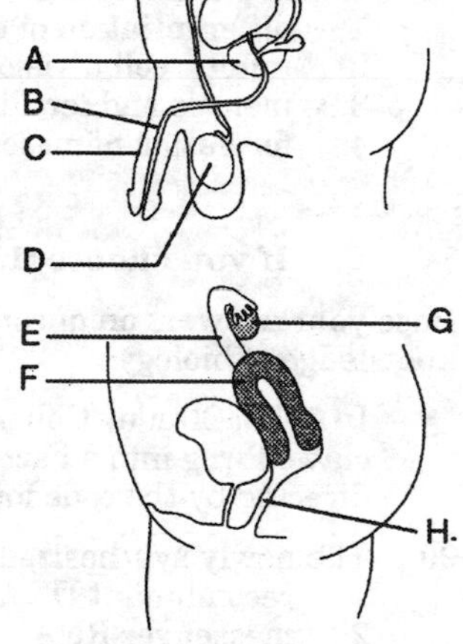

80 In which structure does the corpus luteum develop?
(1) *H* (3) *F*
(2) *G* (4) *D*

81 Which structure produces and secretes a liquid transport medium that is known as semen when it contains sperm?
(1) *A* (3) *F*
(2) *B* (4) *D*

82 In which structure does implantation of the embryo normally occur?
(1) *E* (3) *G*
(2) *F* (4) *H*

83 In which structure does the process of meiosis occur?
(1) *A* and *F* (2) *B* and *H* (3) *C* and *E* (4) *D* and *G*

84 A bluebird reproduces by laying eggs. Which characteristic does the embryo of a bluebird have in common with a human embryo?
1 implantation in the wall of a uterus
2 exchange of materials with the mother through a placenta
3 development within a watery environment inside an amnion
4 a protective shell surrounding a chorion

85 The permanent cessation of the menstrual cycle is known as
1 gestation
2 puberty
3 vascularization
4 menopause

86 Fertilized eggs that develop externally on land generally have more complex structures than those that develop in water because
1 eggs that develop on land cannot easily exchange materials with the environment
2 eggs that develop in water are in less danger from predators
3 eggs that develop in water do not need oxygen
4 a placenta supplies nutrients to eggs that develop in water

Base your answers to questions 87 through 89 on the diagrams at the right and on your knowledge of biology.

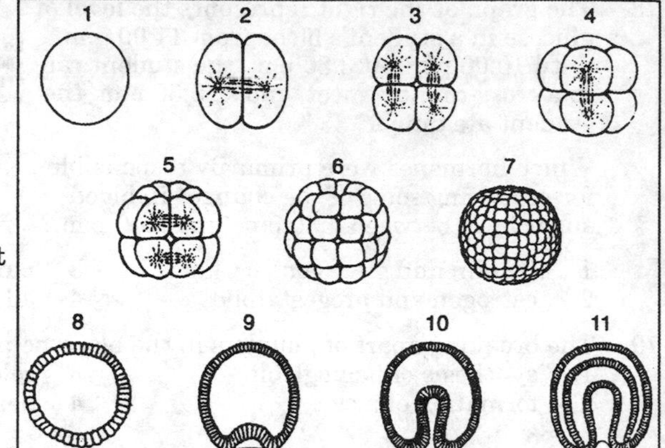

87 Which title would be best for this series of diagrams?
1 Reproduction in Flowering Plants
2 Postnatal Development in Humans
3 Embryonic Development
4 Seed Germination

88 The process of gastrulation is represented in the diagrams numbered
(1) 1 and 2
(2) 3 and 4
(3) 5, 6, 7 and 8
(4) 9, 10, and 11

89 Which phrase best describes the sequence numbered 3 through 7?
1 differentiation of ectoderm
2 mitotic cell division without growth
3 meiosis and fertilization
4 formation of mesoderm by mitosis

Group 4 — Modern Genetics
If you choose this group, be sure to answer question 90–-99.

Base your answers on questions 90 through 93 on the information below and on your knowledge of biology.

In 1973, Stanley Cohen and Herbert Boyer inserted a gene from an African clawed frog into a bacterium. The bacterium then began producing a protein directed by the code found on the inserted frog gene.

90 The newly synthesized genetic material in the bacterium is known as
1 recombinant DNA
2 messenger RNA
3 a gene mutation
4 a multiple allele

91 The procedure used by Cohen and Boyer is known as
1 cloning
2 genetic engineering
3 karyotyping
4 genetic screening

92 Analysis of the DNA from both the frog and the bacterium would reveal that
(1) frog DNA is single stranded, but bacterial DNA is double stranded
(2) frog DNA contains thymine, but bacterial DNA contains uracil
(3) DNA from both organisms is composed of repeating nucleotide units
(4) DNA from both organisms contains the sugar ribose

93 Additional copies of the bacterium containing the frog gene could be produced by
1 asexual reproduction
2 cross-pollination
3 inbreeding
4 grafting

94 The diagram at the right represents a pair of chromosomes. Which diagram best represents the chromatids if *only* crossing-over has occurred?

(1) 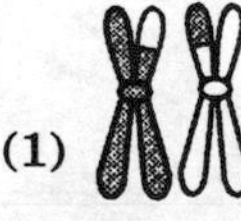
(2)
(3)
(4) 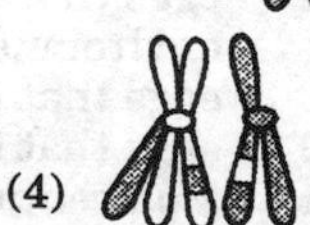

95 Which technique is used to determine whether an adult is a carrier of the gene for sickle-cell anemia?
1 karyotyping
2 amniocentesis
3 urine analysis
4 blood screening

96 What will most likely happen if a population is large and no migration, mutation, or environmental change occurs?
1 Natural selection will increase.
2 Non-random mating will start to occur.
3 The rate of evolution will increase.
4 Gene frequencies will remain constant.

Base your answers to questions 97 and 98 on the diagram at the right, which represents some molecules involved in protein synthesis, and on your knowledge of biology.

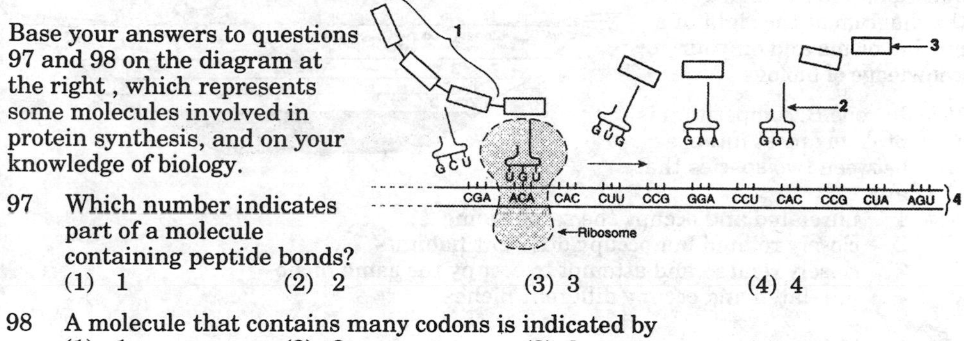

97 Which number indicates part of a molecule containing peptide bonds?
(1) 1 (2) 2 (3) 3 (4) 4

98 A molecule that contains many codons is indicated by
(1) 1 (2) 2 (3) 3 (4) 4

99 A single gene mutation would most likely occur if
1 messenger-RNA molecules temporarily bond to DNA molecules
2 the cytoplasm lacks the amino acids necessary to synthesize a certain polypeptide
3 a base sequence in a DNA molecule is changed
4 transfer-RNA molecules do not line up properly on a messenger-RNA molecule

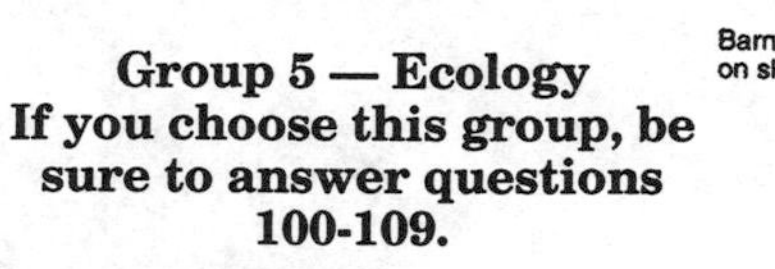

Group 5 — Ecology
If you choose this group, be sure to answer questions 100-109.

Base your answers to questions 100 through 102 on the cnutritional relationships shown at the right and on your knowledge of biology.

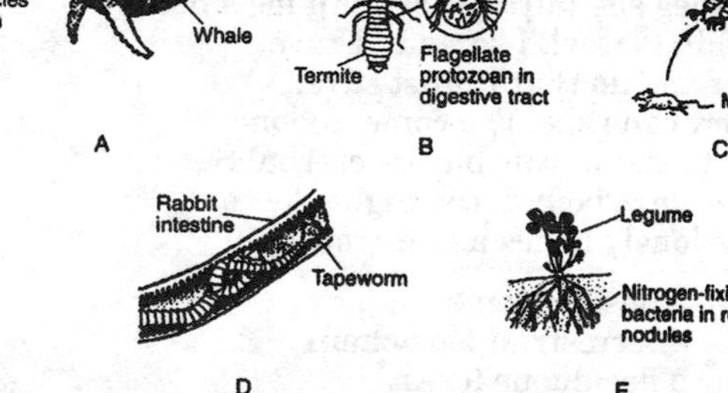

100 Which two diagrams illustrate the same type of nutritional relationship?
(1) *A* and *D*
(2) *B* and *E*
(3) *A* and *C*
(4) *C* and *D*

101 Which organism benefits from its nutritional relationship?
1 whale
2 mouse
3 protozoan
4 rabbit

102 Which diagram shows the relationship that is most similar to that between humans and an athlete's foot fungus?
(1) *E* (2) *B* (3) *C* (4) *D*

103 A climax community is able to exist in a certain geographic region for a long period of time because it
1 provides a habitat for parasites
2 alters the climate of the geographic region
3 attracts many pioneer organisms
4 remains in equilibrium with the environment

104 In the nitrogen cycle, decomposers break down nitrogen compounds and release
1 oxygen gas 2 ammonia 3 urea 4 nitrogen gas

Base your answers to questions 105 through 107 on the diagram at the right of a marine biome and on your knowledge of biology.

105 In zone *B*, competition is probably most intense between two species that are
1 unrelated and occupy the same biome
2 closely related but occupy different habitats
3 closely related and attempt to occupy the same niche
4 unrelated and occupy different niches

106 In which zone does the greatest amount of food production occur?
(1) *A* (2) *B* (3) *C* (4) *D*

107 Autotrophs in zones *A* through *C* depend on the metabolic activities of other organisms for a continuing supply of
1 water
2 carbohydrates
3 respiratory enzymes
4 carbon dioxide

108 Which ecological unit provides the physical setting for the poem below?
The days be hot, the nights be cold,
But cross we must, we rush for gold.
The plants be short, the roots spread wide,
Me leg she hurts, thorn's in me side.
I fall, I crawl, I scream, I rave,
Tiz me life that I must save.
How can it be, I've come undone,
Here 'neath this blazin' eternal Sun?
The days be hot, the nights be cold,
Me lonely bones alone grow old.

1 a desert biome
2 a terrestrial food chain
3 a deciduous forest
4 a coniferous-tree biome

109 Mangrove trees grow in the water on the edge of a subtropical island. In time, grasslike plants will grow on the same spot. Still later, palm trees will grow there. Given enough time (and no natural disasters), all these plants will be gone, and a stable pine forest wills tand where the mangroves once grew. These changes best describe steps involved in
1 the heterotroph hypothesis
2 ecological succession
3 energy cycles
4 the water cycle

Part III

This part consists of five groups. Choose three of these five groups.....[15]

Group 1

If you choose this group be sure to answer questions 110-114.

Base your answers to questions 110 through 113 on the reading passage below and on your knowledge of biology.

"I missed that. What did you say?"

According to the National Center for Health Statistics, one out of 10 Americans has a hearing loss. There are three types of hearing loss: conductive, sensorineural, and mixed. In conductive hearing loss, problems in the outer or middle ear block the transmission of vibrations to the inner ear. Conductive hearing loss can be the result of any number of disorders. The most common disorders are ear infections, excessive ear wax, fluid in the middle ear, and perforated eardrum. This type of hearing loss can usually be treated by medical or surgical procedures.

Sensorineural hearing loss, or "nerve deafness," is most often due to the gradual aging process or long-term exposure to loud noise. However, it can also be caused by high fever, birth defects, and certain drugs.

Some people with impaired hearing have both conductive and sensorineural hearing loss, which is known as mixed hearing loss. Most people with this condition can be helped by either a hearing aid or surgery.

Depending on the symptoms, certain tests can be done to determine the cause and extent of the hearing loss. A standard hearing evaluation may include the following:

- tympanometry, which examines the middle ear, eardrum, and possible blockage of the ear canal
- pure-tone and speech reception testing, which determines the softest level or threshold at which tones and speech are heard
- word discrimination testing, which measures the ability to distinguish words at a comfortable volume

In a recent interview, a rock band saxophone player admitted that over a 6-year period, he developed a 40 percent hearing loss because he neglected to use ear protection during his concert performances. Likewise, the use of personal listening devices, such as headphones, may also cause hearing loss. Your ability to hear is not renewable. It pays to protect your ears from loud noises.

110 Which graph best represents a common relationship between age and nerve deafness?

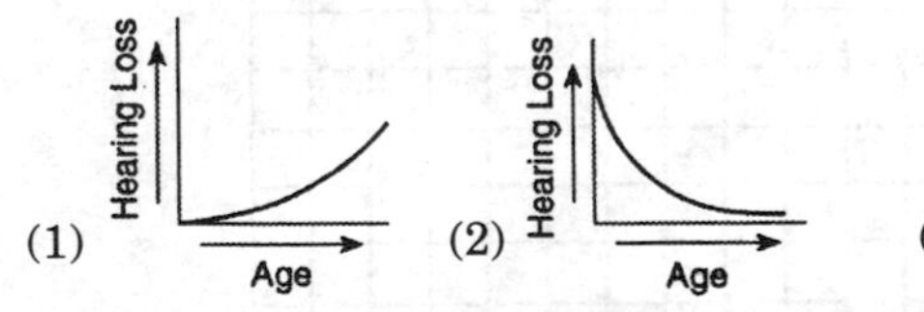

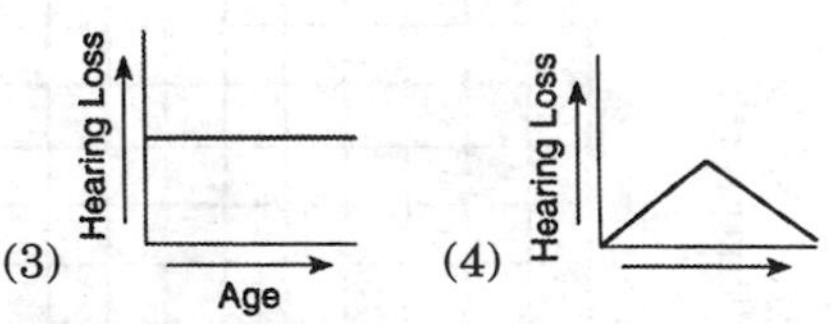

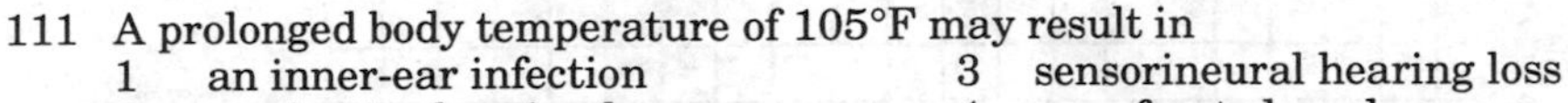

111 A prolonged body temperature of 105°F may result in

1 an inner-ear infection
2 conductive hearing loss
3 sensorineural hearing loss
4 a perforated eardrum

112 Which test is used to determine the presence of excessive wax in the ear canal?

1 sensorineural assessment
2 word discrimination
3 pure-tone and speech reception
4 tympanometry

113 Using one or more complete sentences, explain how a personal listening device may be controlled to decrease damage to the hearing process.

114 The diagram at the right represents a thermometer.

The temperature reading on this thermometer would most likely indicate the temperature
1 of the human body on a very hot summer day
2 at which water freezes
3 at which water boils
4 of a human with a very high fever

Group 2
If you choose this group, be sure to answer questions 115-119.

Base your answers to questions 115 through 119 on the information and data table below and on your knowledge of biology.

A study was made to determine the effect of different salt concentrations on the number of contractions per minute of contractile vacuoles of paramecia. Four beakers of water containing different salt concentrations and equal numbers of paramecia were prepared. All other environmental conditions were kept constant. The paramecia were then observed with a compound microscope, and the contractions of the vacuoles were counted and recorded in the table below.

Directions (115–116): Using the information in the data table, construct a line graph on the grid on your answer paper, following the directions below. The grid on the next page is provided for practice purposes only. Be sure your final answer appears on your answer paper. You may use pen or pencil for your answer.

Data Table

Beaker	Salt Concentration (mg/mL)	Contractions per Minute
A	0.000	5.5
B	0.001	4.0
C	0.010	2.5
D	0.100	1.5

115 Mark an appropriate scale on the axis labeled Contractions per Minute.

116 Plot the data from the data table. Surround each point with a small circle and connect the points.

Example:

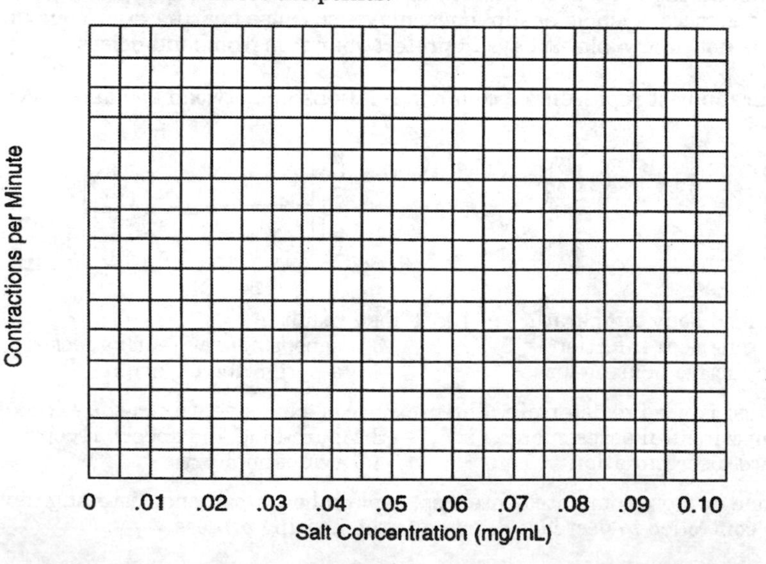

117 According to the data, as the salt concentration increases, the number of contractions per minute changes. What most likely accounts for this change?
 1 a decrease in the water concentration outside the paramecium
 2 a decrease in the salt concentration outside the paramecium
 3 an increased diffusion of salt out of the paramecium
 4 an increased percentage of water in the paramecium

118 According to the information in the data table, which statement is true?
 1 Beaker *B* has a lower salt concentration than beaker *A*.
 2 Beaker *C* has a lower salt concentration than beaker *D*.
 3 The paramecia in beaker *A* respond the least to the water concentration in the beaker.
 4 The paramecia in beaker *D* have nonfunctioning contractile vacuoles.

119 Which safety precaution is recommended when a liquid is being heated in a test tube?
 1 When holding the test tube, keep fingers closest to the open end of the tube.
 2 Direct the flame of the burner into the open end of the test tube.
 3 Stopper the test tube with a rubber stopper.
 4 Wear goggles and a laboratory apron.

Group 3

If you choose this group, be sure to answer questions 120-124.

120 the diagram at the right represents the letter "h" as seen in the low-power field of view of a compound light microscope.

Which diagram best represents the field of view if the slide is not moved and the objective is switched to high power?

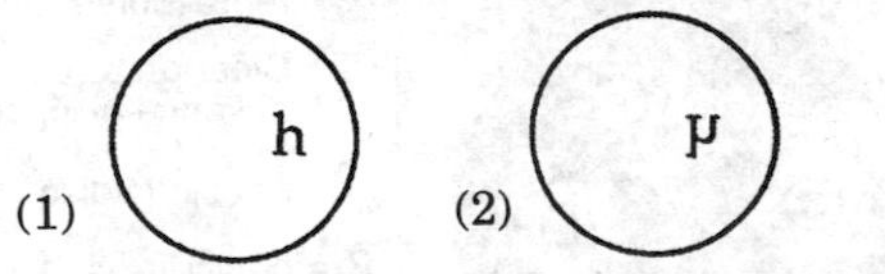

121 The diagram at the right represents cells in a microscopic field of view with a diameter of 1.5 millimeters.

What is the approximate length of a single cell?

(1) 0.5 µm (3) 500 µm
(2) 50 µm (4) 5,000 µm

122 What is the *lowest* possible magnification that can be obtained using the microscope shown at the right?
(1) 20X
(2) 200X
(3) 40X
(4) 800X

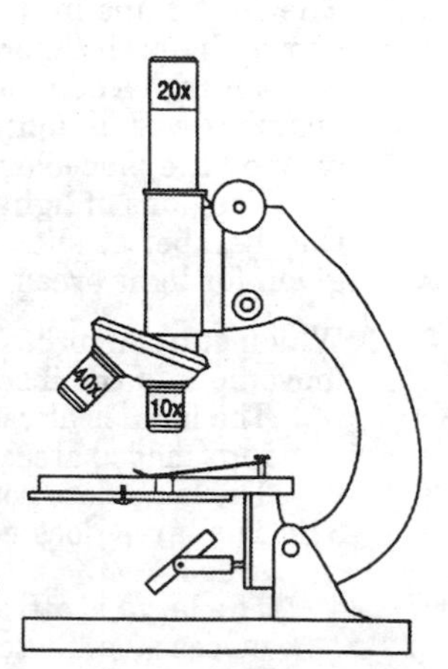

123 While attempting to study a specimen using the low-power objective of a compound light microscope, a student found that the field of view was too dark. Using one or more complete sentences, describe a procedure that would increase the amount of light in the field of view. Write your answer in ink in the space provided on your answer paper.

124 The diagram at the right shows an enlarged microscopic view of two human cheek cells. A copy of this diagram appears on your answer paper.

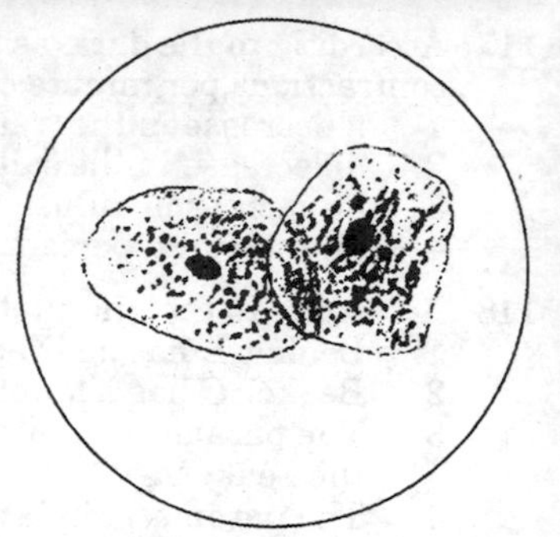

On your answer paper, draw an arrow to a plasma membrane and label it *A*. The point of the arrow should touch the membrane. You may use pen or pencil for your answer.

Group 4
If you choose this group, be sure to answer questions 125-129.

125 In an investigation to determine the effects of environmental pH on the germination of dandelion seeds, 25 dandelion seeds were added to each of five petri dishes. Each dish contained a solution that differed from the others only in its pH, as shown at the right. All other environmental conditions were the same. The dishes were covered and observed for 10 days.

Petri Dish	pH of Solution
1	9
2	8
3	7
4	6
5	5

Using one or more complete sentences, state the variable in this investigation. Write your answer in ink in the space provided on your answer paper.

126 Four discs, each soaked in a different antibiotic, were placed on the surface of a culture plate that had been inoculated with *E. coli* bacteria. The diagram at the right shows the culture plate after it had been incubated for 48 hours.

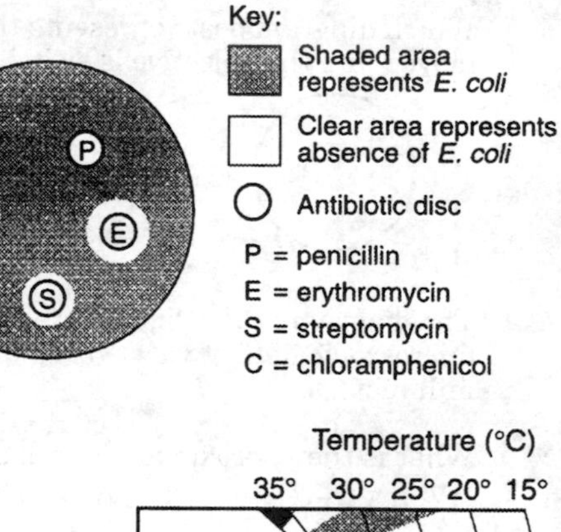

Which antibiotic was most effective in inhibiting the growth of *E. coli*?
1 penicillin
2 erythromycin
3 streptomycin
4 chloramphenicol

127 The dotted line in the diagram at the right shows the path taken by an insect larva when it is placed in a round experiment chamber with a light in the center. Blocks of wood are placed around the light, which cause regions of light and shade within the chamber as shown. Temperatures are given for light areas.

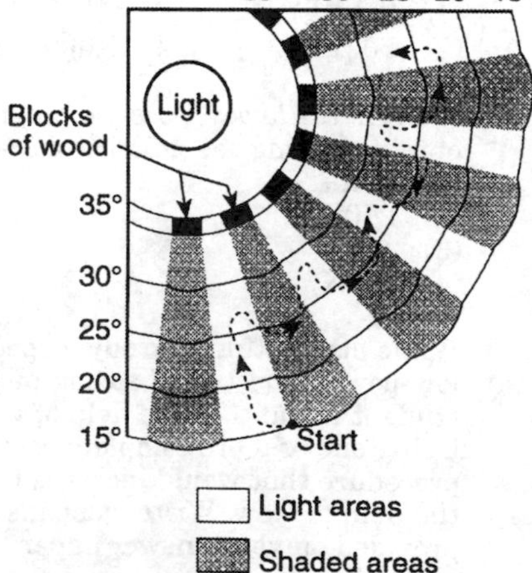

Which statement best describes the movement of the insect larva?
1 The larva is attracted to light but moves into shaded areas when it is too hot.
2 The larva does not move into light areas.
3 The larva does not move into shaded areas.
4 The larva is attracted to light and moves steadily toward the light source.

128 On a slide preparation of a thin slice of potato starch grains can be made more visible by adding
1 Benedict's solution
2 distilled water
3 Lugol's iodine
4 salt solution

129 Bromthymol blue turns yellow in the presence of carbon dioxide. This characteristic makes it possible for bromthymol blue to function as
1 a measure of volume
2 an indicator
3 a catalyst
4 an energy source

Group 5
If you choose this group, be sure to answer questions 130-134.

130 A student performed an experiment involving two strains of microorganisms, strain *A* and strain *B*, cultured at various temperatures for 24 hours. The results of this experiment are shown in the data table at the right.

Data Table

Temperature (°C)	Microorganism Growth (Number of Colonies)	
	Strain A	Strain B
25	10	11
28	10	7
31	11	3
34	12	0

Based on the results of the experiment, the student inferred that strain *A* was more resistant to higher temperatures than strain *B* was. What, if anything, must the student do for this inference to be considered valid?
1 nothing, because this inference is a valid scientific fact
2 repeat this experiment several times and obtain similar results
3 repeat this experiment several times using different variables
4 develop a new hypothesis and test it

Base your answers to questions 131 and 132 on the diagram at the right of some internal structures of an earthworm and on your knowledge of biology.

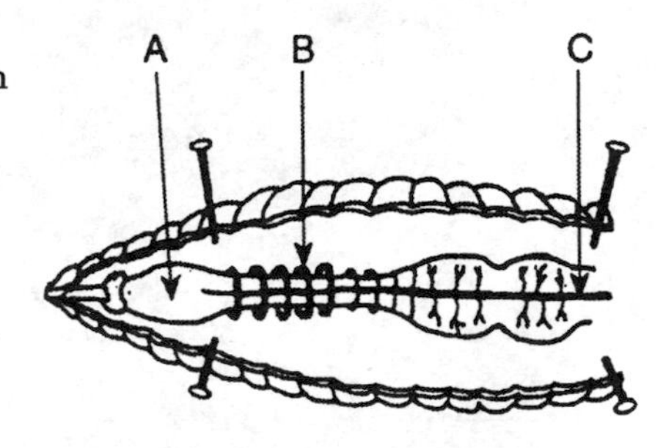

131 Which laboratory equipment should be used to observe the surface details of structure *B*?
1 dissecting microscope
2 compound light microscope
3 ultracentrifuge
4 graduated cylinder

132 Structure A has a diameter of 3 millimeters. What is the approximate diameter of the blood vessel indicated by arrow C?
(1) 2.5 mm (2) 2.0 mm (3) 1.5 mm (4) 0.5 mm

Base your answers to questions 133 and 134 (next page) on the information below.

A student conducting an experiment placed five geranium plants of equal size in environmental chambers. Growing conditions were the same for each plant except that each chamber was illuminated by a different color of light of the same intensity. At the end of 20 days, plant growth was measured.

Directions (133–134): Write your answers in the spaces provided on the separate answer paper. Your answers must be written in ink.

133 Using one or more complete sentences, state a possible hypothesis for this experiment.

134 Using one or more complete sentences, state the control that should be used in this experiment.

Biology Practice Exam 3

Part I

Answer all 59 questions in this part. (65)

Directions (1-59): For each statement or question, select the word or expression that, of those given, best completes the statement or answers the question.

1 Organisms combine simple molecules to form complex molecules by the process of
1 ingestion 3 regulation
2 synthesis 4 hydrolysis

2 In which kingdom are the organisms represented in the cartoon at the right classified?
1 protista
2 monera
3 fungi
plant

3 The development of the cell theory was most directly related to the
1 improvement of the microscope and microscopic techniques
2 use of a five-kingdom classification system
3 development of the gene-chromosome theory
4 discovery of bacteria and viruses

4 Which element is *not* part of a disaccharide?
1 carbon 2 nitrogen 3 oxygen 4 hydrogen

5 Which cell organelle indicated in the diagram at the right controls the synthesis of enzymes?
1 *A*
2 *B*
3 *C*
4 *D*

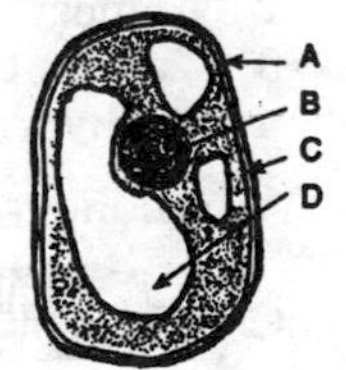

6 A biologist would most likely study all the chemical activities of an organism to obtain information about the
1 number of mutations in the organism
2 reproductive cycle of the organism
3 development of the organism
4 metabolism of the organism

7 Four plants of the same size and type were grown for 24 hours under identical conditions except for the color of the light source. The chart at the right identifies the color of the light each plant was exposed to during the experiment.

Plant	Light Color
A	Red
B	Green
C	Blue
D	White

After 24 hours, the leaves of each plant were tested for sugar. Which plant most likely contained the *smallest* amount of sugar?
1 *A* 2 *B* 3 *C* 4 *D*

8 In order to survive, all organisms must carry out
1 autotrophic nutrition
2 heterotrophic nutrition
3 enzyme-controlled reactions
4 the process of locomotion

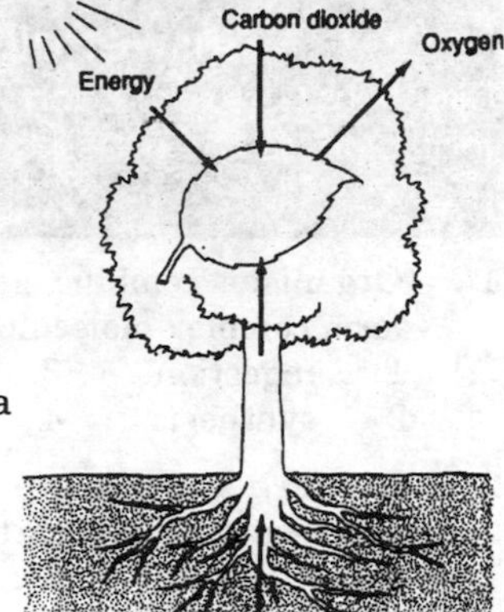

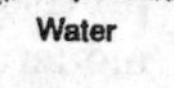

9 Some activities that take place in a plant are represented in the diagram at the right.

The arrows in the diagram are associated with a process that directly involves the use of
1 mitochondria, lenticels, and pistils
2 chloroplasts, stomates, and xylem
3 anthers, phloem, and vacuoles
4 cilia, stamens, and cambium

10 Which process can be performed by organism *A* but not by organism *B*?
1 mitotic cell division
2 aerobic respiration
3 transport of needed materials
4 ingestion of organic molecules

11 The arrows in the diagrams at below right represent the direction of movement of a certain type of molecule through the cell membrane of two different cells. The dots represent the relative concentrations of this molecule. Which processes are illustrated in the diagrams?
1 phagocytosis and diffusion
2 pinocytosis and osmosis
3 active transport and diffusion
4 dehydration synthesis and circulation

Cell A Cell B

12 Which diagram best represents the transport system of an earthworm?

(1) (3)

(2) (4)

13 Which organisms exchange oxygen and carbon dioxide with the environment through thick, moist membranes on their external surfaces?
1 hydras and paramecia
2 amebas and geranium plants
3 blue-green algae and grasshoppers
4 bread molds and maple trees

14 In the diagram below as an impulse travels from structure 1 to structure 5, which sequence does *not* include the secretion of a neurotransmitter?
(1) $1 \rightarrow 4$
(2) $3 \rightarrow 5$
(3) $3 \rightarrow 4$
(4) $1 \rightarrow 2$

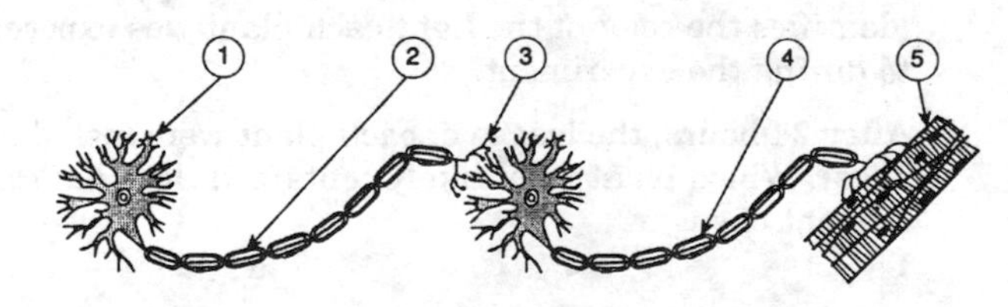

15 The diagram at the right represents root epidermal cells. The major function of the structures labeled *X* is to

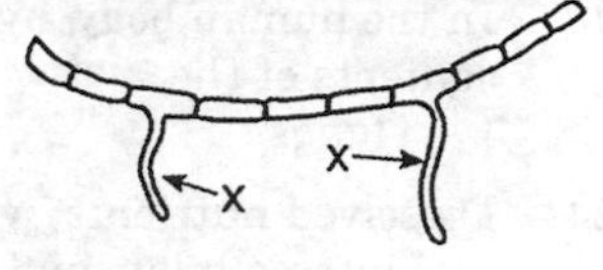

1 transport carbon dioxide from the roots to the leaves
2 move food particles into the phloem
3 absorb complex food particles for transport to the leaves
4 increase the surface area for absorption of water

16 Which process involves the transfer of energy from carbohydrates to ATP molecules?
1 photosynthesis
2 egestion
3 respiration
4 cyclosis

17 When the tentacles of a hydra touch an object, impulses travel throughout the hydra. These impulses travel over
1 definite pathways from receptor to effector
2 random pathways containing fused ganglia
3 modified neurons that form a nerve net
4 specialized neurons that form a central nervous system

18 The diagram at the right includes a microscopic view of a portion of the lower surface of a leaf. The regulatory function of the structures labeled *X* is most closely associated with

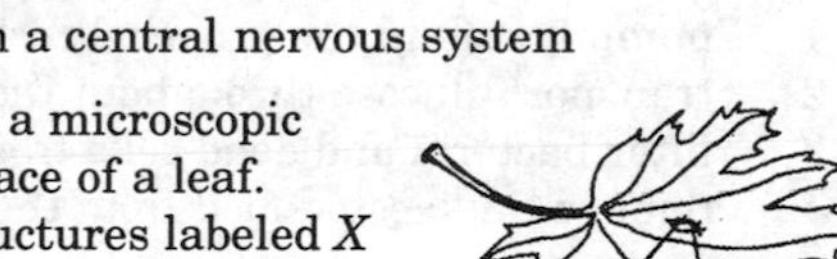

1 auxin secretion
2 transpiration
3 vegetative propagation
4 digestion

19 Mineral salts and uric acid are excreted from the grasshopper by the
1 nephridia and setae
2 neurons and nephrons
3 contractile vacuoles and spiracle openings
4 Malpighian tubules and digestive tract

20 Pituitary growth hormone can affect cells in various parts of the human body because the hormone is secreted directly into
1 glandular ducts
2 muscle tissue
3 the bloodstream
4 the digestive tract

21 The ability to obtain food, avoid polluted environments, and escape from predators help many organisms survive to reproductive age. Which statement accounts for this ability?
1 Evolution has assured that organisms that move survive longer than those that do not.
2 Locomotion has provided certain advantages for many organisms.
3 Reproduction is a life process that requires a sessile state of life.
4 All organisms coordinate life activities by means of a nervous system.

22 Which structure of a paramecium serve the same basic function as the setae of an earthworm?
1 cilia
2 nuclei
3 cell membranes
4 contractile vacuoles

23 In the human body, hydrochloric acid is responsible for the low pH of the contents of the
1 lungs 2 kidney 3 liver 4 stomach

24 Dissolved nutrients, wastes, and oxygen are exchanged between the blood and intercellular fluid through the walls of
1 arteries 2 veins 3 capillaries 4 ventricles

25 In the diagram of the human urinary system at the right, which letter indicates a structure responsible for filtering urea out of the bloodstream?
(1) *A*
(2) *B*
(3) *C*
(4) *D*

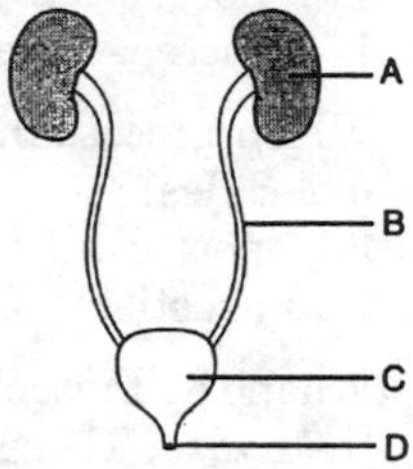

26 The diagram at the right represents the human lymphatic system. A major function of the structures labeled *X* is to
1 pump lymph in the proper direction
2 transport glucose throughout the body
3 filter bacteria and dead cells from the lymph
4 remove undigested food from the blood

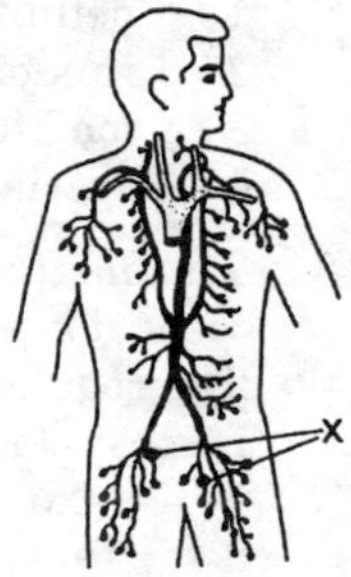

27 The diagram below right represents a demonstration of the breathing process in humans. The balloons represent lungs. The change in the balloons is brought about by
1 a change in air composition outside the bell jar
2 a change in air pressure inside the bell jar
3 an expansion of the balloons, which pulls the rubber sheet into the bell jar
4 a contraction of the balloons, which forces air into the bell jar

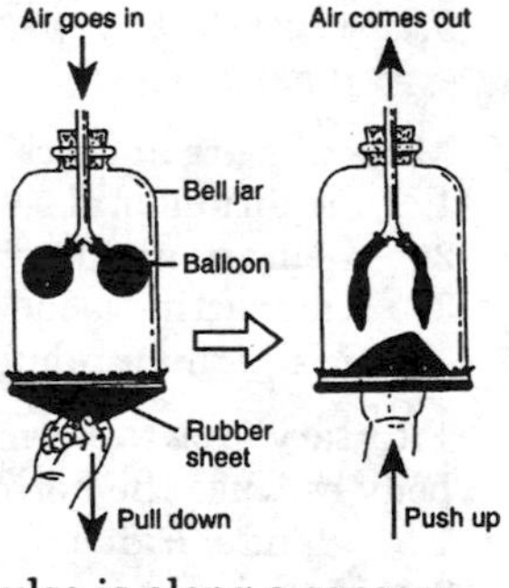

28 In a simple spinal reflex, the pathway for an impulse is along a sensory neuron directly to a motor nueron through
1 an effector 3 an interneuron
2 a receptor 4 the brain

29 Fatigue and a buildup of lactic acid are most closely associated with
1 aerobic respiration in cardiac muscle
2 anaerobic respiration in striated muscle
3 glycogen synthesis in visceral muscle
4 gas exchange in smooth muscle

30 Which diagram best represents mitotic cell division?

(1) 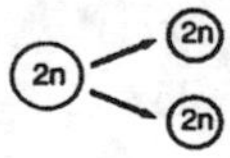 2n → 2n, 2n (2)  n + n → 2n (3) 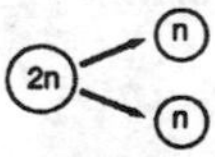 2n → n, n (4) 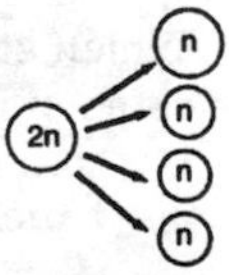 2n → n, n, n, n

31 The separation of homologous pairs of chromosomes during gametogenesis is known as
1 replication 2 synthesis 3 alignment 4 disjunction

32 The normal diploid chromosome number of the house mouse, *Mus musculus*, is 40. How many pairs of homologous chromosomes would a normal zygote of *Mus musculus* contain?
(1) 10 (2) 20 (3) 40 (4) 80

33 In the process of oogenesis in humans, a primary sex cell undergoes divisions that normally produce
1 four monoploid cells
2 one monoploid sperm and three diploid eggs
3 one monoploid egg and three polar bodies
4 four monoploid eggs

34 The diagram at the right includes information on asexual reproduction. Which term belongs in the area labeled *X*?
1 grafting
2 regeneration
3 cutting
4 sporulation

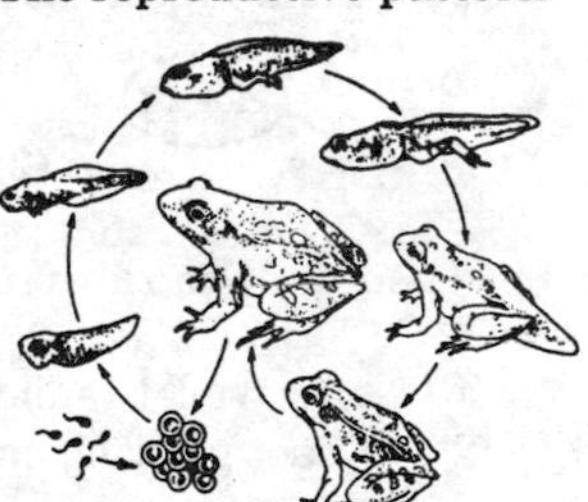

35 The diagram below shows a vertebrate life cycle. The reproductive pattern involved in this cycle can be best described as
1 internal fertilization followed by external development
2 internal fertilization followed by internal development
3 external fertilization followed by external development
4 external fertilization followed by internal development

36 In animals, the process of cleavage is best described as the
1 fertilization of a mature egg cell by an immature sperm cell
2 production of daughter cells having twice the number of chromosomes as the parent cell
3 production of daughter cells having half the number of chromosomes as the parent cell
4 division of cells resulting in the development of an embryo from a zygote

37 Which processes are represented in the diagram of plant structures shown at the right?
1 pollen tube growth, fruit formation, and external fertilization
2 cross-pollination, seed germination, and meiotic cell division
3 self-pollination, pollen germination, and fertilization
4 pollen formation, seed dispersal, and mitotic cell division

38 Mendel developed the basic principles of heredity by
1 examining chromosomes with microscopes
2 using x rays to induce mutations
3 analyzing large numbers of offsprings
4 observing crossing-over during meiosis

39 When two heterozygous tall plants are crossed, some short plants appear in the offspring. The appearance of these short plants illustrates
1 segregation and recombination
2 crossing-over and differentiation
3 intermediate inheritance
4 codominant inheritance

40 In humans, the gene for red hair and the gene for freckles are often inherited together because both genes are located on the same chromosomes. This observation best illustrates the concept of
1 gene linkage
2 independent assortment
3 dominance
4 hybridization

41 A man with a blood type A and a woman with blood type B have a child with blood type O. The genotypes of the man, woman, and child, respectively must be
(1) I^Ai, I^BI^B, ii
(2) I^AI^A, I^Bi, I^Ai
(3) I^AI^A, I^BI^B, I^AI^B
(4) I^Ai, I^Bi, ii

42 Which diagram represents a sperm that can unite with a normal egg to produce a zygote that will develop into a normal human embryo?

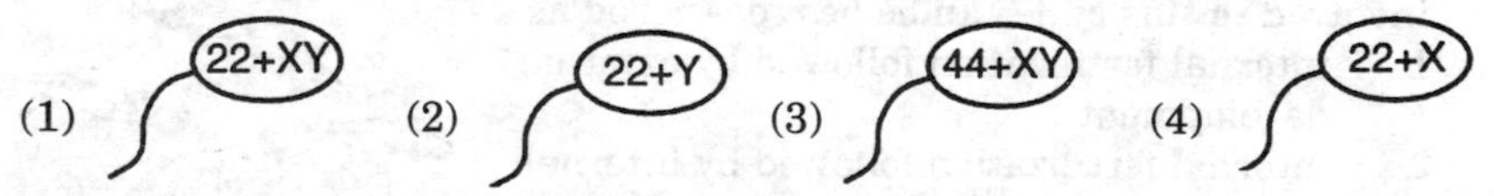

43 The mutation rates in *Drosophila* will most likely increase after exposure to
1 ultraviolet radiation
2 yeast growing on a nutrient medium
3 oxygen gas
4 extremely cold temperatures

44 The sequence of amino acids in horse hemoglobin is very similar to the sequence of amino acids in human hemoglobin. This evidence supporting organic evolution has been drawn from studies in the field of comparative
1 biochemistry 2 cytology 3 anatomy 4 embryology

45 Scientists conducted a study of identical twins who were separated at birth and raised in different homes. They found that in some sets of twins the individuals showed a marked difference in intelligence. The most likely explanation for this difference is that
1 expression of inherited traits can modify the environment
2 intelligence is a sex-linked trait
3 environment can influence the development and expression of inherited traits
4 nondisjunction occurred in the autosomes of one twin but not the other twin

46 The parts of a DNA nucleotide are indicated in the chart at the right by letter *A*, *B*, and *C*. An *X* indicates which chemical elements are present in each part.
Which diagram represents a DNA nucleotide?

DNA Nucleotide Parts	Elements				
	C	O	H	N	P
A		X	X		X
B	X	X	X		
C	X	X	X	X	

(1) A – C – B (2) A – B – C (3) A – C – B (4) B – A – C

47 In Yellowstone National Park, some species of algae and bacteria can survive and reproduce in hot springs at temperatures near the boiling point of water. The ability to survive and reproduce at these temperatures is an example of
1 aggregate formation
2 adaptation
3 artificial selection
4 reproductive isolation

48 The diagrams at the right show embryos of three different vertebrate species. According to one theory, similarities in these embryos suggest common ancestry. As these embryos mature, they will most likely

1 develop new organs according to the nutritional requirements of each organism
2 continue to closely resemble each other as adults
3 show no similarity as adults
4 develop the distinctive characteristics of their species

49 Models I and II in the graph at the right show two different evolutionary pathways. Which evolutionary concepts are best represented by model I and model II?

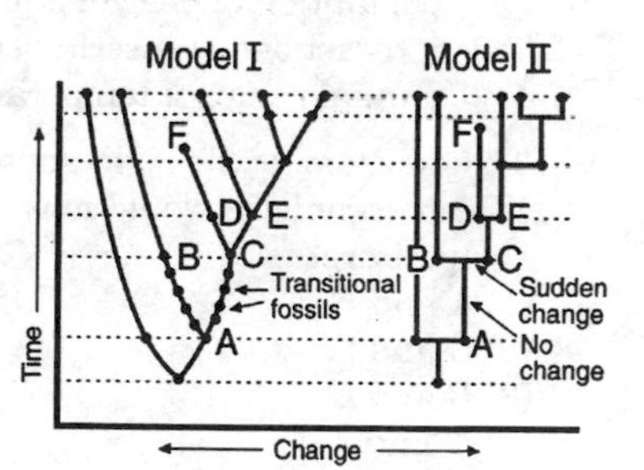

1 Model I represents gradualism; model II represents punctuated equilibrium.
2 Model I represents punctuated equilibrium; model II represents gradualism.
3 Model I represents speciation; model II represents acquired characteristics.
4 Model I represents acquired characteristics: model II represents speciation.

50 According to Charles Darwin, one factor that affects the evolution of a species is
1 variation due to genetic mutations
2 rapid fossil formation
3 survival of the fittest
4 exposure to environmental pollutants

51 Which statement is *not* included as a part of the modern theory of evolution?
1 Sexual reproduction and mutation provide variation among organisms.
2 Traits are transmitted by genes and chromosomes.
3 More offspring are produced than can possibly survive.
4 New organs arise when they are needed.

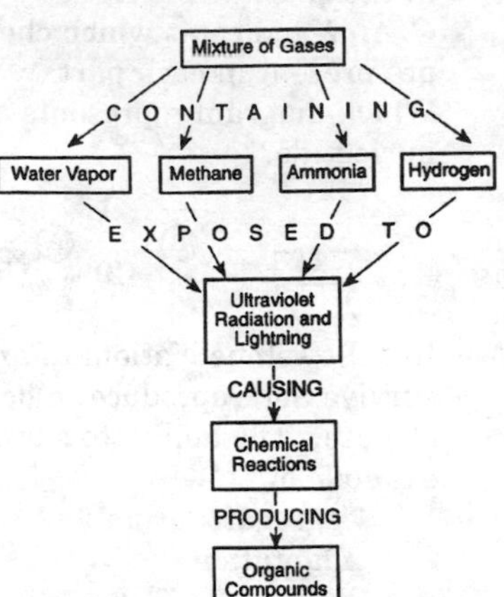

52 Which concept is represented in the diagram at the right?
1 heterotroph hypothesis
2 gene-chromosome theory
3 theory of natural selection
4 use and disuse

53 A pond community is represented in the diagram at the right. Which term includes the interactions between the organisms of this community and the physical factors of their environment?
1 population 3 biotic
2 ecosystem 4 competition

54 An abiotic component of the biosphere is represented by the
1 mushrooms in a cave in the Catskill Mountains
2 population of deer in a forest at the close of the hunting season
3 interactions of insect species in a cornfield
4 annual range of temperatures in a lake in New York State

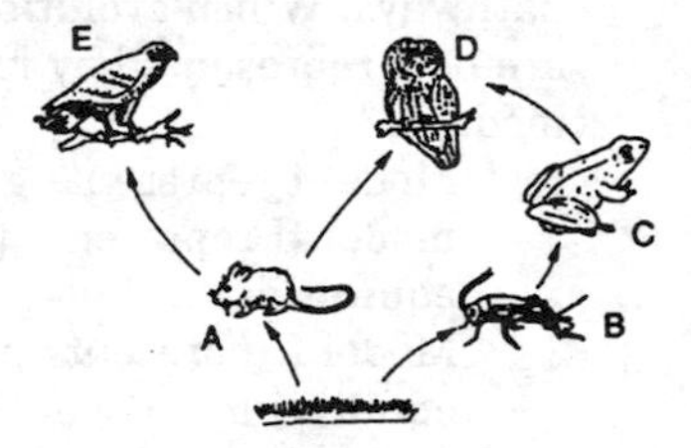

55 The diagram at the right shows a food web. Which organisms would most likely be competitors?
(1) *A* and *C*
(2) *B* and *C*
(3) *B* and *D*
(4) *D* and *E*

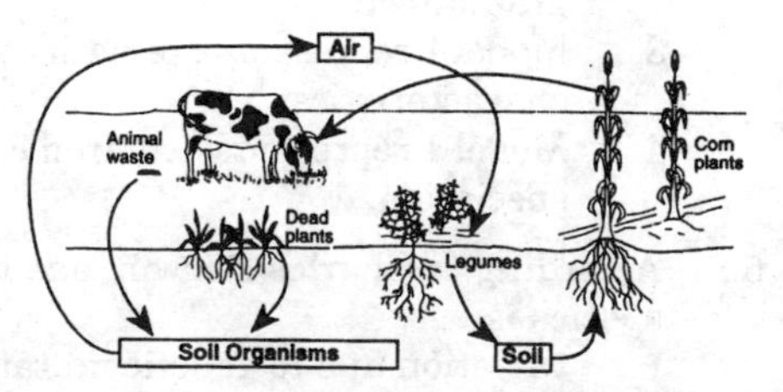

56 A cycling of materials is represented in the diagram at the right. Which statement is supported by evens shown in the diagram?
1 Materials are cycled between living organisms, only.
2 Materials are cycled between heterotrophic organisms, only.
3 Materials are cycled between the living and nonliving components of the environment.
4 Materials are cycled between the physical factors of the environment by the processes of condensation and evaporation.

57 The diagram at the right represents a food chain. The arrows in the diagram indicate the
  1 direction in which organisms move in the environment
  2 direction of energy flow through a series of organisms
  3 order of importance of the various organisms
  4 return of chemical substances to the environment

58 Ecosystems tend to change with time until a stable system is formed. All stable ecosystems are characterized by the presence of
  1 a specific climax vegetation influenced by the climate of the area
  2 trees as the most abundant autotroph in the area
  3 fewer primary consumers than secondary consumers in the area
  4 a larger number of heterotrophs than autotrophs in the area

59 Which activity would most likely control an insect pest and be the *least* harmful to the environment?
  1 spraying area with biocides that affect the insect pest
  2 releasing imported insects that prey on the insect pest
  3 using traps baited with sex hormones to attract the insect pest
  4 eliminating the plants that the insect pest feeds on

Part II

This part consists of five groups, each containing ten questions. Choose two of these five groups. Be sure that you answer all ten questions in each group chosen. (20)

Group 1 – Biochemistry

If you choose this group, be sure to answer questions 60-69.

Base your answers to questions 60 through 62 (on the next page) on diagrams of molecules below and on your knowledge of biology.

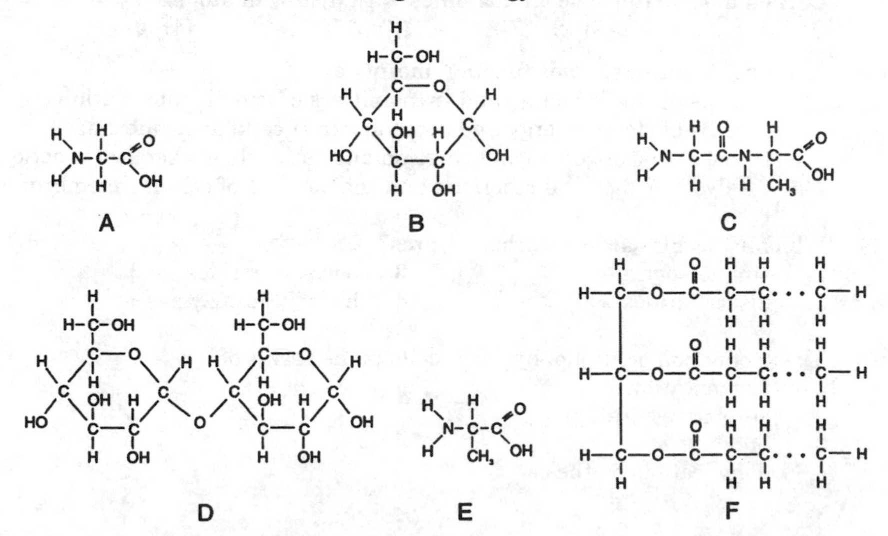

60 Which food contains the highest percentage of the type of molecule represented by *F*?
(1) bread (2) lettuce (3) table sugar (4) butter

61 Amino groups are found in
(1) *A*, *C*, and *E*, only
(2) *B* and *D*
(3) *A*, *C*, *E*, and *F*
(4) *F*, only

62 How many different elements are found in molecule *D*?
(1) 37 (2) 2 (3) 3 (4) 45

63 A chemical reaction is shown below.

maltose + maltase ⟶ maltose-maltase ⟶ glucose + glucose + maltase
A B C D

Which letter indicates an enzyme-substrate complex?
(1) *A* (2) *B* (3) *C* (4) *D*

Directions (64-66): For each statement in questions 64 through 66, select the reaction of photosynthesis, chosen from the list below, that is most closely associated with that statement. Then record its number on the separate answer paper.

Reactions of Photosynthesis

(*1*) Photochemical reactions, only
(*2*) Carbon-fixation reactions, only
(*3*) Both the photochemical and the carbon-fixation reactions
(*4*) Neither the photochemical nor the carbon-fixation reactions

64 Chlorophyll pigments absorb light energy.
(1) *1* (2) *2* (3) *3* (4) *4*

65 The reaction takes place in the chloroplast.
(1) *1* (2) *2* (3) *3* (4) *4*

66 Carbon dioxide from the air combines with hydrogen atoms.
(1) *1* (2) *2* (3) *3* (4) *4*

67 In living organisms, lipids function mainly as
1 sources of stored energy and transmitters of genetic information
2 sources of stored energy and components of cellular membranes
3 transmitters of genetic information and catalysts of chemical reactions
4 catalysts of chemical reactions and components of cellular membranes

68 Which molecules are *not* carbohydrates?
1 simple sugars
2 disaccharides
3 polysaccharides
4 hydrolytic enzymes

69 The production of alcohol by yeast cells is the result of
1 fermentation
2 aerobic respiration
3 budding
4 dehydration synthesis

Group 2 – Human Physiology

If you choose this group, be sure to answer questions 70-79.

Base your answers to questions 70 through 72 on the diagram at the right and on your knowledge of biology. The diagram represents the digestive system of a student who has eaten a sandwich consisting of two slices of bread, chicken, lettuce and mayonnaise.

70 The final reactions for the complete hydrolysis of the bread would occur in organ
(1) 1 (3) 3
(2) 2 (4) 5

71 Which organ produces a substance that would be used in the emulsification of the lipids in the mayonnaise?
(1) 1 (2) 2 (3) 7 (4) 4

72 In which organs would chemical digestion of the chicken take place?
(1) 1 and 3 (2) 1 and 7 (3) 2 and 6 (4) 3 and 5

73 An accumulation of hardened cholesterol deposits is associated with the development of
1 gout 2 gallstones 3 constipation 4 appendicitis

Directions (74-76): In *each* of question 74 through 76, five structures are listed. Select the structure that does *not* belong to the same system as the other structures in the list. [Note: Answers are listed under each question number.]

74
1 thyroid
2 salivary glands
3 adrenal glands
4 islets of Langerhans
5 pituitary

75
1 Bowman's capsule
2 ureter
3 glomerulus
4 urethra
5 esophagus

76
1 aorta
2 right ventricle
3 superior vena cava
4 alveolus
5 pulmonary artery

77 One type of allergic reaction results in constriction of the bronchial tubes, which interferes with the passage of air into and out of the lungs. This type of allergic reaction is most closely associated with
1 asthma 2 emphysema 3 bronchitis 4 meningitis

78 Which response usually occurs after an individual receives a vaccination for influenza virus?
1 Hormones in the blood inhibit reproduction of the virus.
2 Antigens from the vaccine deactivate the virus.
3 Enzymes released from platelets hydrolyze the virus.
4 Antibodies against the virus form in the blood.

79 The diagram below shows a sequence of events that occurs in humans.

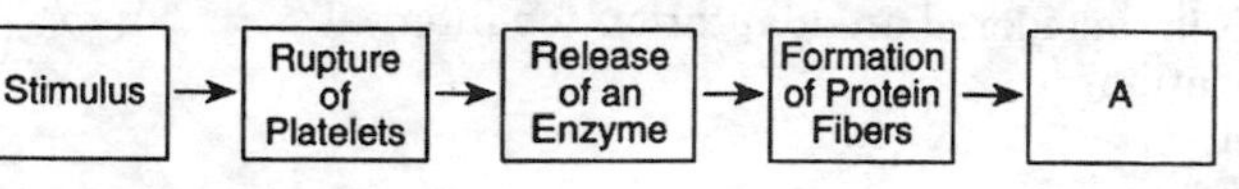

Which information belongs in box *A*?
1 increase in breathing rate
2 decrease in body temperature
3 formation of a clot
4 formation of urea

Group 3 – Reproduction and Development
If you choose this group, be sure to answer questions 80-89.

Base your answers to questions 80 through 81 on the diagram of a developing bird embryo at the right and on your knowledge of biology.

80 In what structure is nitrogenous waste stored?
(1) *A* (3) *E*
(2) *B* (4) *D*

81 Where are the nutrients needed for growth and development stored?
(1) *F* (2) *G* (3) *C* (4) *H*

82 The ovary releases an egg in a process known as
1 fertilization 2 gestation 3 ovulation 4 implantation

83 Which embryonic structure is described as a hollow ball of cells?
1 blastula 2 placenta 3 chorion 4 gastrula

84 The periodic shedding of the uterine lining usually takes place
1 after menopause
2 during pregnancy
3 if fertilization does not occur
4 before the onset of puberty

85 In order to unite with an egg, a sperm cell must travel from
1 ovary to oviduct to uterus
2 vagina to uterus to oviduct
3 vagina to umbilical cord to oviduct
4 ovary to urethra to uterus

86 Which set of environmental factors would most influence the process of germination represented in the diagram at the right?
1 sunlight, temperature, and fertile soil
2 moisture, carbon dioxide, and sunlight
3 oxygen, temperature, and moisture
4 fertile soil, carbon dioxide, and oxygen

Base your answers to questions 87 through 89 on the diagram at the right and on your knowledge of biology.

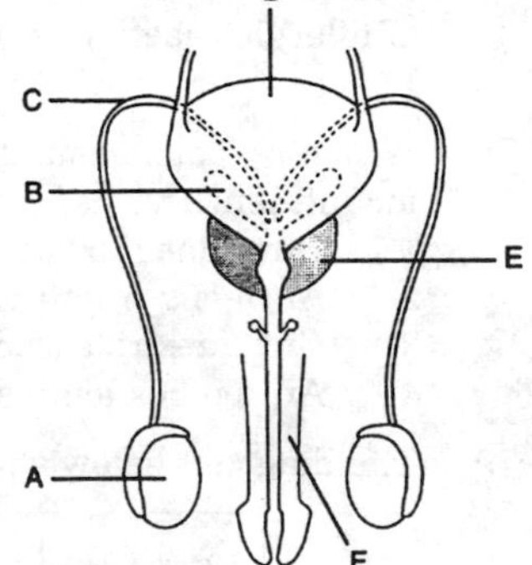

87 Structure *A* is considered part of both the reproductive system and the
1 circulatory system
2 excretory system
3 digestive system
4 endocrine system

88 Which is considered an adaptation for internal fertilization?
(1) *A*
(2) *F*
(3) *C*
(4) *D*

Base your answer to question 89 on the diagram above, the cartoon at the right, and your knowledge of biology.

"Well, Mr. Rosenburg, your lab results look pretty good–although I might suggest your testosterone level is a tad high."

89 Mr. Rosenburg appears to have a problem with secretions produced by which structure in the diagram above?
(1) *A*
(2) *B*
(3) *E*
(4) *D*

Group 4 – Modern Genetics
If you choose this group, be sure to answer questions 90-99.

Base your answers to questions 90 through 93 on the diagram at the right and on your knowledge of biology. The diagram represents molecular structures involved in protein synthesis.

1: G G A T A C C T G / C C T A T G G A C

2: C C U A U G G A C

G G A — Proline; U A C — Methionine; C U G — Aspartic Acid

90 Structure 1 represents
1 part of a polypeptide chain
2 a portion of a DNA molecule
3 a portion of an RNA molecule
4 the building block of proteins

91 The DNA code for aspartic acid is
(1) C-T-G (2) C-C-U (3) C-C-T (4) C-U-G

92 Proline, methionine, and aspartic acid represent three types of
1 fatty acids 2 hormones 3 amino acids 4 enzymes

93 Structure 2 is synthesized in the
1 nucleus 2 vacuole 3 ribosome 4 lysosome

Directions (94-95): For *each* phrase in questions 94 and 95, select the concept, *chosen from the list below*, that is most closely associated with that phrase. Then record its *number* on the separate answer paper.

Concepts
(1) Population (3) Gene frequency
(2) Gene Pool (4) Hardy-Weinberg principle

94 The sum of all heritable genes for traits in a group of interbreeding organisms
(1) *1* (2) *2* (3) *3* (4) *4*

95 The percentage of an allele for a trait in a group of interbreeding organisms
(1) *1* (2) *2* (3) *3* (4) *4*

96 Which event takes place first during DNA replication?
1 A single-stranded RNA molecule is formed.
2 Transfer RNA links to an amino acid.
3 Free nucleotides are bonded together in the correct sequence.
4 The DNA molecules "unzips" along weak hydrogen bonds.

97 Which chemical components may be parts of a molecule of transfer RNA?
1 ribose, phosphate group, uracil base
2 deoxyribose, phosphate group, guanine base
3 glucose, amino group, thymine base
4 maltose, carboxyl group, uracil base

Base your answers to questions 98 and 99 on the diagram at the right and on your knowledge of biology.

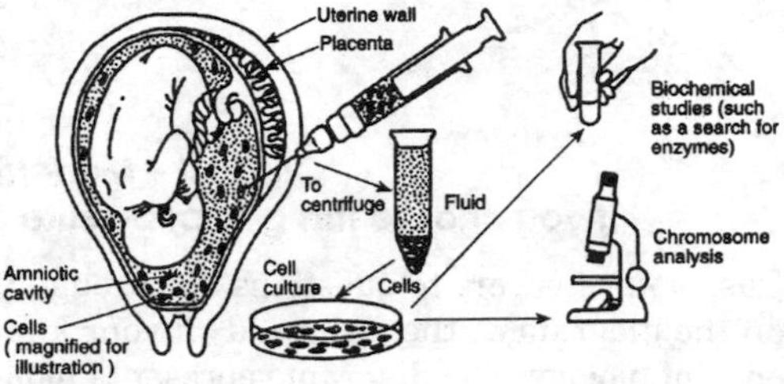

98 The diagram represents a technique that can be used for
1 obtaining a chemical analysis of blood and urine
2 transferring recombinant DNA to a fetus
3 preventing sickle-cell anemia
4 determining the sex of the fetus

99 The chromosome analysis shown could be used to identify the presence of
1 Tay-Sachs in the mother
2 phenylketonuria in both the mother and the fetus
3 Down syndrome in the fetus
4 hemophilia in both the mother and the fetus

Group 5 – Ecology

If you choose this group, be sure to answer questions 100-109.

Base your answers to questions 100 through 102 on the chart of land biomes at the right and on your knowledge of biology.

Land Biome	Climatic Conditions	Climax Flora	Climax Fauna
A	Heavy rainfall; constant warmth	Broad-leafed trees	Snakes, monkeys
Desert	*B*	Succulent plants	Lizards, kangaroo rats
Taiga	Long, severe winters; thawing of subsoil in summer	*C*	Moose, black bear

100 Which biome is represented by *A*?
1 tundra
2 tropical rain forest
3 temperate deciduous forest
4 grassland

101 Which information should be included in box *B*?
1 extreme daily temperature fluctuations; sparse rainfall
2 constant rainfall; high temperatures
3 strong prevailing winds; small variations in temperature
4 permanently frozen subsoil; no precipitation

102 Which information should be included in box *C*?
1 mosses
2 grasses
3 conifers
4 deciduous trees

103 The relationship between fleas and a dog is most similar to the relationship between
1 honeybees and a flower
2 orchids and a tree
3 nitrogen-fixing bacteria and a legume
4 athlete's-foot fungus and a human

104 Which foods are derived from organisms that occupy the level that contains the greatest amount of energy in a biomass pyramid?
1 bread and tomatoes
2 shrimp and rice
3 hamburger and french fries
4 chicken and lettuce

105 The diagram at the right represents a food web.

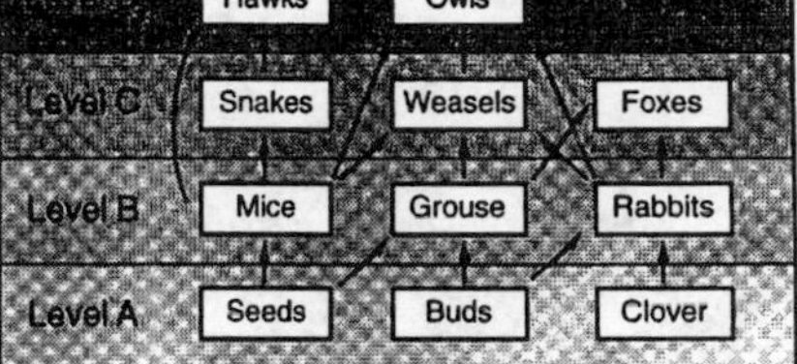

Which level contains the greatest number of consumers?
(1) *A*
(2) *B*
(3) *C*
(4) *D*

106 The diagram at the right represents the nitrogen cycle.
Which letter in the diagram indicates the activity of bacteria of decay?
(1) *W*
(2) *X*
(3) *Y*
(4) *Z*

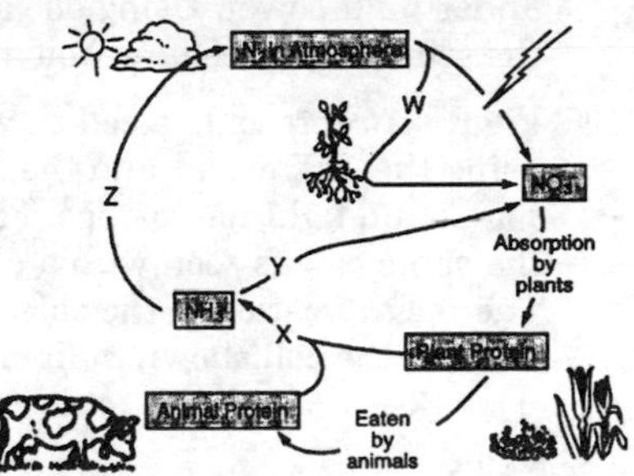

107 Over a period of time, an abandoned field develops into a forest by a series of events known as
1 symbiosis
2 exploitation
3 ecological succession
4 the carbon-hydrogen-oxygen cycle

108 Which factor most directly limits the variety of herbivores living in an area?
1 an amount of nitrogen in the atmosphere
2 number of decomposers present
3 types of plants present
4 kinds of minerals in the water

109 In a marine environment, which factor is *least* effective in determining the kinds and numbers of organisms at different depths?
1 quantity of available oxygen
2 temperature
3 amount of light
4 moisture content

Part III

This part consists of five groups. Choose three of these five groups and record your answers in accordance with the directions given in the question. (15)

Group 1

If you choose this group, be sure to answer questions 110-114 on next page.

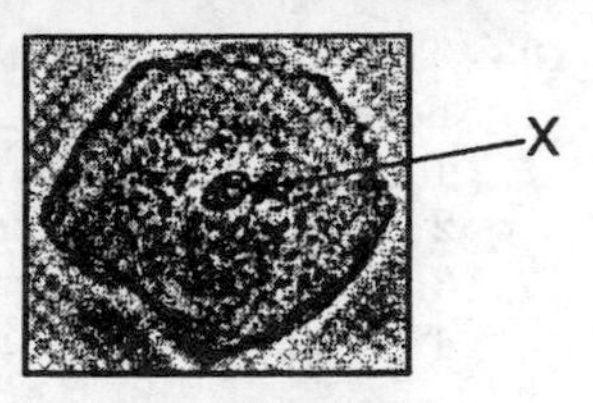

Base your answers to questions 110 through 111 on the photograph at the right and on your knowledge of biology. The photograph shows an animal cell viewed with a compound light microscope.

110 Name the organelle indicated by letter *X*. You may use pen or pencil for your answer.

111 Using one or more complete sentences, state a technique that could have been used to make the organelle indicated by letter *X* more visible. You may use pen or pencil for your answer.

112 A specimen was viewed under the high-power objective of a compound light microscope. Its length was estimated to be 0.75 millimeter. What is the approximate length of the specimen in micrometers?
(1) 0.00075 (2) 0.75 (3) 75 (4) 750

113 In preparing a wet mount of onion cells for observation with a compound light microscope, a student placed a small piece of onion epidermis in a drop of water on a clean slide. Next he added a clean coverslip. He then placed the slide on the stage of the compound light microscope and focused under high power. Using one or more complete sentences, state an error in the student's procedure. You may use pen or pencil for your answer.

114 Figure *A* represents a cell as viewed by a student using the 10X ocular and the 10X objective of a compound light microscope. Figure *B* represents the same cell as seen with a different objective. The magnification of the objective used to observe the cell shown in figure *B* is most likely

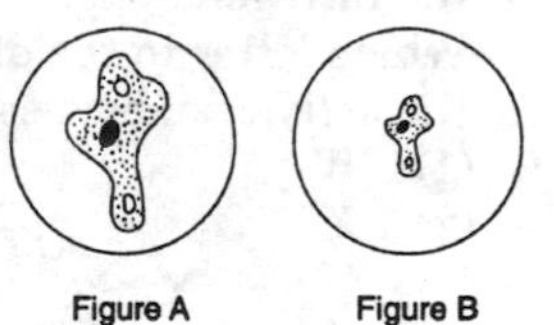

(1) 4X (2) 40X (3) 60X (4) 100X

Group 2

If you choose this group, be sure to answer questions 115-119.

Base your answers to questions 115 through 118 on the information and data table below and on your knowledge of biology.

A field study was conducted to observe a deer population in a given region over time. The deer were counted at different intervals over a period of 40 years. During this period of time, both ranching and hunting increased in the study region. A summary of the data is presented in the table below.

Directions (115-116): Using the information in the data table, construct a line graph on the grid provided *on your answer paper*, following the directions below. The grid on the next page is provided for practice purposes only. Be sure your final answer appears on your answer paper. You may use pen or pencil for your answer.

115 Mark an appropriate scale on each labeled axis.

116 Plot the date for the deer population on the grid. Surround each point with a small circle and connect the points.

Data Table

Year	Deer Population (thousands)
1900	3.0
1910	9.5
1920	65.0
1924	100.0
1926	40.0
1930	25.0
1940	10.0

117 During which 10-year period did the greatest increase in the deer population occur?
(1) 1900-1910 (3) 1920-1930
(2) 1910-1920 (4) 1930-1940

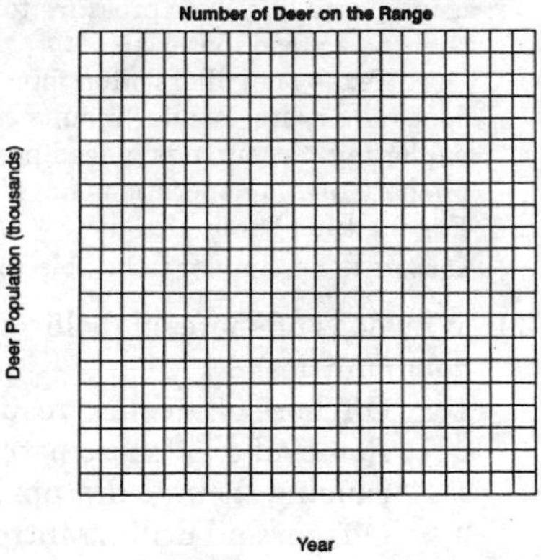

Number of Deer on the Range

Deer Population (thousands)

Year

118 Using one or more complete sentences, state one possible action that could have been used to help maintain a more stable population of deer in the area. You may use pen or pencil for your answer.

119 In attempting to demonstrate the effectiveness of a new vaccine, a scientist performed the experimental procedures below.

I One hundred genetically similar rats were divided into two groups of 50 rats each (group *A* or group *B*).
II Each rat in group *A* was given an injection of the vaccine suspended in a glucose-and-water solution.
III Each rat in group *B* was given an injection of a glucose-and-water solution containing no vaccine.
IV After several weeks, all rats in both groups were exposed to the disease for which the vaccine was developed.

Using one or more complete sentences, state the variable being studied in this experiment. You may use pen or pencil for your answer.

Group 3

If you choose this group, be sure to answer questions 120-124.

Base your answers to questions 120 through 123 on the passage below and on your knowledge of biology.

Oil Leaks Offshore Don't Always Mean Pollution

People have long criticized oil companies for the accidental release of oil into the sea because of its negative impact on marine life. However, over the past decade, a strange new twist has developed. Some scientists have begun to wonder whether drilling in the Gulf of Mexico could threaten marine life by – oddly enough – reducing oil leaks.

Work has recently focused on many natural oil seeps on the Gulf floor. Researchers have discovered a variety of organisms associated with the seeps, including fields of huge tubeworms, giant mussels, and deep-sea crabs. These petroleum-seep communities are isolated from sunlight. Life there must rely on a food chain that begins with energy from constant petrochemical bath of the seeps. For example, bacteria living in the gills of the giant mussels metabolize methane from the seeps and provide food for their hosts.

Soon after their discovery, protection of seep dwellers became a priority. Initially, researchers thought that these organisms were so rare that they should be classified as endangered species. As a result, the Minerals Management Service issued special guidelines for work that might affect chemosynthetic organisms.

Continuing study revealed that seep organisms are not rare after all. Communities dependent on chemosynthesis seem to develop wherever substantial leaks occur on the deep-sea floor. There is little fear now that offshore drilling could accidentally destroy some unique species, although offshore extraction of petroleum could possibly reduce pressure to nearby seeps and rob these deep-sea species of their basic food materials.

Texas oil drillers often found that the extraction of oil on land caused nearby seeps to dry up. Similar results could happen offshore. The loss of a seep by exploiting its source is indeed possible. However, evidence shows that certain offshore drilling operations have not affected nearby petroleum-seep communities. For the time being, the giant mussels and giant tubeworms, as well as the giant oil corporations, appear to be able to continue to feed together happily in the Gulf.

120 Why does offshore oil drilling present a possible threat to petroleum-seep communities?
1 Offshore oil drilling results in an increased number of oil seeps.
2 Removal of offshore petroleum may reduce pressure to oil seeps, causing them to dry up.
3 Offshore oil drilling introduces new species into the petroleum-seep communities.
4 Removal of offshore petroleum cause overpopulation problems within the petroleum-seep communities.

121 The bacteria living within the gill cells of the giant mussels illustrate
1 saprophytism
2 predation
3 an adaptation for survival
4 a competition between species

122 Shortly after the petroleum-seep communities were discovered, researchers became concerned that the petroleum-seep dwellers might
1 decrease the world oil supply
2 multiply and affect other ocean dwellers
3 pollute the sea
4 be endangered species

123 Which phrase best describes the giant tubeworms of the petroleum-seep community?
1 chemosynthetic-dependent organisms
2 deep-sea decomposers
3 heterotrophic monerans
4 photosynthetic marine organisms

124 A solution in a test tube is tested for the presence of glucose using an indicator and a heat source. Using one or more complete sentences, describe a control that should be used in this activity. You may use pen or pencil for your answer.

Group 4
If you choose this group, be sure to answer questions 125-129.

125 The data at the right were obtained during an investigation involving freshwater sunfish.

Water Temperature (°C)	Average Rate of Opening of Gill Covers (openings/minute)
10	15
15	25
18	30
20	38
23	60
25	57
27	25

Which set of labeled axes should be used to present the data most clearly?

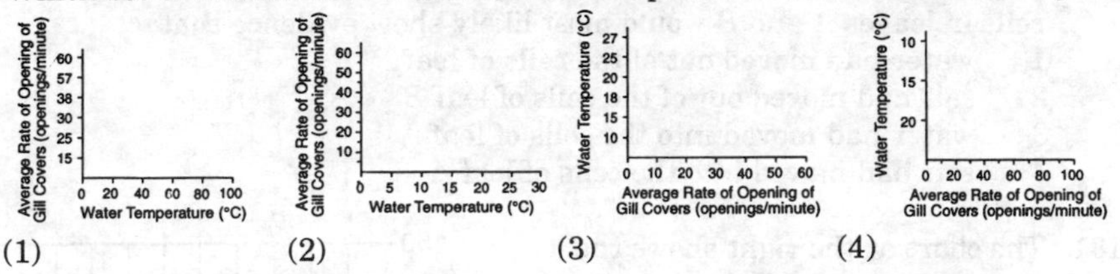

(1) (2) (3) (4)

126 Which graduated cylinder represented below contains a volume of liquid closest to 15 milliliters?

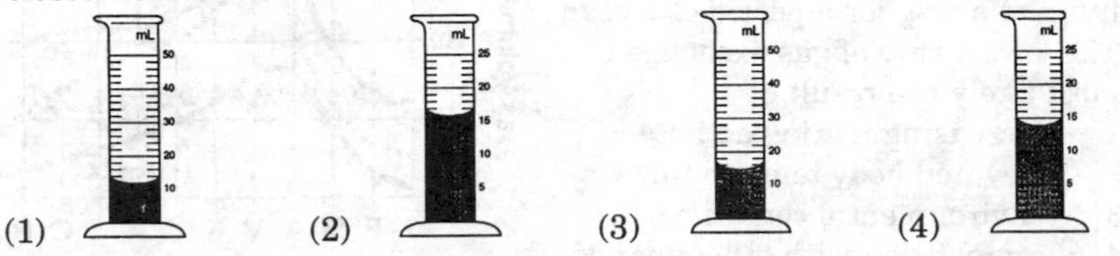

(1) (2) (3) (4)

127 A student prepared the list of steps shown below for performing a laboratory investigation. She omitted one important step for completing the investigation.

Steps to follow in an Experiment

(1) Define a problem.
(2) Develop a hypothesis.
(3) Select suitable lab materials and perform a controlled experiment to test the hypothesis.
(4) Collect, organize, and graph the experimental data.
(5)
(6) Repeat steps 3 through 5.

Using one or more complete sentences, state the procedure that is missing at step 5. You may use pen or pencil for your answer.

128 Which substance should be used in an investigation designed to determine the degree of acidity at which an enzyme works best?

(1) pH paper
(2) glucose solution
(3) Lugol's iodine
(4) methylene blue

129 Using one or more complete sentences, state a scientific purpose of dissecting an organism. You may use pen or pencil for your answer.

Group 5

If you choose this group, be sure to answer questions 130-134 on next page.

130 Diagrams *A* and *B* below represent two different slide preparations of elodea leaves. The tap water used contained 1% salt 99% water, while the salt solution contained 6% salt and 94% water. Elodea cells normally contain 1% salt.

A

Elodea leaf mounted in tap water

B

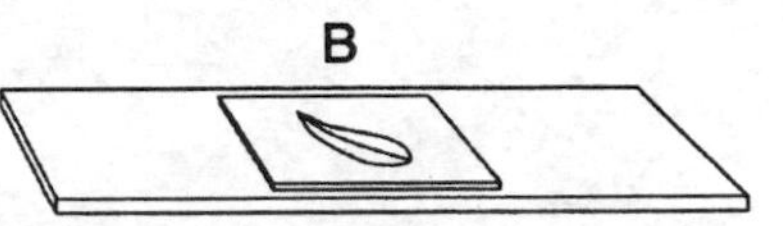

Elodea leaf mounted in 6% salt solution

Ten minutes after the slides were prepared, a microscopic examination of cells in leaves *A* and *B* would most likely show evidence that

1 water had moved out of the cells of leaf *B*
2 salt had moved out of the cells of leaf *B*
3 water had moved into the cells of leaf *A*
4 salt had moved into the cells of leaf *A*

131 The chart at the right shows the amount of oxygen and carbon dioxide exchanged through the skin and lungs of a frog for a period of 1 year. The lowest rate of gas exchange is most likely the result of

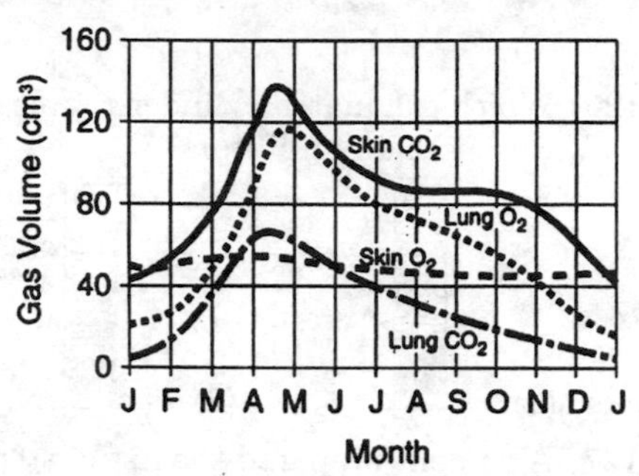

1 increasing mating activity
2 elevated body temperatures
3 environmental conditions
4 competition with other species

132 When stained with certain dyes, nucleoproteins appear black. These dyes would most likely be used to

1 identify specific nucleoproteins within cells
2 stain all types of cell organelles
3 determine the chemical composition
4 determine the chemical composition of nucleoproteins
4 indicate the presence of nucleoproteins within cells

133 The diagram at the right represents some tissue as seen through a compound light microscope.
Which procedure was most likely used to prepare this tissue for viewing with the microscope?

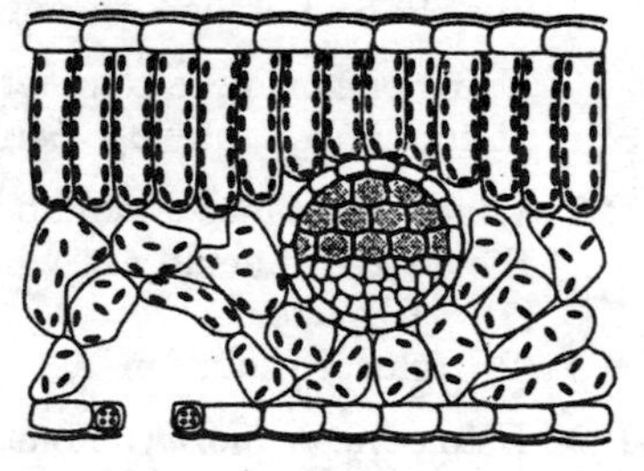

1 A small slice of a root tip was mounted on a slide.
2 A thin cross section of a leaf was mounted on a slide.
3 A vertical cut was made through the body of an earthworm, and a thin slice was mounted on a slide.
4 A section of the exoskeleton of a grasshopper was removed and mounted on a slide.

134 Using one or more complete sentences, describe one safety procedure a student should use when performing a test with Benedict's solution to determine the presence of simple sugar in a food sample. You may use pen or pencil for your answer.